BIOTECHNOLOGY AND CRISPR

Kruger Brentt Publishers

BIOTECHNOLOGY AND CRISPR

Joe Figel
Christopher Clark
Keryn Gedan

Kruger Brentt
Publishers
2025

Kruger Brentt Publishers UK. LTD.
Company Number 9728962

Regd. Office: 68 St Margarets Road, Edgware, Middlesex HA8 9UU

ISBN: 978-1-78715-285-4

For information on all our publications visit our website at http://krugerbrentt.com/

PREFACE

"Biotechnology and CRISPR" represents a groundbreaking exploration of the transformative potential of biotechnology, with a particular focus on the revolutionary gene-editing tool CRISPR-Cas9. This book serves as a comprehensive guide for researchers, students, policymakers, and enthusiasts seeking to understand the profound implications of biotechnology and CRISPR for science, medicine, agriculture, and beyond.

Biotechnology encompasses a wide range of techniques and applications that harness biological systems for practical purposes, from improving crop yields and developing new medicines to combating disease and addressing environmental challenges. Among the most revolutionary advancements in biotechnology is CRISPR-Cas9, a precise and efficient gene-editing tool that has revolutionized genetic engineering and opened up new frontiers in biology and medicine.

As editors of this volume, we have endeavored to compile a diverse array of perspectives, methodologies, and case studies from leading experts and practitioners in the fields of biotechnology and CRISPR. Through a blend of theoretical insights, experimental techniques, and real-world applications, this book covers a wide range of topics, including the history and development of CRISPR-Cas9, its potential applications in human health and agriculture, ethical considerations, regulatory challenges, and future directions.

Our overarching goal in presenting this book is to foster a deeper understanding of the promises and perils of biotechnology and CRISPR, while also promoting dialogue, collaboration, and responsible innovation in these rapidly evolving fields. By showcasing the latest advancements, challenges, and ethical considerations in biotechnology and CRISPR research, we aim to empower readers to critically engage with these technologies and contribute to their ethical and responsible use for the betterment of society.

We extend our sincere gratitude to all the contributors who have generously shared their expertise, experiences, and insights in the fields of biotechnology and CRISPR. It is our hope that this book will serve as a valuable resource and inspiration for researchers, educators, policymakers, and students as they navigate the exciting and complex landscape of biotechnology and CRISPR.

Joe Figel
Christopher Clark
Keryn Gedan

CONTENTS

CHAPTER-1

INTRODUCTION TO BIOTECHNOLOGY

Karl Ereky (1878-1952)

The term biotechnology was coined by Karl Ereky, a Hungarian scientist, in his book entitled Biotechnologie der Fleish-, Fett-und Milcherzeugung im landwirtschaftlichen Grossbetriebe (Biotechnology of Meat, Fat and Milk Production in an Agriculture Large-scale Farm) in 1917.

Biotechnology, the term, is a combination of two words 'bio' and 'technology', - 'bio' means biological systems or processes, and 'technology' refers to methods, systems, and devices used to make useful products from these biological systems. Thus, biotechnology refers to the different technologies that make use of living cells and/or biological molecules to generate useful products for the benefit of mankind.

Mankind has been practicing biotechnology since long. Right from the domestication of sheep and cattle in the Paleolithic age, conservation of plant stocks by the early Egyptian farmers (ancient germplasm conservation), to the classical examples of early fermentation technology in the form of making bread, cheese and wine. However, modern biotechnology is a multidisciplinary subject which involves knowledge sharing between different areas of science such as Cell and Molecular Biology, Microbiology, Genetics, Anatomy and Physiology, Biochemistry, Computer Science and Recombinant DNA technology (rDNA technology).

Biotechnology is the stream of study that integrates biology and technology to develop products and systems that can improve our lives and benefit the health of our planet as a whole. The most significant contribution of biotechnology is the production of therapeutic proteins and other drugs through genetic engineering.

Biotechnology research focuses on discovering the root molecular causes of diseases and intervening precisely at that level to find out the remedies. Modern biotechnology has generated solutions through products and technologies to combat rare diseases, address environmental concerns, develop cleaner energy, and have a safer, cleaner, and more efficient industrial manufacturing process.

The basics of biotechnology refer to the application of technology to use living systems and organisms in developing new solutions for human needs. This process uses different techniques that have created a breakthrough in genetic studies and research.

The study of biotechnology involves analysis of the design and functions of living organisms and provides scientists with the tools and techniques to implement in various commercial, industrial, and healthcare applications for human benefits.

Origin of Modern Biotechnology

Modern biotechnology refers to newer applications in the field of biotechnology involving genetic engineering and cell fusion. It is a common question: when did modern biotechnology begin?

Modern biotechnology is said to begin with the invention of genetic engineering. It is believed that modern biotechnology started in 1973 when scientists first genetically engineered Escherichia coli bacteria with a new gene structure that made them resistant to an antibiotic.

The scientists introduced a process of using proteins called restriction enzymes to produce cuts in the DNA chain and insert foreign DNA. The bacteria would then start using the new genetic information and act in the desired way.

Concept of Biotechnology

The term Biotechnology was first introduced in 1919 by Karloly Ereky which indicates the process of making products from raw materials with the aid of living organisms. Biotechnology is the use of biology to develop new products, methods, and organisms which can be used systematically to improve human health and society.

The concept of biotechnology started long ago with the domestication of plants, and animals and the discovery of fermentation. The field has shown significant development over the last century evolving various methods to manipulate the genetic configuration of living organisms.

The modern concept of biotechnology draws inputs from various disciplines of science and technology such as chemistry, macular biology, genetics, etc. Modern biotechnology begins with the study of genes in plants and animals that determine the characteristics of the organism. Scientists have developed techniques to modify genes to strengthen the characteristics of an organism or create an entirely new one. These modified organisms can be beneficial to humans by producing high-yield crops or increasing resistance to diseases and extreme climatic conditions.

THE EVOLUTION OF BIOTECHNOLOGY

Biotechnology began at least 6,000 years ago with the agricultural revolution. This early era was characterized by exploiting living organisms in their natural forms or modifying their genetic makeup through selective breeding.

Around the same time, humans learned to harness the biological process of fermentation to produce bread, alcohol and cheese. People also began changing the genetic makeup of domesticated plants and animals through selective breeding.

Selective breeding works by breeding parents with desirable characteristics to express or eliminate certain genetic characteristics in their offspring. Over time, species that are selectively bred evolve to be different from their wild ancestors. For instance, during the agricultural revolution, wheat was selectively bred to stay on its stem when harvested instead of falling to the ground like wild wheat. Dogs were selectively bred to be more docile than their wolf ancestors.

However, biotech methods such as selective breeding can take a long time to show changes in species. Biotechnology remained limited to these slow, agricultural methods until the 19th century when biologist Gregor Mendel discovered the basic principles of heredity and genetics.

Also, during that era, scientists Louis Pasteur and Joseph Lister discovered the microbial processes of fermentation. This laid the foundation for biotechnology industries where scientists interact more directly with the molecular and genetic processes of organisms.

Based on the work of these scientists, genetic engineering was developed in 1973. This method is the foundation of modern biotechnology practices and recent advances. It enabled the first direct manipulation of plant and animal genomes, which is the complete set of genes present in a cell.

Over the last 100 hundred years or so, biotechnology emerged with the following discoveries and advancements:

- 1919. Hungarian scientist Karl Ereky coins the term *biotechnology*.
- 1928. Alexander Fleming discovers penicillin, the first true antibiotic.
- 1943. Oswald Avery proves DNA carries genetic information.
- 1953. James Watson and Francis Crick discover the double helix structure of DNA.
- 1960s. Insulin is synthesized to fight diabetes, and vaccines for measles, mumps and rubella are developed.

- 1969. The first synthesis of an enzyme in vitro, or outside the body, is conducted.

1973. Herbert Boyer and Stanley Cohen develop genetic engineering with the first insertion of DNA from one bacteria into another.

1980s. The first biotech drugs to treat cancer are developed.

1890. The United States Supreme Court rules that a "live human-made microorganism is patentable subject matter," meaning GMOs can be intellectual property.

1982. A biotech-developed form of insulin becomes the first genetically engineered product approved by the U.S. Food and Drug Administration (FDA).

1983. The first genetically modified plant is introduced.

1993. GMOs are introduced into agriculture with the FDA approval of growth hormones that produce more milk in cows.

1997. The first mammal is cloned.

1998. The first draft of the Human Genome Project is created, giving scientists access to over 30,000 human genes and facilitating research on treatment of diseases such as cancer and Alzheimer's.

2010. The first synthetic cell is created.

2013. The first bionic eye is created.

2020. MRNA vaccine and monoclonal antibody technology is used to treat the SARS-CoV-2 virus.

Science and traditional knowledge have always gone hand-in-hand in order to produce beneficial results. Greater efforts are being made to document and utilise the Indian traditional knowledge of medicine and biotechnology. People of ancient India had immense knowledge of their environment and properties of plants and animals. The practice of making fermented food such as dahi, idli, kinema and beverages using local biological resources was common in medieval India. The relevance of traditional Indian knowledge in making curd (dahi) has been indicated in few of the patents found in the United States patent database.

Fermentation can be explained as a microbial process in which enzymatically-controlled conversion of organic compounds occurs. Fermentation was practiced for years without any actual knowledge of the processes involved. Fermented dough was discovered by accident when dough was not baked immediately and consequently it underwent fermentation by yeast such as Saccharomyces winlocki. Egypt and Mesopotamia exported bread to Greece and Rome. In efforts to improve the technique, Baker's Yeast was discovered by the Romans, which revolutionised the bread-making technology prevalent then. The Chinese were also using fermentation technology by 4000 B.C., for production of their traditional food items, such as soy sauces and fermented vegetables. Vinegar production was known to the Egyptians by 2000 B.C., by preserving crushed dates for a longer time. The art of preserving animal foods by drying, smoking, and pickling in the brine were popular in pre-historic East and Europe

Beer making may have begun as early as between 6000 and 5000 B.C. using cereal grains such as sorghum, corn, rice, millet, and wheat. Brewing was considered as an art until the fourteenth century A.D. However, early brewers had no practical knowledge about the microbial basis of fermentation. Wine was probably made by accident, when grape juice was contaminated with yeast and other microbes. Between 1850s and 1860s, Louis Pasteur established that yeast and other microbes were responsible for fermentation.

Nineteenth century witnessed an increase in the production scale of fermentation based products such as glycerol, acetone, butanol, lactic acid, citric acid, etc. Industrial fermentation was established during World War I because of Germany's requirement for large amounts of glycerol for explosives. By 1940's, significant improvement was made to techniques involving sterility maintenance, aeration methods, product isolation and purification. World War II was the catalyst which led to the invention of the modern fermenter (vessels used for fermentation), also called bioreactor for mass production of antibiotic penicillin. Today, many chemicals such as antibiotics, amino acids, hormones, pigments and even enzymes are produced with extreme precision in controlled environments of industrial bioreactors.

The foundation of modern biotechnology were laid down with the advancements in science and technology during the eighteenth and nineteenth centuries. Thus with the advent of the first compound microscope, made by Dutch spectacle-maker Zacharias Janssen in 1590, which could magnify about 3×-9×, enabled humans to 'see' things that were not perceived by naked eye.

In 1665, Robert Hooke, a physicist, examined thinly sliced cork and drew rectangular components, which he called *cellulae* (Latin for 'small chambers'). In 1676, Antonie van Leeuwenhoek, a Dutch shopkeeper, saw living organisms in pond water and called them 'animalcules'. During the eighteenth century, the cell theory was developed by German biologists, Matthias Schleiden and Theodor Schwann, who determined that all plant and animal tissues were composed of cells. In 1858, Rudolf Virchow, a German pathologist, concluded that 'all cells arise from pre-existing cells' and that cell is the basic unit of life.

Between 1850 and 1880, Pasteur developed the process of pasteurisation. By 1860, he also concluded that spontaneous generation of organisms did not occur, proving that 'all cells arise from pre-existing cells'. In 1896 Eduard Buchner converted sugar to ethyl alcohol using yeast extracts, showing that biochemical transformations can occur without the use of cells. By 1920s and 1930s, the biochemical reactions of many important metabolic pathways were established.

Genetics and Principles of Heredity was developed by an Austrian monk named Gregor Mendel, beginning in 1857, when he cross-pollinated pea plants to examine traits such as petal color, seed color, and seed texture. In 1869, Johann Friedrich Miescher, a Swiss biochemist, isolated a substance that he called nuclein from the nuclei of white blood cells. The substance contained nucleic acids. In 1882, German cytologist Walter Flemming described thread-like bodies that were visible during cell division, as well as the equal distribution of this material to daughter cells. These thread like bodies were actually chromosomes dividing between the two daughter cells during mitosis.

Many path breaking experiments were conducted during the twentieth century which established the nature of the gene and the chromosome, most important being identification of DNA as the genetic material by the classical Alfred Hershey and Martha Chase experiment in 1952. James Watson and Francis Crick proposed the double helical structure of DNA in 1953. Many experiments followed that determined how the information in the gene is used, such as the manipulation of enzymes involved in DNA replication, and DNA repair.

Modern biotechnology is based on the rDNA technology that has revolutionised biotechnology by allowing scientists to cut and join different pieces of DNA, and place the new recombinant (Chimeric/ hybrid) DNA into a new host. It allows the transfer of gene(s) from one organism to another conferring a novel property. This has revolutionised the age old process of biotechnology with regards to its precision and efficiency with limitless possibilities. Since the advent of rDNA technology, biotechnology has become more advanced and led to advancements in medicine, agriculture, animal science and environmental science.

APPLICATIONS IN MEDICINE, AGRICULTURE, AND INDUSTRY

Modern biotechnology, which is based on rDNA technology, exhibits a wide range of applications. The broad application areas of biotechnology include pharmaceutical and therapeutic research, disease diagnostics, crop improvement, vegetable oil, biofuels, and development of enviromental friendly products (for example biodegradable plastics). Thus, the applications of modern biotechnology mainly focus on the following major areas:

- Medicine and health care
- Crop production and agriculture

- Food processing
- Environmental protection

Medicine and health care

Biotechnology techniques are used in the field of medicine for diagnosis via development of diagnostic tools and kits, which have proved helpful in detecting certain molecules and cellular components which are expressed in diseased conditions. Using rDNA technology, tools of bioinformatics, modern instrumentation and bioprocess technologies, synthetic drug analogs can be predicted and possibly synthesised, which may show improved disease treatment. Production of vaccines and gene therapy are also important applications of biotechnology in the field of medicine.

Biotechnology has a variety of applications in the field of medicine. Some of the biotechnology applications in medicine include the following:

Recombinant Insulin

Insulin is required by diabetic patients to remove excess sugar from the blood. Diabetic patients have a very low level of insulin or no insulin produced by the body. Therefore, they need external insulin to control blood glucose levels.

Later it was discovered that the insulin produced by the pancreas of the pigs can be used by humans. But there were not enough pigs to provide the quantities of insulin required. This led to the cloning of the human insulin gene.

The specific gene sequence that codes for human insulin were introduced in *E.coli bacteria*. The gene sequence altered the genetic composition of the *E.coli* cells. Within 24 hours several *E.coli* bacteria containing the recombinant human insulin gene were produced. The recombinant human insulin was isolated from *E.coli* cells.

Gene Therapy

Gene Therapy holds the most promising answer to the problem of genetic diseases. Gene therapy is used to treat genetic disorders usually by the insertion of a normal gene or correct gene for the defective or inactive gene into an individual with the help of vectors such as retrovirus, adenovirus, and herpes simplex virus.

The normal gene replaces the defective or inactive gene and carries out its functions. The therapy has the highest chances of developing a permanent cure if introduced in the earliest stages of life.

Gene therapy is a genetic engineering technique that may one day be used to cure certain genetic diseases. In its simplest form, it involves the introduction of a non-mutated gene at a random location in the genome to cure a disease by replacing a protein that may be absent in these individuals because of a genetic mutation. The non-mutated gene is usually introduced into diseased cells as part of a vector transmitted by a virus, such as an adenovirus, that can infect the host cell and deliver the foreign DNA into the genome of the targeted cell (Fig. 1). To date, gene therapies have been primarily experimental procedures in humans. A few of these experimental treatments have been successful, but the methods may be important in the future as the factors limiting its success are resolved.

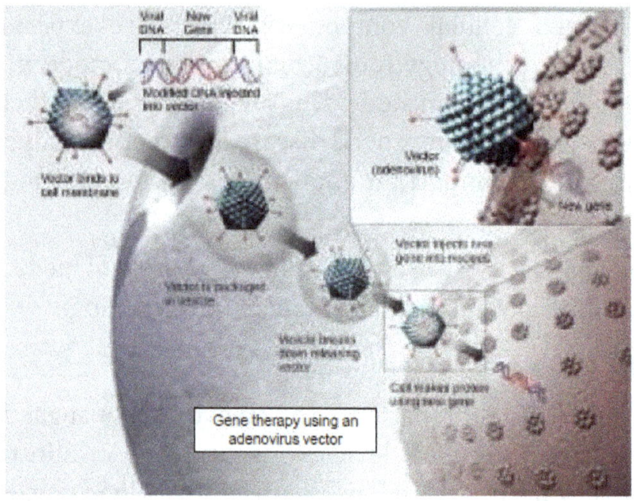

Fig. 1. The steps involved in curing disease with gene therapy using an adenovirus vector.

Production of Vaccines, Antibiotics, and Hormones

Traditional vaccination strategies use weakened or inactive forms of microorganisms or viruses to stimulate the immune system. Modern techniques use specific genes of microorganisms cloned into vectors and mass-produced in bacteria to make large quantities of specific substances to stimulate the immune system. The substance is then used as a vaccine. In some cases, such as the H1N1 flu vaccine, genes cloned from the virus have been used to combat the constantly changing strains of this virus.

Antibiotics kill bacteria and are naturally produced by microorganisms such as fungi; penicillin is perhaps the most well-known example. Antibiotics are produced on a large scale by cultivating and manipulating fungal cells. The fungal cells have typically been genetically modified to improve the yields of the antibiotic compound.

Recombinant DNA technology was used to produce large-scale quantities of the human hormone insulin in *E. coli* as early as 1978. Previously, it was only possible to treat diabetes with pig insulin, which caused allergic reactions in many humans because of differences in the insulin molecule. In addition, human growth hormone (HGH) is used to treat growth disorders in children. The HGH gene was cloned from a cDNA (complementary DNA) library and inserted into *E. coli* cells by cloning it into a bacterial vector.

Transgenic Animals

Although several recombinant proteins used in medicine are successfully produced in bacteria, some proteins need a eukaryotic animal host for proper processing. For this reason, genes have been cloned and expressed in animals such as sheep, goats, chickens, and mice. Animals that have been modified to express recombinant DNA are called transgenic animals.

Several human proteins are expressed in the milk of transgenic sheep and goats. In one commercial example, the FDA has approved a blood anticoagulant protein that is produced in the milk of transgenic goats for use in humans. Mice have been used extensively for expressing and studying the effects of recombinant genes and mutations.

Molecular Diagnosis

Medical diagnosis is another application of biotechnology in the health sector. Many times the pathogen concentration increases by the time the disease is diagnosed. Hence, early diagnosis and knowledge of pathophysiology are essential for an effective cure.

This can be achieved with the help of techniques such as Recombinant DNA Technology, Polymerase Chain Reaction (PCR) and Enzyme-Linked Immunosorbent Assay (ELISA), etc.

Pharmacogenomics

Pharmacogenomics has led to the production of drugs that are best suited to an individual's genetic makeup. It can be applied in diseases such as cancer, depression, HIV, asthma, etc.

Edible Vaccines

Vaccines are obtained by animals and cell cultures. These vaccines contain inactivated pathogens.

The transgenic plants can produce antigens that can be used as edible vaccines. Antigenic proteins from several pathogens can be expressed in plants such as tomato and banana. Transgenic sugarbeet can treat foot and mouth disease of animals, transgenic banana and tomato can cure diseases such as cholera and hepatitis B.

Biotechnology in Agriculture

Biotechnology is best defined as the use of scientific techniques to improve and alter plants, animals, and microorganisms' value. Over the years, biotechnology has found its way to several fields including medicine, agriculture, genetic engineering, etc.

One can define agricultural biotechnology as a set of scientific techniques which can improve plants, micro-organisms, and animals based on DNA and its concepts.

Arguably the use of biotechnology in agriculture is deemed to be more effective than that of agrochemicals. The latter is believed to be responsible for causing environmental distress and is also somewhat unfeasible for farmers.

The following highlight the few ways in which biotechnology has found its way in agriculture:

- **Genetic engineering / rDNA technology:** It is a technology in which one or more genes are modified deliberately in the lab. This is achieved by the process of using recombinant DNA (rDNA) technology, thereby altering the genetic makeup of an organism.
- **Tissue culture:** Tissue culture involves nurturing fragments of plant or animal tissue in a controlled environment where they survive and continue to grow. For this tissue has to be isolated first.
- **Embryo rescue:** It is a form of in-vitro culture technique for plants. Here an immature embryo is nurtured in a controlled environment to ensure its survival. This can help in the preservation of species of seeds that are nearing extinction. This can include heritage seeds, local grains of cultural significance, etc.
- **Somatic hybridisation:** It is a process through which the cellular genome is manipulated through the process of protoplast fusion.
- **Molecular-gene markers:** In genetic engineering, Molecular-gene markers are specific segments of DNA that are associated with a particular location within the genome.
- **Molecular diagnostics:** Molecular diagnostics is a set of techniques used to analyze biological markers in the genome and proteome. It helps in determining how their cells express their genes as proteins
- **Micropropagation:** It is a clonal propagation of plants in a closed vessel under aseptic and controlled conditions.

Role of Biotechnology in Agriculture

The role of biotechnology in agriculture is multifaceted. Some of the most prevalent benefits of biotechnology in agriculture include –

- **Increase in Crop Production:** With better disease control and increased tolerance to drought and flooding, biotechnology leads to a significant increase in crop production. This does not just match the ever-growing demand for food but also helps farmers to lower losses.

- **Better Crop Protection:** The techniques of biotechnology serve as cost-effective solutions to problems about pests. Farmers have been able to transform crops like cotton, corn, and potato to synthesize a protein that tackles issues of pests effectively.
- **Increase in Nutrition Value:** It has also enabled farmers to produce crops with a higher nutritional value and enhanced flavour and texture. For instance, the technology has made it possible to cultivate soybeans with high protein content, beans with more amino acids, and potatoes with starch.
- **Fresher Produce and Better Taste:** It further helps to improve the taste and flavor of crops by enhancing the activity of enzymes present in plants. Also, it helps in keeping the yield fresh for longer.
- **Chemical Tolerance:** Most farmers rely on herbicides to control the growth of weeds which often leads to soil erosion. However, genetically engineered food is resistant to a variety of chemicals, including herbicides; as a result, the scale of soil erosion is significantly low.
- **Disease Resistance:** Viral infections spread by insects are often difficult to contain, and also the use of insecticides tends to pose a threat to both soil and the quality of produce. Nonetheless, genetically modified plants are less susceptible to viral infection and make it easier for farmers to contain crop damage.

Though there is an array of benefits in applying biotechnology in agriculture, it is not entirely devoid of flaws. To elaborate, there are some fears about health, societal and environmental issues.

For example, resistance to antibiotics, resistance to insecticide, growth of superweed, and loss of biodiversity are among the many fears that relate to the application of biotechnology in agriculture. However, one can hope that with the advancement of technology, scientists will find practical solutions to tackle the fears and associated risks effectively.

Transgenic Plants

Manipulating the DNA of plants (creating genetically modified organisms, or GMOs) has helped to create desirable traits such as disease resistance, herbicide, and pest resistance, better nutritional value, and better shelf life (Fig. 2). Plants are the most important source of food for the human population. Farmers developed ways to select for plant varieties with desirable traits long before modern-day biotechnology practices were established.

Fig. 2. Corn, a major agricultural crop used to create products for a variety of industries, is often modified through plant biotechnology.

- **Genetically modified biopesticide:** Bt toxin is produced by a bacterium called *Bacillus thuringiensis* (Bt for short).
    - Bt toxin gene has been cloned from the bacteria and been expressed in plants to provide resistance to insects without the need for insecticides
    - In effect created a bio-pesticide. Examples are Bt cotton, Bt corn, rice, tomato, potato and soyabean etc. Bt Cotton.
    - Some strains of Bacillus thuringiensis produce proteins that kill certain insects such as lepidopterans (tobacco budworm, armyworm), coleopterans (beetles) and dipterans (flies, mosquitoes).
    - B. thuringiensis forms protein crystals during a particular phase of their growth. These crystals contain a toxic insecticidal protein.
- **Pest resistant plants:** Several nematodes parasitise a wide variety of plants and animals including human beings.

- A nematode Meloidogyne incognita infects the roots of tobacco plants and causes a great reduction in yield.
- A novel strategy was adopted to prevent this infestation which was based on the process of RNA interference (RNAi).
- RNAi takes place in all eukaryotic organisms as a method of cellular defense.
- This method involves silencing of a specific mRNA due to a complementary dsRNA molecule that binds to and prevents translation of the mRNA (silencing).

Application of Biotechnology in Industry

Industrial processes can be made more efficient and environment-friendly by using biotechnology.

- Using biotechnology enzymes can be produced in large quantities. This can be used as a biocatalyst for the synthesis of many chemicals
- Genetically modified plants supply different resources like starch, fuel, pharmaceuticals in huge quantity
- Proteases are used in detergents
- Biofertilizers production
- The processed food industry is based on biotechnology and the use of micro-organisms for the fermentation process
- Large scale production of fructose from glucose derived from inexpensive sources
- Corn can be used in place of petroleum to synthesize chemicals
- Biotechnology is useful in getting a better quality of products, e.g. cotton produced through biotechnology is warmer, stronger and wrinkle resistant
- Production of biopesticide. Biopesticides act selectively by killing pathogens such as bacteria, viruses and insects. These are not harmful like chemical pesticides
- Production of aquatic species by using biotechnological tools has greatly enhanced the quality and growth of aquatic organisms

In the energy sector, biotechnology has played an important role in the production of sustainable energy. The use of microorganisms, such as bacteria and algae, in the production of biofuels, such as second-generation ethanol, has the potential to reduce dependence on fossil fuels and mitigate greenhouse gas emissions. In addition, biotechnology is also being used in the development of third-generation solar cells, which seek to improve the efficiency and sustainability of solar energy.

Biotechnology also plays a fundamental role in the search for more sustainable industrial practices and in the preservation of the environment. The use of microorganisms in the biodegradation of pollutants and in the remediation of contaminated soils, the production of biodegradable plastics from renewable sources and the development of biotechnological processes for the treatment of industrial effluents are examples of how biotechnology contributes to environmental sustainability.

The following industries can be driven by biotechnology.

Food and Drink

The first application of biotechnology was in the food and drink industry. Even though the making of bread, cheese, yoghurt, vinegar and alcoholic drinks such as beer and wine dates back to thousands of years, it was Louis Pasteur in the nineteenth century who first showed that fermentation was due to the activity of microorganisms. This necessitated the need for large scale industrial production of yeast and other microorganisms. Biotechnologists have designed and built bioreactors or fermenters which are chambers in which microorganisms are cultured in a liquid or solid medium. Fermentation takes place in the bioreactors and the desired products are either the cells e.g. yeast (biomass) or some enzymes, drugs, alcohol etc. The fermentation of milk produces several diary products which include cheese, butter, yoghurt, sour cream etc.

A new source of food is the Single Cell Protein (SCP). It is protein obtained from large scale growth of microorganisms such as bacteria, yeasts and other fungi and algae. SCP is also a useful source of minerals, vitamins, fats and carbohydrates for human and animal consumption. The advantage of using microorganisms as source of food includes their ability to grow rapidly and occupy less space than conventional crops and animals. They can survive and grow on a wide range of wastes land not suitable for agriculture and are independent of climatic conditions. They are more easily modified genetically and their protein content are relatively high. Examples of SCP products are pruteen and mycoprotein. Yeast cells are also rich in vitamins and can be made into vitamin-rich tablets.

Biofuels

The recent increase in the price of petroleum products have pushed many nations to embark on a desperate search for alternative sources of fuel. The search has focused on using living organisms and biological processes as sources of fuel especially ethanol and biogas.

- **Ethanol** – Since 1975, Brazil has successfully produced and used ethanol as fuel using sugarcane as the raw material. Some Brazilian cars are adapted to run on pure alcohol. Over 11billion litres of ethanol was produced in Brazil in 1985. Other crops used in the production of Bio-fuels include jathropha, grains such as maize, cassava and sorghum. On February 24, 2008 the first ethanol powered airplane belonging to Virgin Atlantic flew from London to Amsterdam on a 20percent bio-fuels. Ethanol is used to power the agricultural spraying aircraft in Brazil. A flight using bio-fuel produced from Jathropha was scheduled for the 4th quarter of 2008 by Air New Zealand.

Vegetable oil, palm oil, castor oil and soybean oil are being used for Biodiesel..

- **Biogas** – Researches have also been carried out on the production of biogas from the fermentation of waste materials by bacteria and other microbes. Biogas is about 50 – 70% methane with traces of nitrogen, hydrogen, and other gasses. Natural gas is about 80% methane. The waste materials usually used include, animal manure, sewage sludge, domestic waste, food waste, paper wastes and plant wastes. Water hyacinth, a plant that causes blockages of waterways have been fermented to produce biogas in the US, India and China.

Microbiological Mining

Microorganisms are becoming important tools in the extraction of metals from their ores. This process is called biotechnology. Bacterial leaching is used throughout the world as an additional technique of extracting metals from ores (mainly copper and uranium). More than 10% of the copper extracted in the USA in 1983 worth over 300 million dollars was through bioleaching. The bacteria convert insoluble metal to soluble metal compounds from which the metal can easily be extracted.

Bacterial leaching is less expensive compared to conventional mining methods which can only be used for ores that are rich in the metal and commercially viable. But low grade ores can be exploited commercially using bacterial leaching.

Another advantage of bioleaching is that the cost, danger and environmental damages associated with conventional deep mining is avoided. The leaching solution is pumped into the ground to dissolve the metal. The solution containing the metal is the pumped out and the metal easily extracted. The damage to land and the wastes created by digging and excavation is prevented.

Medicine

The traditional medical products of biotechnology were antibiotics. Advances in biotechnology especially in genetic engineering have led to the production of new drugs. One typical example is human insulin, the hormone responsible for regulation of blood sugar. Inability to produce this hormone causes the disease called diabetes which is controlled by taking insulin injection. Traditionally, insulin was extracted from the pancreas of cows and pigs, but this differs from human insulin and some patients have allergic reactions to the insulin from cow and pig.

Biotechnologists solved this problem by genetically engineering bacteria to produce the human insulin. This was done by inserting the human gene that produces human insulin into the genome of bacteria. The human gene directs the bacteria to produce human insulin. Human insulin produced by the bacterium Eschericha coli was first marketed in Canada in 1983.

Another example of the application of genetic engineering is in the treatment of dwarfism caused by very low level of human growth hormone in children. Because the growth hormone of different animals works only in the species of origin, the treatment of dwarfism relied on growth hormone extracted from the pituitary glands of dead humans.

This problem was however solved when the human gene for the hormone was inserted into a bacterium which acquired the ability to produce the hormone in large quantity. The hormone Bovine Sometotrophin (BST) which is similar to the human growth hormone (HGT) has been produced using the same method.

Waste and Oil Spillage Management

Micro-organisms have come to the rescue of man in clearing up the wastes generated by his activities. Waste materials such as sewage, domestic wastes, food wastes, paper wastes, spoilt crops etc are becoming raw material for bacteria and yeast for the production of biogas and ethanol respectively as biofuels.

The ecological damage caused to the environment especially aquatic life by oil spillage is a major issue and area of conflict between oil bearing communities and exploration companies. Present methods of cleaning oil spillages through physical barriers are yielding little result. Biotechnologists are attempting to produce genetically engineered bacteria capable of cleaning up oil spills. When sprayed on surfaces polluted by oil, the bacteria is capable of breaking down the four main groups of hydrocarbon present in the oil into harmless soluble compounds.

Biological Detergent

Enzymes which are biological catalysts found in living cells are being used as washing powders. They have the ability of removing stains of biological origin. Examples of such enzymes are proteases which remove stains of protein origin, amylases remove starch stain and lipases digest fat, oil and grease stains. The first successful biological washing powders were introduced in the 1960s.

ETHICAL CONSIDERATIONS IN BIOTECHNOLOGICAL ADVANCEMENTS

Ethical Issues of Biotechnology represent actions that go against moral principles, potentially leading to significant conflicts within society. These issues can be categorized into specific subgroups, including socio-economic, cultural, legal, environmental, and religious aspects. Principles associated with ethics, medical ethics, scientific and technological ethics, justice, human dignity, and related ethical considerations contribute to the multifaceted landscape of ethical issues in biotechnology. These concerns highlight the importance of addressing moral implications and societal impacts in the development and application of biotechnological advancements.

Genetic Modification and Engineering

- **Concerns:** Manipulating the genetic code of organisms, including humans, raises ethical questions about playing "nature's role" and potential unforeseen consequences.
- **Ethical Considerations:** Balancing the benefits of genetic advancements with potential risks, ensuring transparency, and addressing long-term effects on ecosystems and biodiversity.

Cloning

- **Concerns:** Cloning, especially human cloning, sparks debates about the creation of identical individuals and the potential for exploitation or misuse.
- **Ethical Considerations:** Establishing guidelines to prevent unethical cloning practices, addressing issues related to identity and autonomy, and ensuring the well-being of cloned individuals.

Privacy and Genetic Information

- **Concerns:** The increasing availability of genetic information raises concerns about privacy, discrimination, and the potential misuse of personal genetic data.
- **Ethical Considerations:** Establishing robust privacy safeguards, addressing issues of consent and data ownership, and preventing genetic discrimination in employment, insurance, and other areas.

Biopiracy and Intellectual Property

Concerns: Exploiting genetic resources from indigenous communities without fair compensation or consent raises ethical concerns about biopiracy.

- **Ethical Considerations:** Ensuring equitable benefit-sharing arrangements, respecting indigenous knowledge, and establishing guidelines for responsible sourcing and use of genetic resources.
- ***Access to Biotechnological Benefits***
- **Concerns:** Disparities in access to biotechnological advancements may widen existing social and economic inequalities.
- **Ethical Considerations:** Promoting equitable distribution of benefits, addressing affordability, and ensuring that technological advancements contribute to global well-being.

Environmental Impact

- **Concerns:** Introducing genetically modified organisms into the environment may have unintended ecological consequences.
- **Ethical Considerations:** Conducting thorough risk assessments, implementing stringent regulations, and prioritizing environmental sustainability in biotechnological practices.

Dual-Use Technology

- **Concerns:** Biotechnological advancements with dual-use potential for both beneficial and harmful purposes raise ethical dilemmas.
- **Ethical Considerations:** Implementing strict controls and regulations, fostering responsible research practices, and promoting international collaboration to address potential misuse.

Informed Consent in Research

- **Concerns:** Conducting research involving human subjects or genetic information without informed consent poses ethical challenges.
- **Ethical Considerations:** Upholding principles of autonomy and informed consent, ensuring transparency, and protecting the rights and well-being of research participants.

Ethical Issues of Biotechnology

The ethical landscape of biotechnology is complex and multifaceted, encompassing socio-economic, cultural, legal, environmental, and religious dimensions. The manipulation of genetic codes, cloning, gene therapy, and issues related to privacy and intellectual property present significant ethical concerns. The analysis of these concerns underscores the delicate balance required between scientific advancements and ethical considerations.

The potential consequences of genetic modification, such as unforeseen ecological impacts and long-term effects on biodiversity, necessitate rigorous risk assessments and stringent regulations. The ethical dimensions of biotechnology also extend to issues of equity, with disparities in access to biotechnological benefits raising concerns about widening social and economic inequalities. Striking a balance between promoting innovation and ensuring responsible research practices is crucial in addressing dual-use technology concerns.

For research involving human participants and clinical trials, a separate framework by the Central Drugs Standard Control Organization (CDSCO) under the MoHFW is applicable. The Drugs and Cosmetics Act, 1940 is the governing legislation under which Indian Good Clinical Practice Guidelines for Clinical Trials, 2001 and New Drugs and Clinical Trials Rules, 2019 have been notified. CDSCO has mandated that clinical trials should be registered online in Clinical Trials Registry–India (CTR-I), a national public record system for registration of clinical trials. The guidelines also mandate that each research institution establish an Institutional Ethics Committee (IEC). More than 1200 Ethics Committees have already been established by various institutions and registered with CDSCO.

The National Apex Committee for Stem Cell Research and Therapy (NAC-SCRT) is constituted by MoHFW to oversee the activities in the field of stem cell research in India. The committee examines the scientific, technical, ethical, legal and social issues involving stem cell research and therapy. All institutions carrying out research on human stem cells are mandated to constitute an Institutional Committee for Stem Cell Research (IC-SCR) and register it with the NAC-SCRT.

The Indian Council of Medical Research (ICMR) along with Department of Biotechnology (DBT) has also published National Guidelines for Stem Cell Research, 2017. The guidelines restrict genome modification including gene editing by CRISPR-Cas9 technology of stem cells, germ-line stem cells or gamete and human embryos to in-vitro studies only. The guidelines also prohibit culturing of genome modified human embryos beyond 14 days of fertilization. Furthermore, ICMR has also published National Ethical Guidelines for Biomedical and Health Research Involving Human Participants, 2017. These are applicable to all biomedical, social, and behavioural science research for health conducted in India involving human participants, their biological material, and data.

At the global level, India is signatory to the Cartagena Protocol on Biosafety to the Convention on Biological Diversity and India's domestic export control regulations are fully aligned with guidelines of the Australia Group (AG). However, there is need for better coordination and collaboration between countries at regional and global level regarding these guidelines.

Challenges with Current Regulatory Mechanisms

The transnational nature of biotechnology research makes it increasingly difficult for both scientists and companies to adhere to different regulatory norms in different countries. Countries such as Germany, the UK and the US each have their well-developed biotechnology sector with the capability to manufacture cutting-edge laboratory equipment such as gene sequencers, gene synthesisers, advanced experimental kits, and reagents.

Meanwhile, countries like India, Israel and many in Asia, Africa and South America have to rely exclusively on imports of such equipment. Collaborations between scientists of different countries have also increased due to the decentralised nature of biomedical research. Given the bioethical and national security implications of many emerging technologies, it is imperative that a multilateral regulatory system is developed.

Traditionally, export control regimes have regulated technologies with implications on global and national security. Export control regimes are multilateral institutions that seek to prevent proliferation of technologies that may be used to develop nuclear, chemical, or biological weapons.

These institutions have played an important role in regulating cross-border flow of potentially dangerous dual-use technologies. However, there are significant challenges involved in regulating emerging technologies in biotechnology through export controls. First, it is difficult to regulate the global flow of information relating to these technologies due to the open and collaborative nature of scientific research. Many of these techniques use commonly available reagents and laboratory equipment that would be practically impossible to regulate through export controls.

Furthermore, tight export control regulation would only contribute to stifling research and development (R&D) activities as these technologies are fast becoming the cornerstone of biomedical research across various fields. For example, the CRISPR technology is essentially a variant of natural self-defence mechanism in bacteria that has been tweaked for use as a gene editing tool. The biological principle underpinning the technology is public knowledge and a trained biologist would face little difficulty in conducting experiments using CRISPR anywhere in the world. Furthermore, commercially available CRISPR kits have only simplified the experimental process. It makes it harder to formulate a specific criterion to label certain technologies as dual use and recommend their inclusion in the export control lists.

Second, in stark contrast to the period between 1960s-1980s when the first export control regimes were established, cutting-edge R&D activities are no longer the exclusive domain of select few industrialised or advanced economies.

This complicates efforts to regulate emerging technologies using export controls. For example, only a handful of countries or private corporations could afford to invest in nuclear technology—this constraint reduced the risk of widespread proliferation once these countries agreed to cooperate on non-proliferation measures. In biomedical research, however, its trans-national nature presents a unique challenge. The heightened role of non-state actors and rogue individuals in spreading terrorism is another factor complicating regulation through export controls.

Finally, for any export control regulations to be effective, they would need to command support from the key geopolitical players of the day and from countries with significant industrial and scientific base. For example, crucial players in the biotechnology sector such as Russia, China and Israel are not members of the Australia Group (AG). This implies that any efforts at regulation under the aegis of AG would be ineffective. Furthermore, export control regimes are seen by non-members as a means to deny them access to advanced technologies. Therefore, export controls are unlikely to become the framework to regulate currently emerging technologies in the field of biotechnology.

The role of geopolitics

No aspect of contemporary world is immune from the reach of global geopolitics and regulation of emerging technologies is no exception. Historically, developed nations like the US, Russia and the European Union have been the dominant force in institutionalising global regulations. Through their technological and economic prowess, they have ensured that the world adopted their standards of regulation, and often keeping in mind their country's parochial interests. However, as a more multi-polar world takes shape, this equilibrium has started to shift: there is significant lack of trust between governments, especially between biotechnology powerhouses such as the US and China. This trend has been further accelerated by the geopolitical fallout of the Covid-19 pandemic.

Similarly, regional powers such as India and the countries of Association of Southeast Asian Nations (ASEAN) along with other developing countries are also asserting themselves and asking for a seat at the regulatory high-table. This has led to a churn at several international institutions including export control organisations and the UN, especially on matters concerning emerging technologies. One recent example is the success of India in getting its own 5G standard known as '5Gi' approved for final evaluation to become part of global 5G standards set to be approved by International Telecommunications Union (ITU) in 2021.

Given the trans-national nature of biotechnological research, both in terms of cutting-edge research and manufacturing of laboratory equipment, geopolitics would certainly gain prominence during discussions seeking to regulate this field. According to a realist approach to geopolitics, countries with significant edge in the biotechnology industry would prefer regulations that maintain their technological lead while preventing other countries from catching up. To prevent itself from being at the receiving end of this geopolitical dynamic, India should cooperate with its partners and participate in international discussions or working groups as an advocate for its interests along with those of the developing world.

International institutions are under strain, and building a broad consensus on emerging technology regulation would be a massive challenge. There is an urgent need to develop an alternate deliberative approach that is both more inclusive than export controls and less fragmented than individual national regulations.

A Way Forward

The UN remains the most representative forum for discussion involving all relevant stakeholders. Therefore, an Open-ended Working Group (OEWG) created under the resolution of the UN General Assembly (UNGA) can be considered as a potential forum for discussions on this issue. An OEWG set up by UNGA in 2018 is already deliberating another globally contentious issue: the application of international law in cyberspace. It has been able to successfully conduct several rounds of discussions. A similar mechanism can be devised to discuss other globally contentious issues such as the regulatory aspects of emerging technologies in biotechnology.

An OEWG is the least restrictive and most deliberative option to facilitate discussions among civil society members, NGOs, and subject-matter experts in addition to representatives from UN member states. The proposed OEWG can choose to start deliberations based on the background paper released by 'Expert Advisory Committee on Developing Global Standards for Governance and Oversight of Human Genome Editing' set up by the World Health Organization (WHO) in 2019 to provide an overview of governance issues around human genome editing. Similarly, the 2015 report submitted by 'The Independent Advisory Group on Public Health Implications of Synthetic Biology Technology Related to Smallpox' to the Director-General of WHO can act as a basis for broader deliberations on dual-use aspects of synthetic biology.

Discussions at the global level often takes time to materialise into relevant conventions, treaties, or guidelines. It may take several additional years across jurisdictions to develop adequate procedures to regulate emerging technologies. In the meantime, the scientific community and private enterprise should step in to fill in the global regulatory gap. There are several precedents for the same.

In 1975, the Asilomar Conference on Recombinant DNA was convened in California that included lawyers, journalists, government officials, and scientists from different parts of the world. The participants considered the issue of regulation of biotechnology and formulated voluntary guidelines to ensure the safety of recombinant DNA technology which was as revolutionary and novel a technology then, as CRISPR is now. These guidelines served as a template for subsequent guidelines issued by the National Institutes of Health and other regulatory bodies. The initiative by the scientific community to self-regulate in response to a revolutionary yet unpredictable technology can serve as a template for present times.

The global scientific community should consider stringent self-regulation regarding controversial aspects of emerging technologies such as human genome editing using CRISPR until a broader consensus can be developed. There have been some encouraging efforts in this regard: UNESCO's International Bioethics Committee (IBC) in 2015 updated its guidelines on the Human Genome and Human Rights. Workshops and discussions related to bioethics have been organised at International Summit on Human Gene Editing in 2015 and 2018. The He Jiankui incident highlighted the importance and urgency of such initiatives and the need to encourage all members of the global scientific community to adhere to these guidelines. It is significant to note that after international backlash, including at the 2018 International Summit on Human Gene Editing, Chinese regulators proposed legislative changes and issued fresh guidelines to prevent a repeat of such incidents.

The private sector has also tried to self-regulate the use of potential dual-use technology in the absence of multilateral guidelines. For example, several multinational companies providing commercial gene-synthesis services have formed the International Gene Synthesis Consortium (IGSC). IGSC members together represent 80 percent of the world's commercial gene synthesis capacity.

The organisation has published its own Harmonised Screening Protocol (HSP) to screen gene synthesis orders across jurisdictions for sequences that have dual-use potential or are restricted under national or export-control regulations. The protocol also encourages IGSC members to coordinate with and share information with local and national law enforcement and intelligence authorities to prevent the potential misuse of synthetic genes.

This system is based on guidelines issued by the US Department of Health and Human Services (DoHHS) that mandate gene synthesis companies to undertake comprehensive screening of the customer, the sequence ordered to be synthesised and do a follow-up screening to verify the legitimacy of the customer.

Initiatives taken by gene-synthesis companies can become a template for other multinational corporations across countries to formulate a common reporting mechanism for their sector to be used by manufacturers, exporters, and importers of sensitive and potential dual-use technologies to keep a record and verify credentials of customers. They can also work with willing national regulators to further refine this mechanism.

CHAPTER-2

FUNDAMENTALS OF MOLECULAR BIOLOGY

Molecular biology is the study of living things at their molecular level, which controls and makes them up. It is also used to study and understand:

- The molecular pathways within the cells
- How do living things interact with the populations?
- How do proteins and nucleic acids interact with the biomolecules?

The study of molecular biology will establish a strong foundation on the fundamental importance of macromolecular mechanisms such as replication, transcription, translation and other cellular functions. The more commonly used molecular biology techniques include- Polymerase Chain Reaction, Electrophoresis, Restriction Digestion, Blotting, Cloning, etc.

The important topics covered in this subject are nucleic acids – DNA, RNA and protein synthesis in cells. Molecular biology is a branch of biology that is also closely related to other sub-disciplines like biochemistry, cell biology, genetics, and genomics.

For decades, DNA was largely an academic subject and not the source of dinner table conversation in the average household. In 1995 this changed when media coverage of the O.J. Simpson murder trial brought DNA fingerprinting to homes across the world. Two years later, the cloning of Dolly the sheep was headline news. Then, in 2001, scientists announced the rough draft of the human genome sequence. In commenting on this landmark achievement, former US President Clinton likened the “decoding of the book of life” to a medical version of the moon landing. Increasingly, DNA has captivated Hollywood and the general public, excited scientists and science fiction writers alike, inspired artists, and challenged society with emerging ethical issues (Fig. 1).

Fig. 1. DNA in art

Molecular biology is a specialized branch of biology and biochemistry, which is specifically concerned with the study of various biological activities at the molecular level.

Molecular biology includes different biomolecules like amino acids, nucleic acids (DNA and RNA) proteins, carbohydrates, and lipids, along with their compositions, interactions, structure, and functions in the life processes.

The term "Molecular Biology" was coined by an American scientist, Warren Weaver in the year 1938. As per the records, the discovery of molecular biology began in the early 1940s and its fundamental development took place in the year 1953, during the invention of the double-helical structure of the DNA molecule by the two molecular biologists named James Watson and Francis Crick.

DNA STRUCTURE AND REPLICATION

This is also true for viruses, as most of these entities have either RNA or DNA as their genetic material. For instance, some viruses may have RNA as their genetic material, while others have DNA as the genetic material. The Human Immunodeficiency Virus (HIV) contains RNA, which is then converted into DNA after attaching itself to the host cell.

Apart from being responsible for the inheritance of genetic information in all living beings, DNA also plays a crucial role in the production of proteins. Nuclear DNA is the DNA contained within the nucleus of every cell in a eukaryotic organism. It codes for the majority of the organism's genomes while the mitochondrial DNA and plastid DNA handles the rest.

The DNA present in the mitochondria of the cell is termed mitochondrial DNA. It is inherited from the mother to the child. In humans, there are approximately 16,000 base pairs of mitochondrial DNA. Similarly, plastids have their own DNA, and they play an essential role in photosynthesis.

DNA Types

There are three different DNA types:

- **A-DNA:** It is a right-handed double helix similar to the B-DNA form. Dehydrated DNA takes an A form that protects the DNA during extreme conditions such as desiccation. Protein binding also removes the solvent from DNA, and the DNA takes an A form.
- **B-DNA:** This is the most common DNA conformation and is a right-handed helix. The majority of DNA has a B type conformation under normal physiological conditions.
- **Z-DNA:** Z-DNA is a left-handed DNA where the double helix winds to the left in a zig-zag pattern. It was discovered by Andres Wang and Alexander Rich. It is found ahead of the start site of a gene and hence, is believed to play some role in gene regulation.

DNA Structure

The DNA structure can be thought of as a twisted ladder. This structure is described as a double-helix, as illustrated in the figure above. It is a nucleic acid, and all nucleic acids are made up of nucleotides. The DNA molecule is composed of units called nucleotides, and each nucleotide is composed of three different components such as sugar, phosphate groups and nitrogen bases.

The basic building blocks of DNA are nucleotides, which are composed of a sugar group, a phosphate group, and a nitrogen base. The sugar and phosphate groups link the nucleotides together to form each strand of DNA. Adenine (A), Thymine (T), Guanine (G) and Cytosine (C) are four types of nitrogen bases.

These 4 Nitrogenous bases pair together in the following way: A with T, and C with G. These base pairs are essential for the DNA's double helix structure, which resembles a twisted ladder.

The order of the nitrogenous bases determines the genetic code or the DNA's instructions.

Among the three components of DNA structure, sugar is the one which forms the backbone of the DNA molecule. It is also called deoxyribose. The nitrogenous bases of the opposite strands form hydrogen bonds, forming a ladder-like structure.

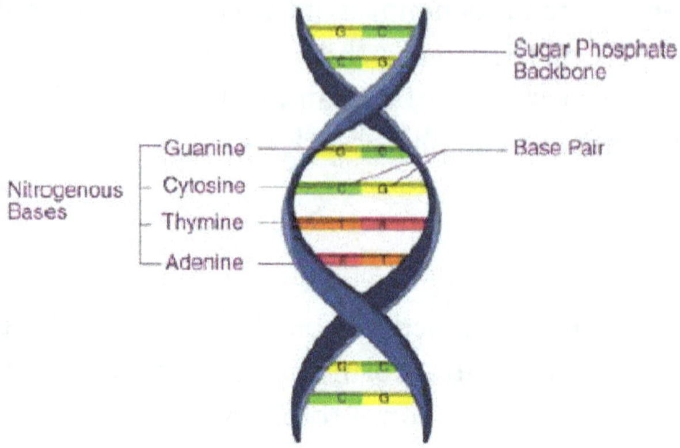

Fig. 2. DNA Structure

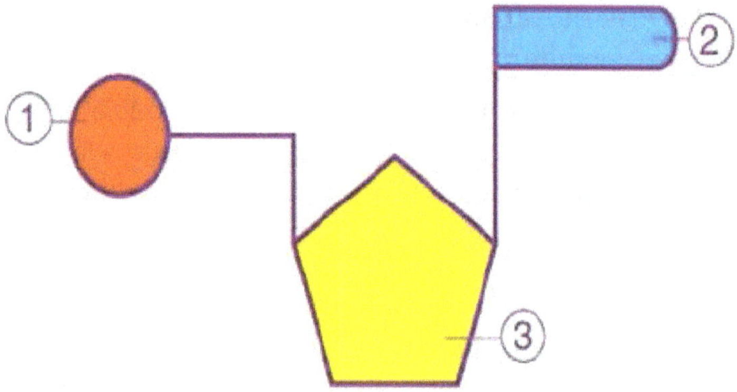

Fig. 3. Components of DNA Structure:
1 Phosphate 2 Nitrogen Base and 3 Deoxyribose sugar

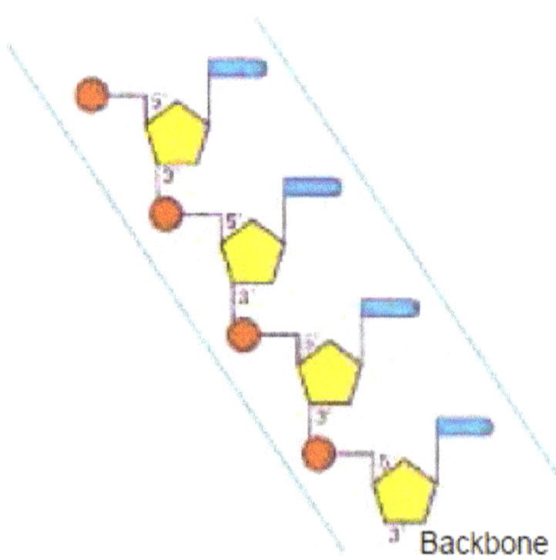

Fig. 4 DNA Structure Backbone

The DNA molecule consists of 4 nitrogen bases, namely adenine (A), thymine (T), cytosine (C) and Guanine (G), which ultimately form the structure of a nucleotide. The A and G are purines, and the C and T are pyrimidines.

The two strands of DNA run in opposite directions. These strands are held together by the hydrogen bond that is present between the two complementary bases. The strands are helically twisted, where each strand forms a right-handed coil, and ten nucleotides make up a single turn.

The pitch of each helix is 3.4 nm. Hence, the distance between two consecutive base pairs (i.e., hydrogen-bonded bases of the opposite strands) is 0.34 nm.

The DNA coils up, forming chromosomes, and each chromosome has a single molecule of DNA in it. Overall, human beings have around twenty-three pairs of chromosomes in the nucleus of cells. DNA also plays an essential role in the process of cell division.

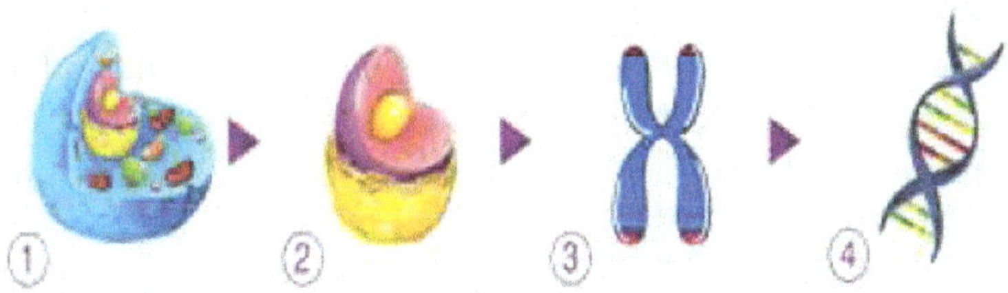

Fig. 5. 1 Cell, 2 Nucleus, 3 Chromosome and 4 DNA

Chargaff's Rule

Erwin Chargaff, a biochemist, discovered that the number of nitrogenous bases in the DNA was present in equal quantities. The amount of A is equal to T, whereas the amount of C is equal to G.

$$A=T; C=G$$

In other words, the DNA of any cell from any organism should have a 1:1 ratio of purine and pyrimidine bases.

DNA REPLICATION

DNA replication is an important process that occurs during cell division. It is also known as semi-conservative replication, during which DNA makes a copy of itself.

DNA replication takes place in three stages:

Step 1: Initiation

The replication of DNA begins at a point known as the origin of replication. The two DNA strands are separated by the DNA helicase. This forms the replication fork.

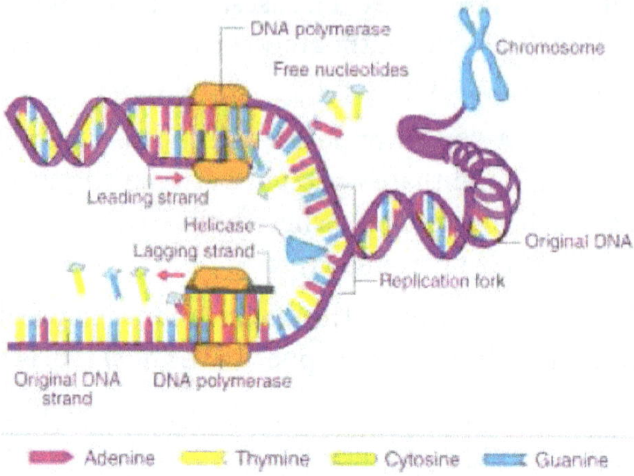

Fig. 6. DNA Replication

Step 2: Elongation

DNA polymerase III reads the nucleotides on the template strand and makes a new strand by adding complementary nucleotides one after the other. For eg., if it reads an Adenine on the template strand, it will add a Thymine on the complementary strand.

While adding nucleotides to the lagging strand, gaps are formed between the strands. These gaps are known as Okazaki fragments. These gaps or nicks are sealed by ligase.

Step 3: Termination

The termination sequence present opposite to the origin of replication terminates the replication process. The TUS protein (terminus utilization substance) binds to terminator sequence and halts DNA polymerase movement. It induces termination.

DNA Function

DNA is the genetic material which carries all the hereditary information. Genes are the small segments of DNA, consisting mostly of 250 – 2 million base pairs. A gene code for a polypeptide molecule, where three nitrogenous bases sequence stands for one amino acid.

Polypeptide chains are further folded in secondary, tertiary and quaternary structures to form different proteins. As every organism contains many genes in its DNA, different types of proteins can be formed. Proteins are the main functional and structural molecules in most organisms. Apart from storing genetic information, DNA is involved in:

- Replication process: Transferring the genetic information from one cell to its daughters and from one generation to the next and equal distribution of DNA during the cell division
- Mutations: The changes which occur in the DNA sequences
- Transcription
- Cellular Metabolism
- DNA Fingerprinting
- Gene Therapy

RNA AND PROTEIN SYNTHESIS

The central dogma describes how the DNA makes its own copies by DNA replication, then codes for the RNA by means of a transcription and RNA codes for the proteins by translation.

For the creation of a new generation, the genetic information stored in the DNA has to be transferred from one generation to another. The process begins with the replication of DNA followed by transcription and translation. Let us have a discussion on translation viz. protein synthesis in cell and the structure of tRNA.

Structure of tRNA

The DNA is transcribed into mRNA on the basis of their complementarity. Translation is the process of conversion of nucleic acid information into amino acids. There is no complementarity between amino acids and mRNA. Hence, translation is not controlled by complementarity but by the genetic code. Since amino acids cannot read this genetic code, they are dependent on an adapter molecule. This adapter molecule is called tRNA (transfer RNA).

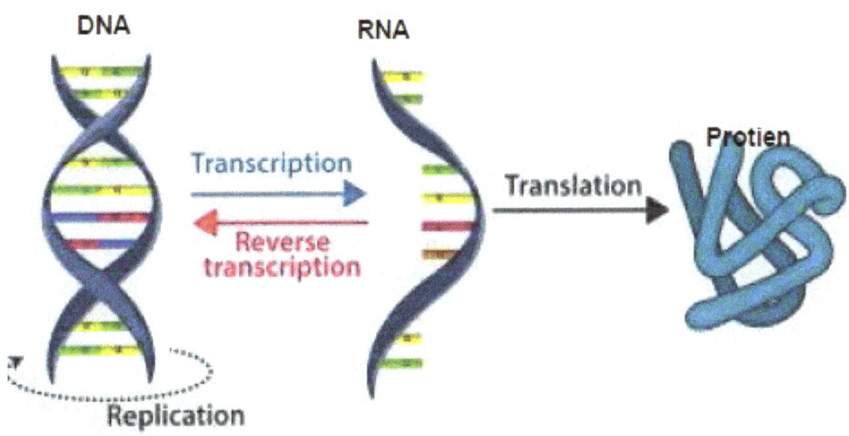

Fig. 7 Central Dogma: DNA to RNA to Protien

Structurally, the tRNA is an inverted L-shaped molecule which has an anticodon loop and amino acid acceptor end. The anticodon loop makes bases complementary to the codes on the mRNA and amino acid end binds to the respective amino acids. Thus, it helps in protein synthesis. Each amino acid has a specific tRNA. Initiator tRNA initiates the translation while stop codons have no tRNA.

Protein Biosynthesis – Translation

The process by which the mRNA codes for a particular protein is known as Translation. In the process, the ribosome translates the mRNA produced from DNA into a chain of specific amino acids. This chain of amino acids leads to protein synthesis.

It is a process where the expense of ATP is required and this energy is given by the charged tRNA. The whole machinery of translation is present in the ribosomes.

The ribosomes consist of a bigger subunit and a smaller subunit. The larger subunit, in turn, consists of two molecules of tRNA that are placed close so that peptide bond could be developed at the expense of sufficient energy. The mRNA enters the smaller subunit which is held by the molecules of tRNA of the complementary codon, that exists in the bigger subunit. Hence, two codons are held by two molecules of tRNA, placed near to each other and a peptide bond is produced among them. When this process repeats, a large chain of amino acids is synthesized. At the stop, codon ribosome releases the amino acid chain.

Key terms	
Term	Meaning
RNA (ribonucleic acid)	Single-stranded nucleic acid that carries out the instructions coded in DNA
Central dogma of biology	The process by which the information in genes flows into proteins: DNA '! RNA '! protein
Polypeptide	A chain of amino acids
Codon	A sequence of three nucleotides that corresponds with a specific amino acid or start/stop signal during translation
Transcription	Process during which a DNA sequence of a gene is copied to make an RNA molecule
Translation	Process during which an mRNA molecule is used to assemble amino acids into polypeptide chains
Mutation	A change in a genetic sequence

Structure of RNA

DNA alone cannot account for the expression of genes. RNA is needed to help carry out the instructions in DNA.

Like DNA, RNA is made up of nucleotide consisting of a 5-carbon sugar ribose, a phosphate group, and a nitrogenous base. However, there are three main differences between DNA and RNA:

- RNA uses the sugar *ribose* instead of *deoxyribose.*
- RNA is generally single-stranded instead of double-stranded.
- RNA contains uracil in place of thymine.

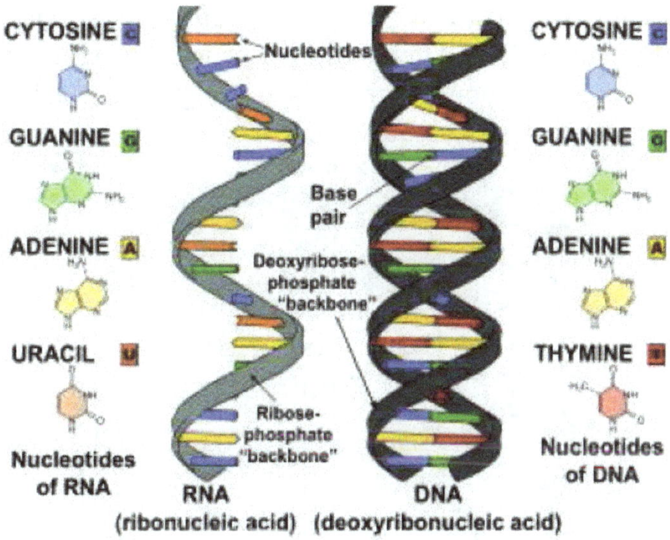

Fig. 8 Image comparing the structure of single-stranded RNA with double-stranded DNA.

These differences help enzymes in the cell to distinguish DNA from RNA.

Types of RNA	
Type	Role
Messenger RNA (mRNA)	Carries information from DNA in the nucleus to ribosomes in the cytoplasm
Ribosomal RNA (rRNA)	Structural component of ribosomes
Transfer RNA (tRNA)	Carries amino acids to the ribosome during translation to help build an amino acid chain

Central dogma of biology

A gene that encodes a polypeptide is expressed in two steps. In this process, information flows from DNA RNA protein, a directional relationship known as the central dogma of molecular biology.

THE GENETIC CODE

The first step in decoding genetic messages is transcription, during which a nucleotide sequence is copied from DNA to RNA. The next step is to join amino acids together to form a protein.

The order in which amino acids are joined together determine the shape, properties, and function of a protein.

The four bases of RNA form a language with just four nucleotide bases: adenine (A), cytosine (C), guanine (G), and uracil (U). The genetic code is read in three-base words called codons. Each codon corresponds to a single amino acid (or signals the starting and stopping points of a sequence).

First letter	Second letter: U	C	A	G	Third letter
U	UUU Phe	UCU Ser	UAU Tyr	UGU Cys	U
	UUC Phe	UCC Ser	UAC Tyr	UGC Cys	C
	UUA Leu	UCA Ser	**UAA Stop**	**UGA Stop**	A
	UUG Leu	UCG Ser	**UAG Stop**	UGG Trp	G
C	CUU Leu	CCU Pro	CAU His	CGU Arg	U
	CUC Leu	CCC Pro	CAC His	CGC Arg	C
	CUA Leu	CCA Pro	CAA Gln	CGA Arg	A
	CUG Leu	CCG Pro	CAG Gln	CGG Arg	G
A	AUU Ile	ACU Thr	AAU Asn	AGU Ser	U
	AUC Ile	ACC Thr	AAC Asn	AGC Ser	C
	AUA Ile	ACA Thr	AAA Lys	AGA Arg	A
	AUG Met	ACG Thr	AAG Lys	AGG Arg	G
G	GUU Val	GCU Ala	GAU Asp	GGU Gly	U
	GUC Val	GCC Ala	GAC Asp	GGC Gly	C
	GUA Val	GCA Ala	GAA Glu	GGA Gly	A
	GUG Val	GCG Ala	GAG Glu	GGG Gly	G

Fig. 9. Genetic code table. Each three-letter sequence of mRNA nucleotides corresponds to a specific amino acid, or to a stop codon. UGA, UAA, and UAG are stop codons. AUG is the codon for methionine, and is also the start codon.

Transcription and translation

The two strands of DNA have the following sequences:

5'-ATGATCTCGTAA-3' 3'-TACTAGAGCATT-5'

Transcription of one of the strands of DNA produces an mRNA that nearly matches the other strand of DNA in sequence. However, due to a biochemical difference between DNA and RNA, the Ts of DNA are replaced with Us in the mRNA. The mRNA sequence is:

5'-AUGAUCUCGUAA-5'

Translation involves reading the mRNA nucleotides in groups of three, each of which specifies and amino acid (or provides a stop signal indicating that translation is finished).

3'-AUG AUC UCG UAA-5'

AUG Methionine AUC Isoleucine UCG Serine UAA "Stop"

Polypeptide sequence: (N-terminus) Methionine-Isoleucine-Serine (C-terminus)

In transcription, a DNA sequence is rewritten, or transcribed, into a similar RNA "alphabet." In eukaryotes, the RNA molecule must undergo processing to become a mature messenger RNA (mRNA).

In translation, the sequence of the mRNA is decoded to specify the amino acid sequence of a polypeptide. The name translation reflects that the nucleotide sequence of the mRNA sequence must be translated into the completely different "language" of amino acids.

Mutations

Sometimes cells make mistakes in copying their genetic information, causing mutations. Mutations can be irrelevant, or they can affect the way proteins are made and genes are expressed.

Substitutions

A *substitution* changes a single base pair by replacing one base for another (Fig. 10).

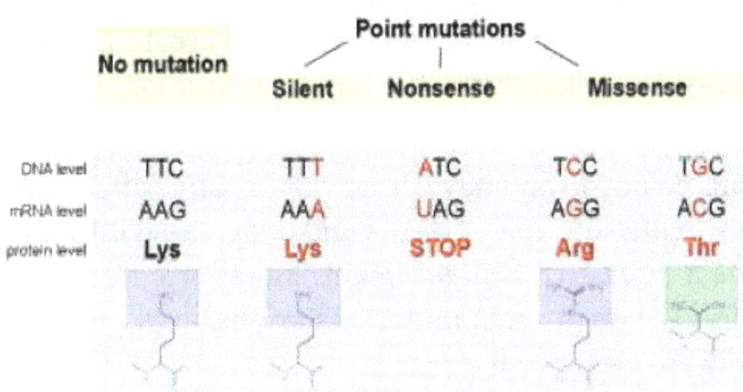

Fig. 10. Substitution changes

There are three kinds of substitution mutations:

- Silent mutations do not affect the sequence of amino acids during translation.
- Nonsense mutations result in a stop codon where an amino acid should be, causing translation to stop prematurely.
- Missense mutations change the amino acid specified by a codon.

Insertions and deletions

An *insertion* occurs when one or more bases are added to a DNA sequence. A *deletion* occurs when one or more bases are removed from a DNA sequence.

Because the genetic code is read in codons (three bases at a time), inserting or deleting bases may change the "reading frame" of the sequence. These types of mutations are called *frameshift mutations*.

A frameshift mutation "shifts" how a sequence of nucleotides is read as triplets (codons) during translation. This may, in turn, alter which amino acids are added to polypeptide. In this example, the original reading frame of a gene encodes an mRNA with codons that specify the amino acid sequence: methionine (Met), isoleucine (Ile), argenine (Arg), and asparagine (Asn). A deletion of the 4th nucleotide (T) shifts the reading frame at the point of the deletion. This produces a new reading frame in the DNA template after the 3rd nucleotide. The mRNA of the new frame bears different codons past the point of the mutation (the first methionine-specifying codon remains unchanged). These codons specify the amino acid sequence: methionine (Met), tyrosine (Tyr), and glycine (Gly).

As this example illustrates, a frameshift mutation changes how nucleotides are interpreted as codons beyond the point of the mutation, and this, in turn, may change the amino acid sequence.

Common mistakes and misconceptions

- Amino acids are not made during protein synthesis. Some students think that the purpose of protein synthesis is to create amino acids. However, amino acids are not being made during translation, they are being used as building blocks to make proteins.
- Mutations do not always have drastic or negative effects. Often people hear the term "mutation" in the media and understand it to mean that a person will have a disease or disfigurement. Mutations are the source of genetic variety, so although some mutations are harmful, most are unnoticeable, and many are even good!
- Insertions and deletions that are multiples of three nucleotides will not cause frameshift mutations. Rather, one or more amino acids will just be added to or deleted from the protein. Insertions and deletions that are not multiples of three nucleotides, however, can dramatically alter the amino acid sequence of the protein.

GENETIC CODE AND GENE EXPRESSION

Genetic Code

Synthesizing a protein with the help of the information in RNA is similar to translating a language to another. A four-letter language is translated to 20 letter language during protein synthesis. There should be a specific relationship among the four bases of DNA and sequence of 20 amino acids in the protein.

The genetic code can be defined as the set of certain rules using which the living cells translate the information encoded within genetic material (DNA or mRNA sequences). The ***ribosomes*** are responsible to accomplish the process of translation. They link the amino acids in an mRNA-specified (messenger RNA) order using tRNA (transfer RNA) molecules to carry amino acids and to read the mRNA three nucleotides at a time.

Genetic Code Table

The complete set of relationships among *amino acids* and *codons* is said to be a genetic code which is often summarized in a table.

It can be seen that many amino acids are shown in the table by more than one codon. For example, there are six ways to write leucine in mRNA language.

Note: *A **codon** is a sequence of three nucleotides which together form a unit of genetic code in a DNA or RNA molecule.*

A key point of the genetic code is its universal nature. This indicates that virtually all species with minor exceptions use the genetic code for protein synthesis.

In other words, genetic code is defined as the nucleotide sequence of the base on DNA which is translated into a sequence of amino acids of the protein to be synthesized.

FIRST LETTER	SECOND LETTER: U	C	A	G	THIRD LETTER
U	UUU Phe	UCU Ser	UAU Tyr	UGU Cys	U
	UUC Phe	UCC Ser	UAC Tyr	UGC Cys	C
	UUA Leu	UCA Ser	UAA Stop	UGA Stop	A
	UUG Leu	UCG Ser	UAG Stop	UGG Trp	G
C	CUU Leu	CCU Pro	CAU His	CGU Arg	U
	CUC Leu	CCC Pro	CAC His	CGC Arg	C
	CUA Leu	CCA Pro	CAA Gln	CGA Arg	A
	CUG Leu	CCG Pro	CAG Gln	CGG Arg	G
A	AUU Ile	ACU Thr	AAU Asn	AGU Ser	U
	AUC Ile	ACC Thr	AAC Asn	AGC Ser	C
	AUA Ile	ACA Thr	AAA Lys	AGA Arg	A
	AUG Met	ACG Thr	AAG Lys	AGG Arg	G
G	GUU Val	GCU Ala	GAU Asp	GGU Gly	U
	GUC Val	GCC Ala	GAC Asp	GGC Gly	C
	GUA Val	GCA Ala	GAA Glu	GGA Gly	A
	GUG Val	GCG Ala	GAG Glu	GGG Gly	G

Fig. 11. Genetic Code Table

Properties of Genetic Code

- Triplet code
- Non-ambiguous and Universal
- Degenerate code
- Nonoverlapping code
- Commaless
- Start and Stop Codons
- Polarity

These properties of genetic code are explained below.

Triplet code

A codon or a code word is defined as a group of bases that specify an amino acid. There is strong evidence, which proves that a sequence of three nucleotides codes for an amino acid in the protein, i.e., the code is a ***triplet***.

The four bases of nucleotide i.e, (A, G, C, and U) are used to produce three-base codons. The 64 codons involve sense codons (that specify amino acids). Hence, there are 64 codons for 20 amino acids since every codon for one amino acid means that there exist more than code for the same amino acid.

Commaless code

No room for punctuation in between which indicates that every codon is adjacent to the previous one without any nucleotides between them.

Nonoverlapping code

The code is read sequentially in a group of three and a nucleotide which becomes a part of triplet never becomes part of the next triplet.

For example

5'-UCU-3' codes for Serine

5'-AUG-3' codes for methionine

Polarity

Each triplet is read from 5' '! 3' direction and the beginning base is 5' followed by the base in the middle then the last base which is 3'. This implies that the codons have a ***fixed polarity*** and if the codon is read in the reverse direction, the base sequence of the codon would reverse and would specify two different proteins.

Degenerate code

Every amino acid except tryptophan (UGG) and methionine (AUG) is coded by various codons, i.e, a few codons are synonyms and this aspect is known as the degeneracy of genetic code. For instance, UGA codes for tryptophan in yeast mitochondria.

Start and Stop Codons

Generally, *AUG codon* is the initiating or start codon. The polypeptide chain starts either with eukaryotes (methionine) or prokaryotes (N- formylmethionine).

On the other hand, *UAG, UAA* and *UGA* are called as termination codons or stop codons. These are not read by any tRNA molecules and they never code for any amino acids.

Non-ambiguous and Universal

The genetic code is non-ambiguous which means a specific codon will only code for a particular amino acid. Also, the same genetic code is seen valid for all the organisms i.e. they are universal.

Exceptions to the Code

The genetic code is universal since similar codons are assigned to identical amino acids along with similar START and STOP signals in the majority of genes in microorganisms and plants. However, a few exceptions have been discovered and most of these include assigning one or two of the STOP codons to an amino acid.

Apart from this, both the codons ***GUG*** and ***AUG*** may code for methionine as a starting codon, although GUG is meant for valine. This breaks the property of non-ambiguousness. Thus, it can be said that few codes often differs from the universal code or non-ambiguous code.

Gene Expression

Gene expression is the process by which the information encoded in a gene is turned into a function. This mostly occurs via the transcription of RNA molecules that code for proteins or non-coding RNA molecules that serve other functions. Gene expression be thought of as an “on/off switch” to control when and where RNA molecules and proteins are made and as a “volume control” to determine how much of those products are made. The process of gene expression is carefully regulated, changing substantially under different conditions and cell types. The RNA and protein products of many genes serve to regulate the expression of other genes. Where, when, and how much a gene is expressed can also assessed by measuring the functional activity of a gene product or observing a phenotype associated with a gene.

The genetic code codes for proteins. The information of the DNA is 'translated' into a chain of amino acids that forms a protein. These proteins form the building blocks for structures within the cells and ultimately the whole body. Proteins also form enzymes and other chemicals that perform various functions in the body. Each gene can code for different proteins and thus the number of proteins known to exist in the cells is more than the number of genes.

All genes are not expressed or do not code for any protein. This could be organ specific for example a liver cell expresses different genes than kidney cells.

The environment also plays a role in determining the ultimate trait. The phenotype of an organism thus depends on the interaction of genetics with the environment. The environment for example, has a role in effects of the human genetic disease phenylketonuria. The mutation that causes phenylketonuria disrupts the ability of the body to break down the amino acid phenylalanine. This leads to toxic build-up of an intermediate molecule leading to mental retardation and seizures. Persons with phenylketonuria mutation on a strict diet that avoids this amino acid may remain normal and healthy.

Regulation of gene expression

Not all genes in a cell are active in protein production at any given time. Gene action can be switched on or off in response to the cell's stage of development and external environment. In multicellular organisms, different kinds of cells express different parts of the genome. In other words, a skin cell and a muscle cell contain exactly the same genes, but the differences in structure and function of these cells result from the selective expression and repression of certain genes.

In prokaryotes and eukaryotes, most gene-control systems are positive, meaning that a gene will not be transcribed unless it is activated by a regulatory protein. However, some bacterial genes show negative control. In this case the gene is transcribed continuously unless it is switched off by a regulatory protein. An example of negative control in prokaryotes involves three adjacent genes used in the metabolism of the sugar lactose by *E. coli*. The part of the chromosome containing the genes concerned is divided into two regions, one that includes the structural genes (i.e., those genes that together code for protein structure) and another that is a regulatory region. This overall unit is called an operon. If lactose is not present in a cell, transcription of the genes that code for the lactose-processing enzymes—â-galactosidase, permease, and transacetylase—is turned off. This is achieved by a protein called the lac repressor, which is produced by the repressor gene and binds to a region of the operon called the operator. Such binding prevents RNA polymerase, which initially binds at the adjacent promoter, from moving into the coding region. If lactose enters the cell, it binds to the lac repressor and induces a change of shape in the repressor so that it can no longer bind to the DNA at the operon. Consequently, the RNA polymerase is able to travel from the promoter down the three adjacent protein-coding regions, making one continuous transcript. This three-gene transcript is subsequently translated into three separate proteins.

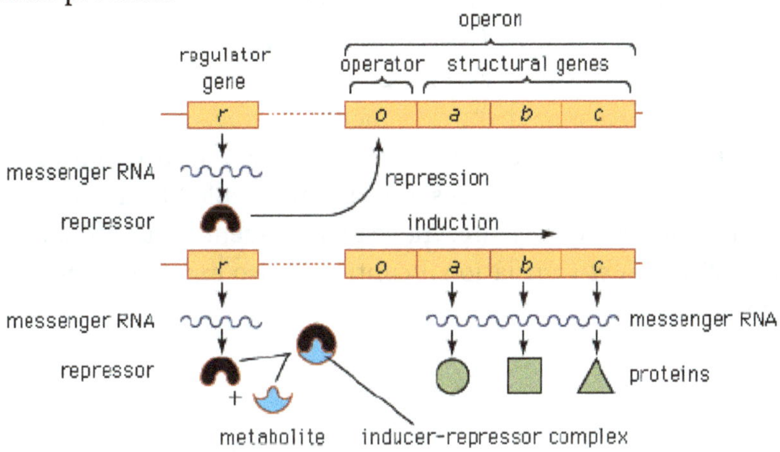

Fig. 12. Model of the operon and its relation to the regulator gene.

Although the operon model has proved a useful model of gene regulation in bacteria, different regulatory mechanisms are employed in eukaryotes. First, there are no operons in eukaryotes, and each gene is regulated independently. Furthermore, the series of events associated with gene expression in higher organisms is much more complex than in prokaryotes and involves multiple levels of regulation.

In order for a gene to produce a functional protein, a complex series of steps must occur. Some type of signal must initiate the transcription of the appropriate region along the DNA, and, finally, an active protein must be made and sent to the appropriate location to perform its specific task. Regulation can be exerted at many different places along this pathway. The fundamental level of control is the rate of transcription. Transcription itself is also a complex process with many different components, and each one is a potential point of control. Regulatory proteins called activators or enhancers are needed for the transcription of genes at a specific time or in a certain cell. Thus, control is positive (not negative as in the lac operon) in that these proteins are necessary for the promotion of transcription. Activators bind to specific regions of the DNA in the upstream regulatory region, some very distant from the binding of the initiation complex.

Following the transcription of DNA into RNA, a process of editing and splicing takes place in which noncoding nucleotide sequences called introns are excised from the primary transcript, resulting in functional mRNA. For most genes this is a routine step in the production of mRNA, but in some genes there are alternative ways to splice the primary transcript, resulting in different mRNAs, which in turn result in different proteins.

Some genes are controlled at the translational and post-translational levels. One type of translational control is the storage of uncapped mRNA to meet future demands for protein synthesis. In other cases, control is exerted through the stability or instability of mRNA. The rate of translation of some mRNAs can also be regulated. Post-translationally, certain proteins (e.g., insulin) are synthesized in an inactive form and must be chemically modified to become active. Other proteins are targeted to specific locations inside the cell (e.g., mitochondria) by means of highly specific amino acid sequences at their ends, called leader sequences; when the protein reaches its correct site, the leader segment is cut off, and the protein begins to function. Post-translational control is also exerted through mRNA and protein degradation.

CHAPTER-3

TOOLS AND TECHNIQUES IN BIOTECHNOLOGY

Biotechnology deals with techniques of using live organisms or enzymes from organisms to produce products and processes useful to humans. In this sense, making curd, bread or wine, which are all microbe-mediated processes, could also be thought as a form of biotechnology. However, it is used in a restricted sense today, to refer to such of those processes which use genetically modified organisms to achieve the same on a larger scale.

Further, many other processes/techniques are also included under biotechnology. For example, in vitro fertilisation leading to a 'test-tube' baby, synthesising a gene and using it, developing a DNA vaccine or correcting a defective gene, are all part of biotechnology

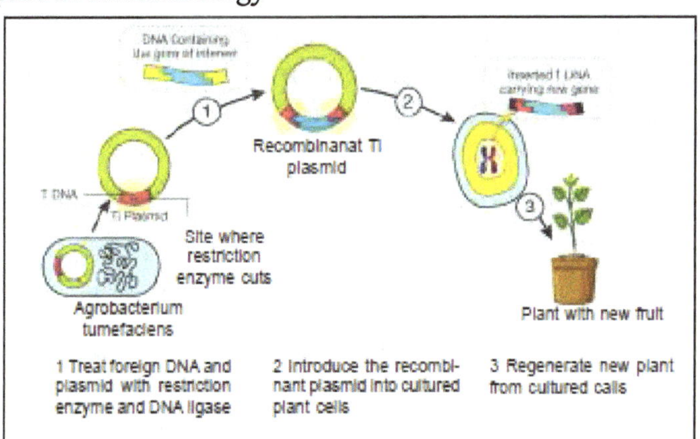

Fig. 1. Recombinant DNA Technology

RECOMBINANT DNA TECHNOLOGY

Recombinant DNA technology alters the phenotype of an organism (host) through a genetically altered vector. This cloning vector is introduced and integrated into the genome of the organism. So, basically, the process involves the introduction of a foreign piece of DNA into the genome which contains our gene of interest. The gene which is introduced is the recombinant gene and the technique is called the recombinant DNA technology. Here we will learn about key tools of recombinant DNA technology.

Tools of Recombinant DNA technology

Inserting the desired gene into the genome of the host is not as easy as it sounds. It involves the selection of the desired gene for administration into the host followed by a selection of the perfect vector with which the gene has to be integrated and recombinant DNA formed. This recombinant DNA then has to be introduced into the host. And at last, it has to be maintained in the host and carried forward to the offspring. Recombinant DNA technology can be complete and achieved with the help of some elemental tools. The different tools used for the purpose are discussed below:

Restriction Enzymes

The restriction enzymes – help to cut, the polymerases- help to synthesize and the ligases- help to bind.

The restriction enzymes used in recombinant DNA technology play a major role in determining the location at which the desired gene is inserted into the vector genome. They are two types, namely endonucleases and exonucleases. The endonucleases cut within the DNA strand whereas the exonucleases cut the nucleotides from the ends of the DNA strands. The restriction endonucleases are sequence-specific which is usually palindrome sequences and cut the DNA at specific points. They scrutinize the length of DNA and make the cut at the specific site called the restriction site. This gives rise to sticky ends in the sequence. The desired genes and the vectors are cut by the same restriction enzymes to obtain the complementary sticky notes, thus making the work of the ligases easy to bind the desired gene to the vector.

Vectors

The vectors help in carrying and integrating the desired gene. These form a very important part of the tools of recombinant DNA technology as they are the ultimate vehicles that carry forward the desired gene into the host organism. Plasmids and bacteriophages are the most common vectors in recombinant DNA technology.

Host Organism

Host organism is the organism into which the recombinant DNA is introduced. The host is the ultimate tool of recombinant DNA technology which takes in the vector engineered with the desired DNA with the help of the enzymes. There are a number of ways in which this recombinant DNA's are inserted into the host, namely – microinjection, biolistics or gene gun, alternate cooling and heating, use of calcium ions, etc.

Cloning Vectors are a type of vector that can be used to make copies

We need vectors to transport the cut DNA into a host organism, just how a mosquito works as a vector for transmitting the malarial infection into the human body. One of the most essential biotech tools is vectors.

Plasmids are useful vectors because they can reproduce independently of chromosomal DNA control in bacterial organisms. The vectors currently in use are designed to facilitate the linking of foreign DNA and to allow for the selection of recombinants over non-recombinants. To allow cloning, a vector must have the following properties:

Replication's Source (ori)

This is where replication starts. This sequence induces a fragment of DNA to replicate in the host cell when it is linked to it. The copy number of linked DNA is likewise controlled by this sequence. In order to retrieve huge amounts of DNA, the target DNA must be cloned into a vector with a high-copy-number 'ori'

Marker with a Variety of Options

A selectable marker must also be included in the vector, allowing recombinants to be chosen over non-recombinants. Genes that confer antibiotic resistance, such as ampicillin, kanamycin, and chloramphenicol, are useful selection markers in E. coli. The recombinants are straightforward to choose since normal E. coli cells do not have these resistance genes.

Sites for Cloning

The vector must have a recognition site for a certain restriction enzyme in order to attach the foreign DNA to it. Multiple recognition sites result in multiple DNA fragments, which makes cloning more difficult. Multiple antibiotic resistance genes can be found in a vector. A restriction site in one of the antibiotic resistance genes is ligated to the foreign DNA.

Consider an E. coli cloning vector with ampicillin and tetracycline resistance genes. The plasmid loses its tetracycline resistance after inserting foreign DNA into a recognition site inside the tetracycline resistance gene. By plating on ampicillin-containing media, it can still be selected for non-recombinants.

We can now distinguish between recombinants and non-recombinants by moving the ones that grow in the ampicillin medium to a tetracycline medium.

Recombinants grow in ampicillin but not tetracycline media, whereas non-recombinants grow in both.

Alternative markers that distinguish recombinants from non-recombinants based on color production are already available. This entails inserting foreign DNA into the DNA sequence of an enzyme, such as -galactosidase, in order to render it inactive. 'Insertional inactivation' is the technical term for this. Recombinants do not produce color when reacting with a substrate, whereas non-recombinants do.

'Biotechnology is the application of natural science to organisms, cells, sections of cells, and molecular analogs in the development of products and services.' The following are the two fundamental techniques that gave rise to modern biotechnology:

Genetic Engineering entails techniques that alter the chemistry of genetic material so that when it is put into a host, the phenotypic of the host is altered.

Maintaining a sterile environment to allow the growth of only the necessary germ in high quantities for the production of vaccines, antibiotics, and other pharmaceuticals.

GMOs

In-plant and animal breeding, traditional hybridization techniques introduce both beneficial and unwanted genes. Recombinant DNA technology, gene cloning, and other genetic engineering approaches solve this difficulty by allowing us to introduce only the genes we want. The following are the steps involved in genetically altering an organism:

- DNA that contains the desired genes must be identified.
- DNA is introduced into the host organism.
- Maintaining the inserted DNA in the host genome and passing it on to descendants.

POLYMERASE CHAIN REACTION (PCR)

Polymerase chain reaction (PCR) is a common laboratory technique used to make many copies (millions or billions!) of a particular region of DNA. This DNA region can be anything the experimenter is interested in. For example, it might be a gene whose function a researcher wants to understand, or a genetic marker used by forensic scientists to match crime scene DNA with suspects.

Typically, the goal of PCR is to make enough of the target DNA region that it can be analyzed or used in some other way. For instance, DNA amplified by PCR may be sent for sequencing, visualized by gel electrophoresis, or cloned into a plasmid for further experiments.

PCR is used in many areas of biology and medicine, including molecular biology research, medical diagnostics, and even some branches of ecology.

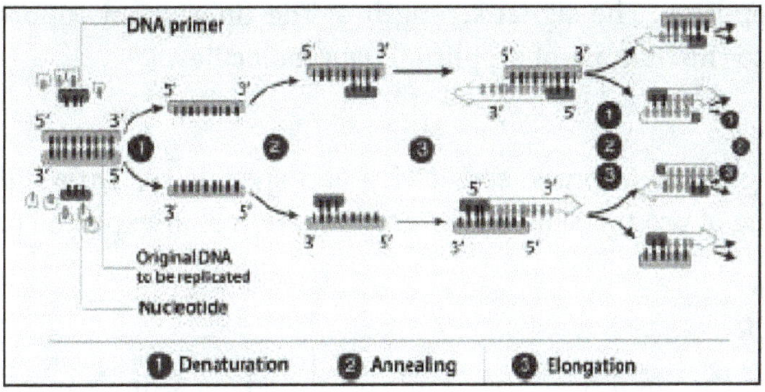

Fig. 2. Polymerase chain reaction (PCR)

Principle of PCR

The PCR technique is based on the enzymatic replication of DNA. In PCR, a short segment of DNA is amplified using primer mediated enzymes. DNA Polymerase synthesises new strands of DNA complementary to the template DNA. The DNA polymerase can add a nucleotide to the pre-existing 3'-OH group only. Therefore, a primer is required. Thus, more nucleotides are added to the 3' prime end of the DNA polymerase.

Components of PCR

Components Of PCR constitutes the following:

- **DNA Template** – The DNA of interest from the sample.
- **DNA Polymerase** – Taq Polymerase is used. It is thermostable and does not denature at very high temperatures.
- **Oligonucleotide Primers** - These are the short stretches of single-stranded DNA complementary to the 3' ends of sense and anti-sense strands.
- **Deoxyribonucleotide triphosphate**– These provide energy for polymerization and are the building blocks for the synthesis of DNA. These are single units of bases.
- **Buffer System**– Magnesium and Potassium provide optimum conditions for DNA denaturation and renaturation. It is also important for fidelity, polymerase activity, and stability.

Types of PCR

PCR is of the following types:

Real-time PCR

In this type, the DNA amplification is detected in real-time with the help of a fluorescent reporter. The signal strength of the fluorescent reporter is directly proportional to the number of amplified DNA molecules.

Nested PCR

This was designed to improve sensitivity and specificity. They reduce the non-specific binding of products due to the amplification of unexpected primer binding sites.

Multiplex PCR

This is used for the amplification of multiple targets in a single PCR experiment. It amplifies many different DNA sequences simultaneously.

Quantitative PCR

It uses the DNA amplification linearity to detect, characterize and quantify a known sequence in a sample.

Arbitrary Primed PCR

It is a DNA fingerprinting technique based on PCR. It uses primers the DNA sequence of which is chosen arbitrarily.

PCR Steps

The PCR involves three major cyclic reactions:

Denaturation

Denaturation occurs when the reaction mixture is heated to 94! for about 0.5 to 2 minutes. This breaks the hydrogen bonds between the two strands of DNA and converts it into a single-stranded DNA.

The single strands now act as a template for the production of new strands of DNA. The temperature should be provided for a longer time to ensure the separation of the two strands.

Annealing

The reaction temperature is lowered to 54-60! for around 20-40 seconds. Here, the primers bind to their complementary sequences on the template DNA.

Primers are single-strand sequences of DNA or RNA around 20 to 30 bases in length.

They serve as the starting point for the synthesis of DNA.

The two separated strands run in the opposite direction and consequently there are two primers- a forward primer and a reverse primer.

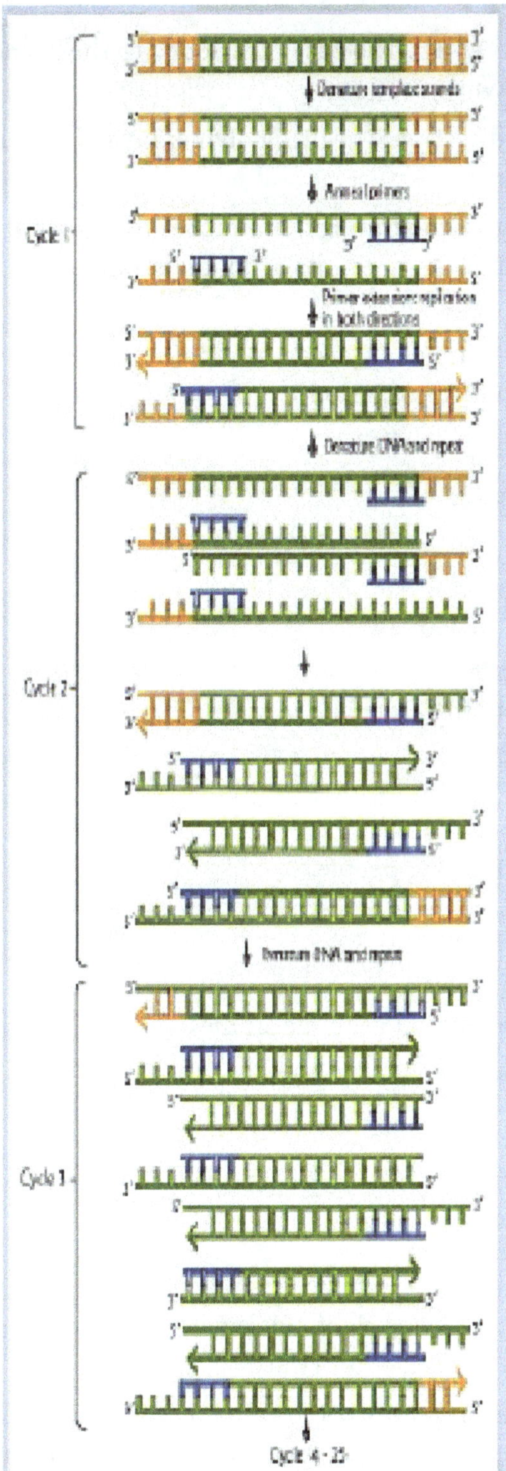

Fig. 3 Polymerase chain reaction (PCR).PCR is an in vitro DNA replication method. The starting material is a doublestranded DNA. The target sequence to be amplified is indicated in green. Large numbers of two primers (blue) are added, each with a sequence complementary to that found in one strand at the end of the region to be amplified. A thermostable DNA polymerase (e.g. Taq polymerase) and dNTPs are also added. In the first cycle, heating to 95°C denatures the double-stranded DNA and subsequent cooling to 55–65°C then allows the primers to anneal to their complementary sequences in the target DNA. Taq polymerase extends each primer from 52 to 32 , generating newly synthesized strands in both directions, which extend to the end of the template strands. The extension is performed at 72°C. In the second cycle, the original and newly made DNA strands are denatured at 95°C and primers are annealed to their complementary sequences at 55–65°C. Each annealed primer again is extended by Taq polymerase. In the third cycle, two doublestrand DNA molecules are generated exactly equal to the target sequence. These two are doubled in the fourth cycle and are doubled again with each successive cycle.

Elongation

At this step, the temperature is raised to 72-80!. The bases are added to the 3' end of the primer by the Taq polymerase enzyme.

This elongates the DNA in the 5' to 3' direction. The DNA polymerase adds about 1000bp/minute under optimum conditions.

Taq Polymerase can tolerate very high temperatures. It attaches to the primer and adds DNA bases to the single strand. As a result, a double-stranded DNA molecule is obtained.

These three steps are repeated 20-40 times in order to obtain a number of sequences of DNA of interest in a very short time period.

Applications of PCR

The following are the applications of PCR :

Medicine

- Testing of genetic disease mutations.
- Monitoring the gene in gene therapy.
- Detecting disease-causing genes in the parents.

Forensic Science

- Used as a tool in genetic fingerprinting.
- Identifying the criminal from millions of people.
- Paternity tests

Research and Genetics

- Compare the genome of two organisms in genomic studies.
- In the phylogenetic analysis of DNA from any source such as fossils.
- Analysis of gene expression.
- Gene Mapping

DNA SEQUENCING TECHNOLOGIES

DNA sequencing is the process of determining the sequence of nucleotide bases (As, Ts, Cs, and Gs) in a piece of DNA. Today, with the right equipment and materials, sequencing a short piece of DNA is relatively straightforward.

Sequencing an entire genome (all of an organism's DNA) remains a complex task. It requires breaking the DNA of the genome into many smaller pieces, sequencing the pieces, and assembling the sequences into a single long "consensus." However, thanks to new methods that have been developed over the past two decades, genome sequencing is now much faster and less expensive than it was during the Human Genome Project.

Sanger sequencing: The chain termination method

Regions of DNA up to about base pairs in length are routinely sequenced using a method called Sanger sequencing or the chain termination method. Sanger sequencing was developed by the British biochemist Fred Sanger and his colleagues in 1977.

In the Human Genome Project, Sanger sequencing was used to determine the sequences of many relatively small fragments of human DNA. (These fragments weren't necessarily bp or less, but researchers were able to "walk" along each fragment using multiple rounds of Sanger sequencing.) The fragments were aligned based on overlapping portions to assemble the sequences of larger regions of DNA and, eventually, entire chromosomes.

Although genomes are now typically sequenced using other methods that are faster and less expensive, Sanger sequencing is still in wide use for the sequencing of individual pieces of DNA, such as fragments used in DNA cloning or generated through polymerase chain reaction (PCR).

Ingredients for Sanger sequencing

Sanger sequencing involves making many copies of a target DNA region. Its ingredients are similar to those needed for DNA replication in an organism, or for polymerase chain reaction (PCR), which copies DNA *in vitro*. They include:

- A DNA polymerase enzyme
- A primer, which is a short piece of single-stranded DNA that binds to the template DNA and acts as a "starter" for the polymerase
- The four DNA nucleotides (dATP, dTTP, dCTP, dGTP)
- The template DNA to be sequenced

However, a Sanger sequencing reaction also contains a unique ingredient:

- Dideoxy, or chain-terminating, versions of all four nucleotides (ddATP, ddTTP, ddCTP, ddGTP), each labeled with a different color of dye

Both images given above have a 5 carbon ring shaped molecule with a base attached to the first carbon and a phosphate group attached to the fifth carbon. The top image has a two hydrogens attached to the number 3 carbon and is labeled Dideoxynucleotide, ddNTP. The bottom image has one hydrogen and an OH attached to the number 3 carbon and is labeled Deoxynucleotide, dNTP.

Dideoxy nucleotides are similar to regular, or deoxy, nucleotides, but with one key difference: they lack a hydroxyl group on the 3' carbon of the sugar ring. In a regular nucleotide, the 3' hydroxyl group acts as a "hook," allowing a new nucleotide to be added to an existing chain.

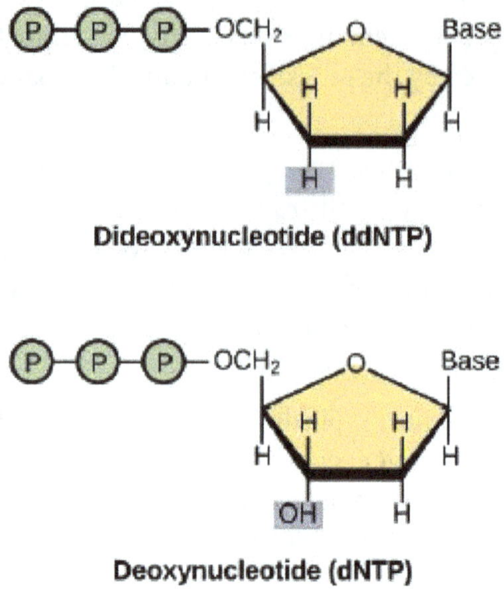

Fig. 4. Two chemical structures of a nucleotide are shown.

Once a dideoxy nucleotide has been added to the chain, there is no hydroxyl available and no further nucleotides can be added. The chain ends with the dideoxy nucleotide, which is marked with a particular color of dye depending on the base (A, T, C or G) that it carries.

Method of Sanger sequencing

The DNA sample to be sequenced is combined in a tube with primer, DNA polymerase, and DNA nucleotides (dATP, dTTP, dGTP, and dCTP). The four dye-labeled, chain-terminating dideoxy nucleotides are added as well, but in much smaller amounts than the ordinary nucleotides.

The mixture is first heated to denature the template DNA (separate the strands), then cooled so that the primer can bind to the single-stranded template. Once the primer has bound, the temperature is raised again, allowing DNA polymerase to synthesize new DNA starting from the primer. DNA polymerase will continue adding nucleotides to the chain until it happens to add a dideoxy nucleotide instead of a normal one. At that point, no further nucleotides can be added, so the strand will end with the dideoxy nucleotide.

This process is repeated in a number of cycles. By the time the cycling is complete, it's virtually guaranteed that a dideoxy nucleotide will have been incorporated at every single position of the target DNA in at least one reaction. That is, the tube will contain fragments of different lengths, ending at each of the nucleotide positions in the original DNA (see figure below). The ends of the fragments will be labeled with dyes that indicate their final nucleotide.

At the top of the diagram (Fig. 5) is a horizontal green line with 17 small vertical lines extending from the horizontal line. The left side of the line is labeled 3 prime, the right side of the line is labeled 5 prime, and the line is titled template. Above the green line is a horizontal gray line with 9 small vertical lines extending from it; the left side of the gray line is labeled 5 prime and the right side is labeled 3 prime and the line is titled primer. An arrow points from the primer to a series of horizontal lines that are increasing in length. The arrow is joined by an arrow from a key that read dNTPs.

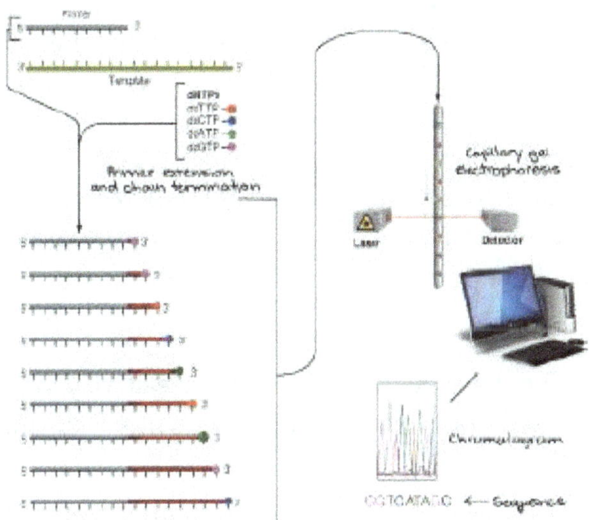

Fig. 5 A diagram with images showing the Sanger DNA sequencing method.

The symbols for the key are ddTTP is red circle, ddCTP is blue circle ddATP is green circle and ddGTP is purple circle. At the point where the arrow from the key joins the arrow from the primer, the arrow is labeled primer extension and chain termination. At the tip of the arrow are 9 parallel lines, each increasing in length. Each line begins with the gray primer and then is joined by a red line segment with a vertical line attached that ends with a colored circle. The first line has one vertical line segment and a purple circle. The second line has 2 red vertical line segments and a purple circle. Each horizontal line continues to extend by 1 vertical line segment adding to the 3 prime end of the segment, and at the end of each horizontal line segment is either a red, blue, green or purple circle.

All 9 horizontal lines are enclosed within a bracket that has an arrow pointing to a tube shaped structure that has different colored bands on the tube and has the label capillary gel electrophoresis. The tube has an arrow pointing to a box labeled laser that shows a laser line running through the tube and connected to a box labeled detector. The detector is hooked up to a computer.

From the computer there is an arrow pointing to a box that has an image of purple, red, green and blue lines on a graph and the box is labeled chromatogram. Under the box is a label sequence that reads G, G, T, C, A, T, A, G, C. The Gs are colored purple, the Ts are colored red, the Cs are colored blue and the As are colored green.

After the reaction is done, the fragments are run through a long, thin tube containing a gel matrix in a process called capillary gel electrophoresis. Short fragments move quickly through the pores of the gel, while long fragments move more slowly. As each fragment crosses the "finish line" at the end of the tube, it's illuminated by a laser, allowing the attached dye to be detected.

The smallest fragment (ending just one nucleotide after the primer) crosses the finish line first, followed by the next-smallest fragment (ending two nucleotides after the primer), and so forth. Thus, from the colors of dyes registered one after another on the detector, the sequence of the original piece of DNA can be built up one nucleotide at a time. The data recorded by the detector consist of a series of peaks in fluorescence intensity, as shown in the chromatogram above. The DNA sequence is read from the peaks in the chromatogram.

Uses and limitations

Sanger sequencing gives high-quality sequence for relatively long stretches of DNA (up to about base pairs). It's typically used to sequence individual pieces of DNA, such as bacterial plasmids or DNA copied in PCR.

However, Sanger sequencing is expensive and inefficient for larger-scale projects, such as the sequencing of an entire genome or metagenome (the "collective genome" of a microbial community). For tasks such as these, new, large-scale sequencing techniques are faster and less expensive.

Next-generation sequencing

The name may sound like Star Trek, but that's really what it's called! The most recent set of DNA sequencing technologies are collectively referred to as next-generation sequencing.

There are a variety of next-generation sequencing techniques that use different technologies. However, most share a common set of features that distinguish them from Sanger sequencing:

- Highly parallel: many sequencing reactions take place at the same time
- Micro scale: reactions are tiny and many can be done at once on a chip
- Fast: because reactions are done in parallel, results are ready much faster
- Low-cost: sequencing a genome is cheaper than with Sanger sequencing
- Shorter length: reads typically range from nucleotides in length

Conceptually, next-generation sequencing is kind of like running a very large number of tiny Sanger sequencing reactions in parallel. Thanks to this parallelization and small scale, large quantities of DNA can be sequenced much more quickly and cheaply with next-generation methods than with Sanger sequencing. For example, in 2001, the cost of sequencing a human genome was almost.

Why does fast and inexpensive sequencing matter? The ability to routinely sequence genomes opens new possibilities for biology research and biomedical applications. For example, low-cost sequencing is a step towards personalized medicine - that is, medical treatment tailored to an individual's needs, based on the gene variants in his or her genome.

Automated DNA sequencing

In 1986, Leroy Hood and Lloyd Smith automated Sanger's method. In this new sequencing technology, radioactive markers are replaced with fluorescent ones. Each ddNTP terminator is tagged with a different color of fluorophore: red, green, blue, or yellow. Thus, instead of having to run four separate sequencing reactions, the reactions can be combined into one tube. The first automated sequencer made use of a polyacrylamide gel to resolve the samples, a laser to excite the dye molecules as they reached a detector near the end of the gel, and a computer to read the results as a DNA sequence.

In this system each automated sequencer was able to produce 4800 bases of sequence per day. The current automated systems replace the old-style gel with arrays of tiny capilliaries, each of which acts as a "lane." A pump loads special capillaries with a polymer that serves as the separation matrix. DNA samples in a 96-well plate are loaded into the capillary array by a short burst of electrophoresis, called "electrokinetic injection." The capillary array is immersed in running buffer and the DNA fragments then migrate through the capillary matrix by size, smallest to largest.

As the DNA fragments reach the detection window, a laser beam excites the dye molecules causing them to fluoresce. Emitted light from 96 capillaries is collected at once, spectrally separated into the four colors and focused onto a CCD camera. Computer software interprets the pattern of peaks to produce a graph of fluorescence intensity versus time (electropherogram), which is then converted to the DNA sequence (Fig. 6). With this system, as many as 2 million bases can be sequenced per day.

RECOMBINANT DNA TECHNOLOGY

Introduction

The technology used for producing artificial DNA through the combination of different genetic materials (DNA) from different sources is referred to as Recombinant DNA Technology. Recombinant DNA technology is popularly known as genetic engineering.

The recombinant DNA technology emerged with the discovery of restriction enzymes in the year 1968 by Swiss microbiologist Werner Arber,

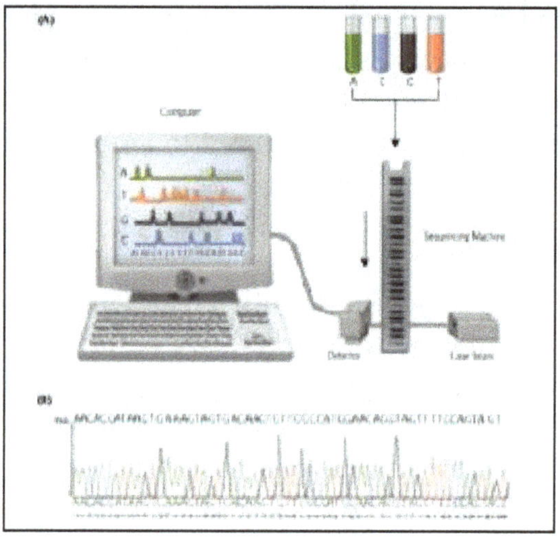

Fig. 6 Automated DNA sequencing. (A) The diagram depicts the steps involved in automated DNA sequencing. The products from each tube will emit a different color fluorescence when excited by light. (B) Sample computer printout of an automated DNA sequencing experiment.

Inserting the desired gene into the genome of the host is not as easy as it sounds. It involves the selection of the desired gene for administration into the host followed by a selection of the perfect vector with which the gene has to be integrated and recombinant DNA formed.

Thus the recombinant DNA has to be introduced into the host. And at last, it has to be maintained in the host and carried forward to the offspring.

The cornerstone of most molecular biology technologies is the gene. To facilitate the study of genes, they can be isolated and amplified. One method of isolation and amplification of a gene of interest is to clone the gene by inserting it into another DNA molecule that serves as a vehicle or vector that can be replicated in living cells. When these two DNAs of different origin are combined, the result is a recombinant DNA molecule. Although genetic processes such as crossing-over technically produce recombinant DNA, the term is generally reserved for DNA molecules produced by joining segments derived from different biological sources. The recombinant DNA molecule is placed in a host cell, either prokaryotic or eukaryotic. The host cell then replicates (producing a clone), and the vector with its foreign piece of DNA also replicates. The foreign DNA thus becomes amplified in number, and following its amplification can be purified for further analysis.

Historical perspective

In the early 1960s, before the advent of gene cloning, studies of genes often relied on indirect or fortuitous discoveries, such as the ability of bacteriophages to incorporate bacterial genes into their genomes. For example, a strain of phage phi 80 with the lac operator incorporated into its genome was used to demonstrate that the Lac repressor binds specifically to this DNA sequence. The synthesis of many disparate experimental observations into recombinant DNA technology occurred between 1972 and 1975, through the efforts of several research groups working primarily on bacteriophage lambda (λ).

Insights from bacteriophage lambda (λ) cohesive sites

In 1962, Allan Campbell noted that the linear genome of bacteriophage l forms a circle upon entering the host bacterial cell, and a recombination (breaking and rejoining) event inserts the phage DNA into the host chromosome. Reversal of the recombination event leads to normal excision of the phage DNA. Rare excision events at different places can result in the incorporation of nearby bacterial DNA sequences (Fig. 7).

Further analysis revealed that phage ë had short regions of single-stranded DNA whose base sequences were complementary to each other at each end of its linear genome. These single-stranded regions were called “cohesive” (cos) sites. Complementary base pairing of the cos sites allowed the linear genome to become a circle within the host bacterium.

The idea of joining DNA segments by “cohesive sites” became the guiding principle for the development of genetic engineering. With the molecular characterization of restriction and modification systems in bacteria, it soon became apparent that the ideal engineering tools for making cohesive sites on specific DNA pieces were already available in the form of restriction endonucleases.

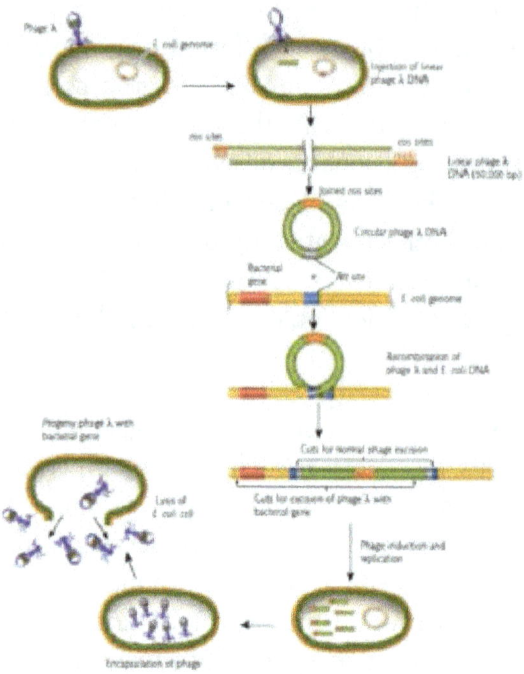

Fig. 7 Bacteriophage lambda (λ) cohesive sites

Insights from bacterial restriction and modification systems

Early on, Salvador Luria and other phage workers were intrigued by a phenomenon termed "restriction and modification." Phages grown in one bacterial host often failed to grow in different bacterial strains ("restriction"). However, some rare progeny phages were able to escape this restriction. Once produced in the restrictive host they had become "modified" in some way so that they now grew normally in this host. The entire cycle could be repeated, indicating that the modification was not an irreversible change. For example, phage ë grown on the C strain of *Escherichia coli* (l.C) were restricted in the K-12 strain (the standard strain for most molecular work). However, the rare phage ë that managed to grow in the K-12 strain now had "K" modification (l.K). These phages grew normally on both C and K-12; however, after growth on C, the phage l with "C" modification (l.C) was again restricted in K-12. Thus, the K-12 strain was able to mark its own resident DNA for preservation, but could eliminate invading DNA from another distantly related strain. In 1962, the molecular basis of restriction and modification was defined by Werner Arber and co-workers.

Tools of Recombinant DNA Technology

The enzymes which include the restriction enzymes help to cut, the polymerases- help to synthesize and the ligases- help to bind. The restriction enzymes used in recombinant DNA technology play a major role in determining the location at which the desired gene is inserted into the vector genome. They are two types, namely Endonucleases and Exonucleases.

The Endonucleases cut within the DNA strand whereas the Exonucleases remove the nucleotides from the ends of the strands. The restriction endonucleases are sequence-specific which are usually palindrome sequences and cut the DNA at specific points. They scrutinize the length of DNA and make the cut at the specific site called the restriction site. This gives rise to sticky ends in the sequence. The desired genes and the vectors are cut by the same restriction enzymes to obtain the complementary sticky notes, thus making the work of the ligases easy to bind the desired gene to the vector.

The vectors - help in carrying and integrating the desired gene. These form a very important part of the tools of recombinant DNA technology as they are the ultimate vehicles that carry forward the desired gene into the host organism. Plasmids and bacteriophages are the most common vectors in recombinant DNA technology that are used as they have a very high copy number. The vectors are made up of an origin of replication- This is a sequence of nucleotides from where the replication starts, a selectable marker – constitute genes which show resistance to certain antibiotics like ampicillin; and cloning sites – the sites recognized by the restriction enzymes where desired DNAs are inserted.

Host organism – into which the recombinant DNA is introduced. The host is the ultimate tool of recombinant DNA technology which takes in the vector engineered with the desired DNA with the help of the enzymes.

There are a number of ways in which these recombinant DNAs are inserted into the host, namely – microinjection, biolistics or gene gun, alternate cooling and heating, use of calcium ions, etc.

Process of Recombinant DNA Technology

The complete process of recombinant DNA technology includes multiple steps, maintained in a specific sequence to generate the desired product.

Step-1. Isolation of Genetic Material.

The first and the initial step in Recombinant DNA technology is to isolate the desired DNA in its pure form i.e. free from other macromolecules.

Step-2.Cutting the gene at the recognition sites.

The restriction enzymes play a major role in determining the location at which the desired gene is inserted into the vector genome. These reactions are called 'restriction enzyme digestions'.

Step-3. Amplifying the gene copies through Polymerase chain reaction (PCR).

It is a process to amplify a single copy of DNA into thousands to millions of copies once the proper gene of interest has been cut using restriction enzymes.

Step-4. Ligation of DNA Molecules.

In this step of Ligation, the joining of the two pieces - a cut fragment of DNA and the vector together with the help of the enzyme DNA ligase.

Step-5. Insertion of Recombinant DNA Into Host.

In this step, the recombinant DNA is introduced into a recipient host cell. This process is termed as Transformation. Once the recombinant DNA is inserted into the host cell, it gets multiplied and is expressed in the form of the manufactured protein under optimal conditions.

As mentioned in Tools of recombinant DNA technology, there are various ways in which this can be achieved. The effectively transformed cells/organisms carry forward the recombinant gene to the offspring.

Application of Recombinant DNA Technology

- DNA technology is also used to detect the presence of HIV in a person.
- Gene Therapy - It is used as an attempt to correct the gene defects which give rise to heredity diseases.
- Clinical diagnosis - ELISA is an example where the application of recombinant
- Recombinant DNA technology is widely used in Agriculture to produce genetically-modified organisms such as Flavr Savr tomatoes, golden rice rich in proteins, and Bt-cotton to protect the plant against ball worms and a lot more.
- In the field of medicines, Recombinant DNA technology is used for the production of Insulin.

DNA Cloning

A clone is a cluster of individual entities or cells that are descended from one progenitor. Clones are genetically identical as the cell simply replicates producing identical daughter cells every time. Scientists are able to generate multiple copies of a single fragment of DNA, a gene which can be used to create identical copies constituting a DNA clone. DNA cloning takes place through the insertion of DNA fragments into a tiny DNA molecule. This molecule is made to replicate within a living cell, for instance, a bacterium. The tiny replicating molecule is known as the carrier of the DNA vector.

Yeast cells, viruses, and Plasmids are the most commonly used vectors. Plasmids are circular DNA molecules that are introduced from bacteria. They are not part of the main cellular genome. It carries genes, which provide the host cell with beneficial properties such as mating ability, and drug resistance. They can be conveniently manipulated as they are small enough and they are capable of carrying extra DNA which is weaved into them.

Applications of Gene Cloning

Listed below are the applications of gene cloning:

- Gene Cloning plays an important role in the medicinal field. It is used in the production of hormones, vitamins and antibiotics.
- Gene cloning finds its applications in the agricultural field. Nitrogen fixation is carried out by cyanobacteria wherein desired genes can be used to enhance the productivity of crops and improvement of health. This practice reduces the use of fertilizers hence chemical-free produce is generated
- It can be applied to the science of identifying and detecting a clone containing a particular gene which can be manipulated by growing in a controlled environment
- It is used in gene therapy where a faulty gene is replaced by the insertion of a healthy gene. Medical ailments such as leukaemia and sickle cell anaemia can be treated with this principle.

CRISPR-CAS9: A REVOLUTIONARY GENE EDITING TOOL

Genome editing (also called gene editing) is a group of technologies that give scientists the ability to change an organism's DNA. These technologies allow genetic material to be added, removed, or altered at particular locations in the genome. Several approaches to genome editing have been developed. A well-known one is called CRISPR-Cas9, which is short for clustered regularly interspaced short palindromic repeats and CRISPR-associated protein 9. The CRISPR-Cas9 system has generated a lot of excitement in the scientific community because it is faster, cheaper, more accurate, and more efficient than other genome editing methods.

CRISPR-Cas9 was adapted from a naturally occurring genome editing system that bacteria use as an immune defense. When infected with viruses, bacteria capture small pieces of the viruses' DNA and insert them into their own DNA in a particular pattern to create segments known as CRISPR arrays. The CRISPR arrays allow the bacteria to "remember" the viruses (or closely related ones). If the viruses attack again, the bacteria produce RNA segments from the CRISPR arrays that recognize and attach to specific regions of the viruses' DNA. The bacteria then use Cas9 or a similar enzyme to cut the DNA apart, which disables the virus.

Researchers adapted this immune defense system to edit DNA. They create a small piece of RNA with a short "guide" sequence that attaches (binds) to a specific target sequence in a cell's DNA, much like the RNA segments bacteria produce from the CRISPR array. This guide RNA also attaches to the Cas9 enzyme. When introduced into cells, the guide RNA recognizes the intended DNA sequence, and the Cas9 enzyme cuts the DNA at the targeted location, mirroring the process in bacteria. Although Cas9 is the enzyme that is used most often, other enzymes (for example Cpf1) can also be used. Once the DNA is cut, researchers use the cell's own DNA repair machinery to add or delete pieces of genetic material, or to make changes to the DNA by replacing an existing segment with a customized DNA sequence.

Genome editing is of great interest in the prevention and treatment of human diseases. Currently, genome editing is used in cells and animal models in research labs to understand diseases. Scientists are still working to determine whether this approach is safe and effective for use in people. It is being explored in research and clinical trials for a wide variety of diseases, including single-gene disorders such as cystic fibrosis, hemophilia, and sickle cell disease. It also holds promise for the treatment and prevention of more complex diseases, such as cancer, heart disease, mental illness, and human immuno-deficiency virus (HIV) infection.

Ethical concerns arise when genome editing, using technologies such as CRISPR-Cas9, is used to alter human genomes. Most of the changes introduced with genome editing are limited to somatic cells, which are cells other than egg and sperm cells (germline cells). These changes are isolated to only certain tissues and are not passed from one generation to the next. However, changes made to genes in egg or sperm cells or to the genes of an embryo could be passed to future generations. Germline cell and embryo genome editing bring up a number of ethical challenges, including whether it would be permissible to use this technology to enhance normal human traits (such as height or intelligence). Based on concerns about ethics and safety, germline cell and embryo genome editing are currently illegal in the United States and many other countries.

Mechanism of the classical CRISPR/Cas9 system

The CRISPR/Cas9 system is a heritable adaptive antiviral immune system of prokaryotes that targets infectious invading viruses and bacteriophages and uses RNA-guided nucleases to cut foreign genetic components. It contains two compartments, one for Cas9 endonuclease and one for single-stranded guide RNA (sgRNA).

The sgRNA directs the Cas9 endonuclease to cleave both DNA strands of the target gene in a sequence-specific manner. DNA cleavage occurs at a sequence 3 base pairs upstream of an "NGG" protospacer adjacent motif (PAM).

The genome DNA is repaired by double-strand break (DNA-DSB) repair mechanisms after the cleavage. Therefore, the utilization of the CRISPR/Cas9 gene editing system achieves genome modifications by the introduction of small insertions or deletions (indels) through the relatively error-prone non-homologous end-joining (NHEJ) or the high-fidelity homology-directed repair (HDR) (Fig. 8).

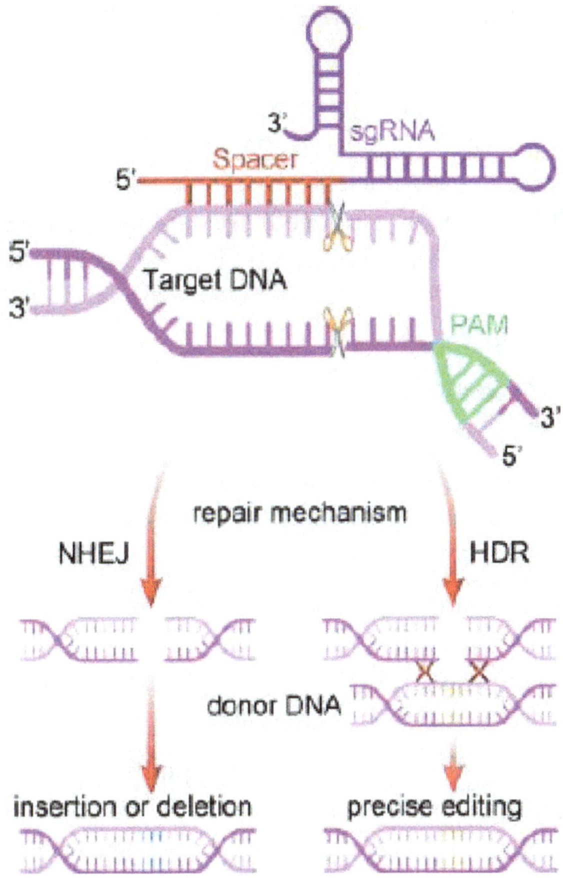

Fig. 8 Mechanism of the CRISPR/Cas9 gene editing system. The single guide RNA (sgRNA) directs the Cas9 nuclease to a complementary sequence in the genome where Cas9 will induce a double strand break (DSB). The target genomic locus must be followed by a 52 -NGG-32 motif (protospacer adjacent motif, PAM) for Cas9 to function. DSB are repaired by non-homologous end joining (NHEJ), or by homology directed repair (HDR) in the presence of a DNA repair template, which can be exploited to introduce precise genetic modifications or exogenous sequences

Development of the CRISPR/Cas9 system and related gene editing tools

In 1987, Japanese scientists discovered some unknown tandem repeats in the *Escherichia coli* genome but did not explore their biological significance.

In 2002, these sequences were named as Clustered Regularly Interspaced Short Palindromic Repeats (CRISPR), but the significance of these sequences remained unknown.

In 2005, three research teams independently discovered the role of CRISPR loci in adaptive immunity, then speculated that protospacer-adjacent motifs (PAMs) might direct the type II Cas9 nuclease to cleave DNA. In 2007, Barrangou et al. proved that the CRISPR system is indeed an adaptive immune system and found that phage gene sequences incorporated by bacteria could change the bacteria's resistance to phages.

In 2008, Brouns et al. found that non-coding RNA transcribed from the CRISPR proto-interregional sequence could guide Cas protein to the target-specific part of DNA to play a defensive role.

In 2011, Deltcheva et al. revealed that trans-coding crRNA (tracrRNA) was involved in the pre-crRNA processing and maturation process and their study revealed new pathways for crRNA maturation.

In 2012, in vitro experiments demonstrated that mature crRNA formed a special double-stranded RNA structure with tracrRNA through base complementary pairing, thus directing Cas9 protein to cause double-stranded fracture on the target DNA.

In 2013, Cong and Mali et al. applied the type II Cas system to the cutting of DNA in mammalian cells, which paved the way for the application of the CRISPR/Cas9 system in genome modification. In the same year, the Cas9 protein mutant dCas9 (endonuclease-deficient Cas9), which had lost nuclease activity, was first developed.

Subsequently, the CRISPR activation (CRISPRa) (Fig. 9a) and interference (CRISPRi) (Fig. 9b) tools were developed by fusing the dCas9 protein with transcription regulators that activate or inhibit gene transcription.

To overcome the problem of unexpected disruptions observed in the CRISPR/Cas9 gene editing system, Komor et al. fused APOBEC (cytosine deaminase) with CRISPR/Cas9. This modified Cas9 makes the C ® T (or G ® A) conversion under the guidance of gRNA without causing DSB. This base editor could effectively correct a variety of point mutations in the genome (Fig. 9c). Furthermore, the adenine bases Editor (ABE) that converts A - T Base pairs to G - C Base pairs was developed.

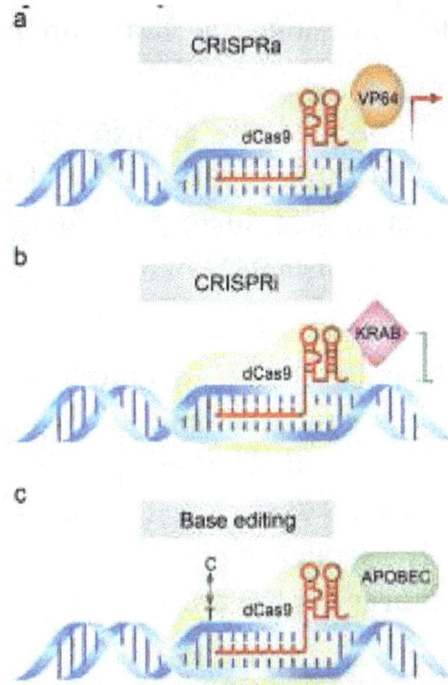

Fig. 9 Schematic diagram of dCas9-based gene editing tools. a CRISPRa: Fusion of deactivated Cas9 (dCas9) with activation domain VP64 induces the expression of gene of interest. b CRISPRi: Fusion of dCas9 with repressor domain KRAB leads to inhibition of gene of interest. c Base editing: Fusion of dCas9 with adenosine deaminase or cytidine deaminase enable introduction of point mutation in the genome

The team further improved the single-base editing and the CRISPR/Cas9 systems, thus reducing greatly the miss rate of the single-base editor and improving the target range of spCas9. Interestingly, Gilpatrick et al. used Cas9 and the adapter ligation to develop a nanopore Cas9-targeted sequence (nCATS) for third-generation nanopore sequencing through modifying the target DNA region and change the structure of the target genome, which can read long fragments at low cost.

Recently, a caged RNA strategy was developed that allows Cas9 to bind to DNA but does not cleave before light-induced activation. This method was named very fast CRISPR (vfCRISPR) and it creates double-strand breaks (DSBs) at the submicrometer and second scales. The vfCRISPR is very accurate and can edit only one allele at a time, providing a basis for investigating complex genetic traits. Collectively, these improvements allowed for the transformation of the CRISPR/Cas9 editing tool from a blunt instrument to a precision instrument.

CRISPR/Cas9 variations

Streptococcus pyogenes Cas9 (SpCas9) was the first to be used outside prokaryotic cells and reprogrammed for genome editing in mammalian cells. SpCas9 targeting of

DNA relies on the 20-nucleotide-long spacer and on the PAM 52 -NGG (N represents any nucleotide).

In 2015, researchers obtained the SpCas9-VRER variant with NGA and the SpCas9-VRER variant with NGCG recognition through the error-prone PCR strategy. In the same year, Nishimasu et al. identified the gene editing role of SaCas9 in mammalian cells; it recognizes the PAM of NNGRRT (N refers to A, T, C, G; R refers to A and G) and has excellent cutting activity and targeting accuracy.

In 2018, researchers built xCas9 3.7 variants that could identify NGG, NG, GAA, and GAT. Furthermore, a more active SpCas9-NG variant was constructed in the Nureki laboratory, and the identified PAM sequence was extended to NG. Recently, researchers developed a SpCas9 variant SpRY that was almost completely liberated from the constraints of the sequence of PAM. These advancements expanded the application scope and accuracy of the CRISPR/Cas9 system in genome editing.

Applications and implications

- CRISPR-Cas9 has a lot of potential as a tool for treating a range of medical conditions that have a genetic component, including cancer, hepatitis B or even high cholesterol.
- Many of the proposed applications involve editing the genomes of somatic (non-reproductive) cells but there has been a lot of interest in and debate about the potential to edit germline (reproductive) cells.
- Because any changes made in germline cells will be passed on from generation to generation it has important ethical implications.
- Carrying out gene editing in germline cells is currently illegal in the UK and most other countries.
- By contrast, the use of CRISPR-Cas9 and other gene editing technologies in somatic cells is uncontroversial. Indeed they have already been used to treat human disease on a small number of exceptional and/or life-threatening cases.

A sperm and egg cell. Carrying out gene editing in germline cells is currently illegal in the UK. Image credit: Shutterstock

Therapeutic applications

CRISPR also raises the prospect of 'correcting' a disease-causing mutation. For example, in people with transthyretin (TTR) amyloidosis (ATTR), fibrils of misfolded TTR slowly accumulate in tissues, especially the heart and nerves, causing cardiomyopathy and polyneuropathy respectively. The lifespan of ATTR patients is typically two to six years after developing symptoms in those with cardiomyopathy and four to 17 years in those with polyneuropathy without cardiac involvement.

Recent interim results from a clinical study suggest that a CRISPR-Cas9 called NTLA-2001 may treat ATTR. In six patients with hereditary ATTR and polyneuropathy, using NTLA-2001 to knock out the underlying gene reduced serum TTR levels by means of 52% and 87% 28 days after CRISPR-Cas9 administration at the two doses tested. The FDA has granted orphan drug designation to NTLA-2001 for ATTR.

CRISPR-Cas9 could help treat several other genetic diseases including sickle cell anaemia, hereditary retinal disorders, beta-thalassemia, cystic fibrosis and Duchenne muscular dystrophy. CRISPR-Cas9 could also lead to new treatments for some cancers and cardiovascular conditions, Huntington's disease and certain viral infections, including HIV and SARS-CoV-2.

For instance, CRISPR could inactivate the protein CCR-5, an important receptor for HIV. This would reduce the chances that the retrovirus will infect T lymphocytes. Recently, researchers reported that a CRISPR-based system using an enzyme called Cas13d can diagnose SARS-CoV-2 in less than an hour. The technology could be included in lateral flow tests, may be more appropriate for resource-poor settings than current diagnostic methods and could help identify emerging pathogens.

"The therapeutic applications of CRISPR are starting to reach clinical trials and can be broadly divided into two camps," Professor Qasim says. "One, where cells are collected from patients or donors, engineered using CRISPR, and then infused into patients. For example, T cells can be modified to prevent alloreactivity [strong immune responses against the foreign protein] in cancer therapy and, thus, be used without matching. Blood stem cells have been modified to promote haemoglobin production to treat sickle cell disease or thalassemia."

In people with beta-thalassaemia, for example, low or absent beta-haemoglobin leads to abnormal red blood cells and, in turn, severe anaemia. Gene-edited autologous haematopoietic stem cells can deliver fetal haemoglobin in erythrocytes to people with beta-thalassaemia, correcting the defect.

"The other approach involves direct applications into the body, either into an organ, such as the eye, or as an infusion into the bloodstream," Professor Qasim adds. "These are more challenging, because of bio-distribution issues and the risk of triggering immune responses." As a foreign protein, bacterial Cas9 can trigger immunological responses, for example.

"One of the first arenas to test gene editing in humans has been immune cell therapy, in particular chimeric antigen receptor T [CAR-T] cells," notes Professor Qasim. Essentially, CAR-T cell therapy involves using apheresis to separate T cells from the blood. These are genetically engineered to express a receptor, such as that binding to a protein expressed by certain leukaemias and lymphomas.

Researchers have used CRISPR to disarm T cells so they don't attack their host but still attack the cancer. Cells can also be made invisible to parts of the host immune system or can evade the effects of certain drugs and so persist longer," Professor Qasim says. Currently, CAR-T cells are made individually for each patient using the patient's T cells. This makes the treatment time-consuming and expensive. "Research is ongoing, but we think it will be possible to manufacture banks of CAR-T cells that can be kept frozen until required and then used without matching.

"Such 'universal' cells could ultimately provide less toxic and more effective alternatives to current CAR-T cells," Professor Qasim says. "We have trials underway testing the approach in B cell leukaemias and will extend to acute myeloid leukaemia and T cell cancers shortly. Indeed, almost every disorder caused by a single gene and many types of blood cancer could be amenable to genetic correction. The tools are evolving and will need to be applied and investigated systematically, and long-term strategies are needed to track side-effects."

Future of CRISPR-Cas9

- It is likely to be many years before CRISPR-Cas9 is used routinely in humans.
- Much research is still focusing on its use in animal models or isolated human cells, with the aim to eventually use the technology to routinely treat diseases in humans.
- There is a lot of work focusing on eliminating 'off-target' effects, where the CRISPR-Cas9 system cuts at a different gene to the one that was intended to be edited.

Better targeting of CRISPR-Cas9

- In most cases the guide RNA consists of a specific sequence of 20 bases. These are complementary to the target sequence in the gene to be edited. However, not all 20 bases need to match for the guide RNA to be able to bind.
- The problem with this is that a sequence with, for example, 19 of the 20 complementary bases may exist somewhere completely different in the genome. This means there is potential for the guide RNA to bind there instead of or as well as at the target sequence.
- The Cas9 enzyme will then cut at the wrong site and end up introducing a mutation in the wrong location. While this mutation may not matter at all to the individual, it could affect a crucial gene or another important part of the genome.

- Scientists are keen to find a way to ensure that the CRISPR-Cas9 binds and cuts accurately. Two ways this may be achieved are through:
- the design of better, more specific guide RNAs using our knowledge of the DNA sequence of the genome and the 'off-target' behaviour of different versions of the Cas9-gRNA complex.
- The use of a Cas9 enzyme that will only cut a single strand of the target DNA rather than the double strand. This means that two Cas9 enzymes and two guide RNAs have to be in the same place for the cut to be made. This reduces the probability of the cut being made in the wrong place.

CHAPTER-4

CRISPR-CAS9: THE MOLECULAR SCISSORS

CRISPR stands for Clustered Regularly Interspaced Short Palindromic Repeats. It is a component of bacterial immune systems that can cut DNA, and has been repurposed as a gene editing tool. It acts as a precise pair of molecular scissors that can cut a target DNA sequence, directed by a customizable guide.

The system is made up of two key parts: a CRISPR-associated (Cas) nuclease, which binds and cuts DNA, and a guide RNA sequence (gRNA), which directs the Cas nuclease to its target. It was discovered in bacterial immune systems, where it cuts the DNA of invading viruses, called bacteriophage, and disables them. Once the molecular mechanism for its DNA-cleaving ability was discovered, it was quickly developed as a tool for editing genomes.

CRISPR is important because it allows scientists to rewrite the genetic code in almost any organism. It is simpler, cheaper, and more precise than previous gene editing techniques. Moreover, it has a range of real-world applications, including curing genetic disease and creating drought-resistant crops.

The History and Discovery Of CRISPR

Today, CRISPR is well-known as a precise gene editing tool, but it took many years for scientists to figure out what it was and how to harness its potential. Let's discuss the inventors of this gene editing tool, the scientists who championed this technology, and the history behind these breakthroughs.

CRISPR was discovered by Dr. Jennifer Doudna (UC Berkeley, seen on right in image above) and Dr. Emmanuelle Charpentier (Max Planck Unit for the Science of Pathogens, Berlin, left). Their groundbreaking paper, revealing that the CRISPR-Cas9 bacterial immune system could be repurposed as a gene editing tool, was published in the journal Science in 2012.

It wasn't until 2020 - well after it had been adopted in labs around the world - that Doudna and Charpentier won the Nobel prize in Chemistry for their discovery, becoming the first all-female team to do so.

Other significant contributors include Feng Zhang at the Broad Institute of MIT and Harvard in Cambridge, Massachusetts, who pioneered the use of CRISPR in eukaryotic cells and discovered novel Cas variants, George Church at Harvard Medical School in Boston, Massachusetts, who was among the first to demonstrate its use in human cells, and biochemist Virginijus Siksnys at Vilnius University in Lithuania, who independently discovered the ability of CRISPR to edit genes in other organisms. For more information on these early pioneers, you can check out our blog about prominent CRISPR scientists.

History of CRISPR

While Doudna and Charpentier were the first to adapt CRISPR-Cas9 as a gene editing tool, the history goes back a little further than their 2012 publication. In 1993, Dr. Fransisco Mojica, a scientist at the University of Alicante in Spain, identified repetitive palindromic segments of DNA interspaced with other fragments of genetic material in bacterial genomes. Dr. Mojico gave these regions the name CRISPR, and proposed that they are a component of the bacterial immune system. In 2007, a team of scientists led by Dr. Philippe Horvath experimentally demonstrated Mojica's theory.

Sincc its adaptation by Dr. Doudna and Dr. Charpentier, this versatile gene editing technology has progressed rapidly. It has been adapted for many different purposes, including RNA editing, base and prime editing, live imaging, and diagnostics. It has been used to edit DNA in a variety of organisms, including humans.

In 2019, the first CRISPR clinical trials began, harvesting cells from patients with sickle cell disease (SCD) and editing them in vitro before infusing them back into the body - a method known as cell therapy. After the success of SCD cell therapy trials, a CRISPR treatment was injected directly into human patients for the first time in 2020. This technique is known as gene therapy, and was used to treat hereditary blindness.

CRISPR/Cas9 is an RNA-guided immune system that protect Archaea and Bacteria cells against genome invading viruses and plasmids. The puzzling CRISPR DNA segments were first observed in 1980, but their function was not well understood for almost two decades. In general, when an unknown virus attacks, bacteria incorporate viral DNA sequences into the repetitive sequences of their genome. Thus, when the bacteria again encounter that virus, bacteria use the DNA in these repetitive sequences to synthesize RNAs that recognize the matching viral sequences.

Next, a protein attached to one of these RNAs cut up the viral DNA. Two forms of gRNAs are used to facilitate Cas9-mediated DNA cleavage, these are a Crispr RNA (crRNAs) array and trans-activating crRNA (tracrRNA) (Fig. 1). This crRNA in association with tracrRNA guide the Cas9 protein to a region called the 'protospacer adjacent motif' (PAM), a three-nucleotide sequence (NGG) juxtaposed to the DNA complementary region in the DNA (Figure 1). In the presence of an appropriate homologous donor, the CRISPR/Cas9 system may be used to generate defined modifications and insertions at a targeted locus through the homologous DNA repair process. In the absence of a donor, single DSBs generated by CRISPR/Cas9 are repaired through the error-prone non-homologous end joining (NHEJ), which is prone to insertion or deletion mutations.

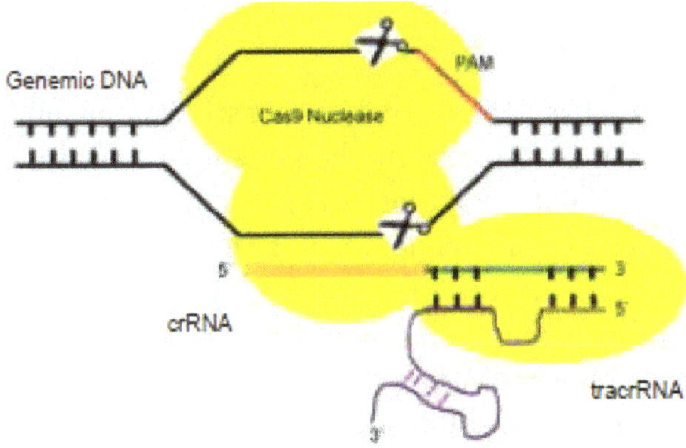

Fig. 1 The components of CRISPR/Cas9 System.
Figure showing the three components: 1. crRNA specifically targets genomic DNA site. 2. tracrRNA- that forms a complex with crRNA. 3. Cas9 Nuclease

Recently, several research groups have been adapted CRISPR/Cas9 system for site-specific genome editing in diverse cell types and organisms such as zebrafish, Drosophila, mice and human cells. A recent report also demonstrated the disruption of a Green Fluorescent Protein transgene in mice using the CRISPR/Cas9 system. Yang et al. created reporter and conditional mutant mice by co-injection of zygotes with Cas9 mRNA and different guide RNAs (gRNAs) as well as DNA vectors of different sizes. This group generated mice carrying a fluorescent reporter construct with different genes. They also analyzed potential off-targets of five gRNAs in gene-modified mice and ESc lines and identified off-target mutations are rare. In a recent study, Zheng et al. used two gRNAs coupled with Cas9 and efficiently created DNA deletions of 10kb. Targeted deletion with CRISPR/Cas9 observed to be independent of the transcriptional status of the targeted locus.

Tools for genome editing seem to be improving faster than software technologies. Recently developed genomic editing technologies have the potential to generate different transgenic animals because of their ability to modulate genes.

CRISPR/Cas9 systems are found to be comparable to ZFN and TALEN regarding activity but cheaper and ease to use. Mashiko et al. generated mutant mice by injecting humanized Cas9 (hCas) mRNA and single guide RNA into fertilized eggs. The same group also produced mutant mice by injecting circular plasmids expressing Cas9 instead of Cas9 mRNA and sgRNA into mouse zygotes.

Interestingly, multiple gRNAs can be used to edit multiple sites within the genome, demonstrating the one-step generation of animals carrying multiple genes mutations. Zheng et al. showed that CRISPR/Cas-mediated gene editing could disrupt multiple genes in mouse ES cells with 80% efficiency. Thus, CRISPR/Cas technology greatly accelerate the *in vivo* study of functionally redundant genes and of epistatic gene interactions.

Palindromes in the genome

"Able was I ere I saw Elba": this rather bombastic statement is a palindrome sentence, in other words it reads exactly the same forwards as it does backwards. The beginning of the CRISPR revolution was marked by the discovery of a large number of repeated palindromic sequences in a region of bacterial DNA. In these sequences, the letters of the genetic code, the four base molecules adenine, cytosine, thymine and guanine, are ordered such that they have the same order as the second complementary DNA-strand – in this case read in opposite direction. This is the property that gives CRISPR (Clustered Regularly Interspaced Short Palindromic Repeats) its tongue-twisting name.

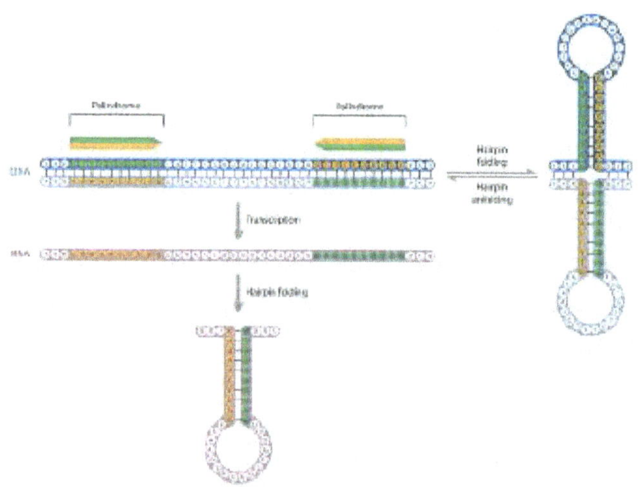

Fig. 2 Palindrome sequence in the DNA of the bacterium Streptococcus agalactiae. Parts of the letter sequence of one strand correspond to those of the other strand in the reverse order. However, the palindrome is not perfect. It also contains a non-palindromic sequence. The DNA can form hairpin structures using a broken palindrome like this one.

Unlike word palindromes like 'civic' and 'tenet', which have a meaning, the palindromes in the genetics dictionary do not make sense and cannot be translated into functional proteins. Nevertheless, they are not entirely meaningless. DNA-cutting proteins frequently use palindrome sequences as recognition sequences, at which they cut the DNA molecule. These sequences can be four, six or eight base pairs long, although some cutting proteins require 20 or more base pairs.

Out of the palindromic sequences of the CRISPR region RNA molecules are transcribed, which adopt a very stable arrangement (secondary structure). They range between 23 and 47 base pairs in length. Variable regions of similar length can be found between these sequences. They originated from the genome of foreign DNA that penetrated the bacterial cell, and are also known as spacer DNA.

The CRISPR region includes a promoter which ensures that the CRISPR region can be read and translated into the CRISPR-RNA (crRNA). Other genes known as CRISPR-associated genes (Cas) are located adjacent to it. These genes provide the blueprint for the Cas proteins - namely, the enzymes that cut the DNA strand. The CRISPR and spacer sequences are followed by a region for an RNA molecule known as tracrRNA, which guides the cutting molecules and the crRNA to their target locations on the virus DNA.

CRISPR Mechanism

The CRISPR system is the basis of adaptive immunity in bacteria and archaea. It utilizes Cas nucleases, which are enzymes that can bind and create double-stranded breaks (DSBs) in DNA. When a bacterium is infected by a virus, it uses a Cas nuclease to snip off a piece of viral DNA known as a protospacer. This fragment is stored in the bacterial genome with fragments from other viruses that have previously infected the cell - an immune memory. These viral spacer fragments are placed between repeated palindromic sequences, and this arrangement of spacers and palindromic repeats is what gives CRISPR its name.

Upon reinfection with the same virus, the bacterium can recognize and destroy it with Cas9. Cas9 activity relies on a CRISPR RNA (crRNA) and a trans-activating CRISPR RNA (tracrRNA). The crRNA is complementary to the viral spacer that was stored after the original infection, while the tracrRNA serves as a scaffold; these two RNAs form a complex known as a guide RNA (gRNA). Think of the Cas9 as scissors, and the gRNA as the hand that's directing them to cut.

Before cutting, the Cas9 acts as a search tool, checking the viral DNA for the protospacer adjacent motif (PAM), a short sequence downstream of the target site. When it recognizes PAM, Cas9 checks the region upstream - if it locates the target provided by the gRNA, it will create a double-stranded break (DSB). DSBs incapacitate the virus because viruses lack their own DNA repair mechanisms.

Once Dr. Doudna and Dr. Charpentier revealed the molecular mechanism that enabled the natural CRISPR-Cas9 system to cut DNA, the next question was obvious: if Cas9 is given a different guide RNA sequence, could it be used to create cuts at any desired location in an organism's genome? Their theory was correct, and the results were groundbreaking.

The natural gRNA complex was engineered into a chimeric single guide RNA (sgRNA), offering a simple and cost-effective method of genetic manipulation. Researchers only needed to provide a different guide RNA - which can be generated with relative ease - and Cas9 could be used to create cuts at a range of target sites in the DNA of any organism (relying on the presence of the correct PAM sequence).

CRISPR-Cas9 gene editing works by creating double-stranded breaks in the DNA and then taking advantage of cellular DNA repair pathways. While there are several DNA repair pathways, the key ones used for gene editing are non-homologous end joining (NHEJ) and homology-directed repair (HDR). NHEJ is exploited to render genes non-functional, while HDR is exploited to insert new genes or fragments of genetic material.

CRISPR Methods and Techniques

If Cas9 creates a DSB, it will most likely be repaired by NHEJ. However, NHEJ is error-prone, and it usually results in insertions and deletions (indels) in the region being repaired. When indels occur within the coding region of a gene and result in a frameshift mutation, the gene becomes non-functional. This is known as a gene knockout (KO).

Gene knockouts are used in a range of research areas, including functional genomics, pathway analysis, drug discovery and screening, and disease modeling. Using multiple guide RNAs that target various regions of the gene ensures high efficiency gene knockout, a method that is gaining popularity.

CRISPR knock-in

In the presence of a DSB induced by Cas9, cells can also repair themselves via HDR, and this pathway offers an opportunity for researchers to insert a new piece of DNA or an entire gene. This method is known as a gene knock-in.

To achieve a gene knock-in, there must be a DNA template for repair known as a donor template. This donor template consists of the sequence or gene of interest flanked by regions of homology that match the area on either side of the cut. The donor is delivered to the cells being engineered, along with the other editing components of Cas9 and sgRNA.

Gene knock-ins have been a key breakthrough in biotechnology, including the production of recombinant proteins, increasing the viability of immortalized cell lines, and precision disease modeling. Perhaps most importantly, CRISPR knock-in can be used in cell and gene therapies to correct genetic mutations that cause human disease.

Compared with gene knockout, gene knock-ins are more challenging. This is because HDR is a less common repair pathway than NHEJ, only occurring at certain stages of the cell cycle.The low frequency of HDR typically results in low knock-in efficiencies. However, scientists have been creating methods to overcome this obstacle, including detailed experimental optimization, cell cycle synchronization, and treatments that either boost HDR or disable NHEJ for knock-in experiments.

CRISPRa & CRISPRi

CRISPR-Cas9 can be used to delete (KO) or insert (KI) genes, but with slight modifications, it can also be used to regulate the expression of genes. This is known as CRISPR activation (CRISPRa) and CRISPR interference (CRISPRi). CRISPRa is used to increase (upregulate) the expression of a gene, while CRISPRi can reduce (downregulate) the expression of a gene.

Both of these technologies function by fusing an engineered variant of Cas9 known as catalytically dead Cas9 (dCas9) with transcriptional effectors to modulate target gene expression. dCas9 cannot cut the DNA, and so it simply leads the transcriptional effectors to the gene of interest.

These gene modulators have many research applications, including developmental biology, infectious disease, disease progression, functional genomics, and screening for genetic elements that mediate drug resistance.

CRISPR screens

CRISPR-Cas9 technology has enabled highly accurate, large-scale screening studies for drug discovery research, allowing us to more easily unravel the relationships between genotype and phenotype. CRISPR screens typically involve generating a large sgRNA library targeting different genes, making a wide range of edits to a particular cell line, then analyzing the effects of the edits on the phenotype of the cells.

By performing knockouts of multiple genes in a healthy cell line, researchers are able to identify which genes are involved in disease pathogenesis, while the same method in a diseased cell line can identify which genes are ideal targets for drugs. CRISPR screens provide more consistent results with less off-target effects than the RNAi screens that were previously employed for drug discovery. The multi-guide sgRNA knockout approach used by Synthego makes these screens even more reliable for target identification and validation.

Base editing and prime editing

Some of the most recently developed CRISPR methods are base editing and prime editing. These technologies work on the same principle, however at a more precise scale, inducing single nucleotide substitutions. Importantly, base editing and prime editing do not induce DSBs in the target DNA.

Base editing uses either a catalytically dead Cas9 (dCas9) or a nickase Cas9 (nCas9). dCas9 is incapable of cutting DNA, while nCas9 produces 'nicks', or single-stranded breaks (SSBs) in the DNA. By fusing either dCas9 or nCas9 to a DNA modifying enzyme, researchers can alter specific nucleotides. One of the limitations of base editing is that they cannot be used to alter every possible nucleotide, and this is one of the factors that led to the development of prime editing.

Prime editing involves fusing nCas9 to an engineered reverse transcriptase and a prime editing guide RNA (pegRNA). The pegRNA contains two sections: one that guides to the region of interest, and another that contains the desired substitution/s for repair after the single-stranded cut has been generated. After one strand has been altered by the prime editor, the complementary strand can also be corrected - an additional gRNA and nCas9 will create a nick in the strand and it will be repaired using the previously edited strand as a template. Prime editing is predicted to be capable of treating 89% of genetic mutations in humans.

Important Applications of CRISPR Technology

In just a few short years, CRISPR has had a massive impact on scientific research, contributing to breakthroughs in medicine and biotechnology. Let's take a closer look at some of the key applications of this technology.

Cell and gene therapies

CRISPR is poised to revolutionize medicine, with the potential to cure a range of genetic diseases, including neurodegenerative disease, blood disorders, cancer, and ocular disorders. As we mentioned earlier, the first trial of a CRISPR cell therapy was performed in 2019, treating patients with sickle cell disease. The treatment restored fetal hemoglobin, eliminating the need for a functional copy of adult hemoglobin.

In 2021, a significant CRISPR trial for transthyretin amyloidosis, a neurodegenerative disease, showed very promising results. It is also revolutionizing pediatric cancer treatment, described in this podcast interview with Shondra Miller from St Jude's.

CRISPR can also be used to generate chimeric antigen receptor (CAR) T cells, a form of immunotherapy used to treat cancer. The T cells are extracted from patients and engineered to express chimeric antigen receptors before being re-injected into the body. The receptors allow the T cells to more efficiently target and destroy the specific type of cancer the patient suffers from.

While we are still in the early years of clinical trials, this technology could be used to treat thousands of genetic conditions in the future, including breast and ovarian cancer linked to BRCA mutations, Huntington's disease, Tay-Sachs, beta-thalassemia, cystic fibrosis, and early-onset Alzheimer's. For all the latest medical developments and clinical trials using this technology to cure a range of human diseases, you can take a look at the CRISPR Medicine News website.

Diagnostics

During the COVID 19 pandemic, CRISPR was used as both a potential therapeutics tool and as a diagnostic tool for the coronavirus. The SHERLOCK™ CRISPR SARS-CoV-2 test kit was granted Emergency Use Authorization from the federal authorities to be used in laboratory settings. You can learn more about SHERLOCK and the more recently developed STOPCovid diagnostic test here and in this podcast.

Mammoth Biosciences has also developed a CRISPR-based Covid-19 diagnostic method, known as DETECTR. Like SHERLOCK and STOPCovid, DETECTR utilizes Cas9's search function to detect genetic material from the virus, employing naturally occurring Cas nucleases, like Cas12 and Cas13. For more information on DETECTR, you can listen to this interview with Trevor Martin, CEO of Mammoth Bioscience.

Similar diagnostics utilizing the search function of Cas9 have also been engineered to identify other diseases, both infectious and genetic. Early in 2021, Dr. Kiana Aran of Cardea Bio published a study which combined three different Nobel Prize-winning technologies - graphene, transistors, and CRISPR - into a tiny chip that can detect pathogenic single nucleotide polymorphisms (SNPs). Since 50% of disease-causing mutations in humans are SNPs, this is a significant breakthrough in medical diagnostics.

Agriculture

Gene editing technology has huge potential in agriculture, and experts suggest that CRISPR-modified foods will be available within 5-10 years. This is primarily because it can be used to create crops that are disease-resistant and drought-resistant. For example, scientists from the University of Berkeley and Innovative Genomics Institute have partnered with Mars, Inc. to create disease-resistant cacao plants.

It can also be used to prolong the shelf-life of other perishable foods, reducing food waste and allowing access to healthy foods at relatively low cost. For more information on these applications, you can read our overview of CRISPR's use in agriculture.

Bioenergy

As one of the leading alternatives to fossil fuels, bioenergy has been under the spotlight for a while now. However, there are several hurdles to producing biofuels at scale. By using CRISPR, scientists have recently been able to make some significant advances in this area.

For example, KO of multiple transcription factors that control production of lipids in algae has led to a huge increase in lipid production for generating biodiesel. Similarly, gene editing can improve the tolerance of yeast to harsh conditions during the production of biofuels. It has also increased editing efficiencies in bacterial species that are used to produce ethanol. For more details, you can check out this blog on how CRISPR is helping the biofuel industry.

CRISPR in Disease Models and Animal Models

The CRISPR system has enabled a faster and cheaper production of animal models compared with traditional gene-targeting methods, and can provide much more precise models of disease. Let's take a look at gene editing progress with respect to cell and animal models.

Stem cells and organoids

CRISPR editing has been a game-changer for stem cell research. Stem cells are pluripotent, meaning that they can differentiate into any type of cell. This makes them incredibly valuable in medical research, however harvesting stem cells from embryos is highly controversial. The advent of induced pluripotent stem cells (iPSCs) was a major breakthrough, because they are adult cells (like skin or blood cells) that have been reprogrammed to become pluripotent.

While human iPSCs avoid the ethical concerns of embryonic stem cells, they remain challenging to work with - not only is it difficult to reprogram and grow iPSCs, but they are also quite resistant to genetic manipulation. CRISPR has not only made reprogramming iPSCs easier, but has also delivered much more efficient results than previous gene editing technologies which were originally used to engineer iPSCs.

CRISPR-edited iPSCs have been successfully developed as cell therapies, like the CTX001 trial for sickle cell disease we mentioned earlier, and large-scale disease modeling studies, such as the 2021 Inducible Pluripotent Stem Cell Neurodegeneration Initiative (iNDI). Recently, CRISPR was used to generate triple-knockout, 'off-the-shelf' CAR T cells which are hypoimmunogenic, meaning that they can be used to treat any patient - a massive breakthrough for T cell immunotherapy.

CRISPR-Cas9 edited iPSCs can also be used to generate organoids - tiny, three-dimensional models of human organs that can be grown in the lab. Organoids are more relevant to the study of human disease than cell cultures that are maintained as monolayers or suspensions, because they provide more of the complexity of real human organs and can therefore more accurately mimic disease progression and response to treatment. Recently, edited iPSCs were used to generate brain organoids to study how the human brain evolved.

Primary cells

Primary human cells are incredibly valuable for disease modeling as they provide more accurate biological data than immortalized cell lines, and gene-edited primary cells can be used for both disease modeling and cell therapy. However, successful culturing and genetic manipulation of primary cells have been significant challenges in biomedical research.

In recent years, scientists have provided solutions to increase CRISPR editing efficiency in primary cells, including synchronizing the cell cycle, the use of chemically modified sgRNAs, creating ribonucleoprotein (RNP) complexes for delivery, and optimizing delivery methods based on the specific cell type. These advances have led to many successful cell therapies, including T cell immunotherapies.

CRISPR animal models

Fig. 3

Traditional animal models of human disease are useful, but they have several limitations, including a lack of translational capacity. With CRISPR, however, creating transgenic animals that more accurately recapitulate human disease states has become significantly easier. CRISPR editing in animals is much more efficient than previous genome engineering technologies - it can generate almost any required edit, as well as having the capacity to generate multiple edits within the same organism.

CRISPR can be used to generate 'humanization knock-ins' in animals like mice - deleting a particular gene or region of DNA in the animal and replacing it with the human version. This level of accuracy is key to understanding and treating human disease. For example, gene-edited animal models of Duchenne Muscular Dystrophy have recently led to significant breakthroughs in treating this disease, including the development of a potential gene therapy, experimentally demonstrated in dogs. It has also been used to create large animal models of neurodegenerative disorders which may lead to clinical trials in the near future.

CRISPR Protocols: Learn How To Use CRISPR

Now that you're more familiar with the basics of CRISPR and its applications, you're probably curious about how to use it in your own research. In this section, we'll take you through the main steps of designing your own experiment. A more detailed chapter-by-chapter guide for designing your gene editing experiment is also available.

Guide RNA design

Guide RNA design is one of the most crucial aspects of designing a successful CRISPR experiment. The choice of sgRNA will very much depend on the type of edit you require, whether it be KO, KI, base or prime editing, CRISPRi, or CRISPRa.

Synthego's CRISPR Design Tool is a free resource available to all researchers globally, enabling the design and selection of guide RNA sequences to target any gene across over 120,000 genomes. Synthego also has an in-depth guide to choosing the right gRNA for your experiment.

Choosing the right Cas9 variant

The choice of Cas nuclease is also important for the success of your experiment. Cas9 was the first and most commonly used Cas nuclease, originally isolated from the bacterial strain Streptococcus pyogenes (SpCas9). However, there are now many different naturally occurring and engineered Cas variants available for different purposes.

For example, engineered Cas9 nickases are used to create single-stranded breaks and high-fidelity Cas9 variants have been engineered to reduce off-target edits, while naturally-occurring Cas 13 is used for its diagnostic potential.

Transfection protocols

There is no universal transfection protocol for all experiments, because delivery of editing components depends on the cell and organism you're targeting, as well as the type of editing cargo being delivered. There are various transfection protocols available, and you can also check out our in-depth blog on delivery methods.

Researchers at Synthego have determined that the best method for introducing gene editing components into the majority of cell types involves the nucleofection of ribonucleoproteins (RNP) complexes consisting of chemically modified synthetic sgRNA and purified recombinant Cas9.

Analyzing your CRISPR experiment

The last step in evaluating your experiment is picking an analytics tool to confirm that editing was successful. Like every other step, this must be tailored to the experiment. The gold standard is to use Next Generation sequencing (NGS), however this is not always in every lab's budget, especially if you need to sequence frequently or have many samples to run.

Synthego's Inference of CRISPR Edits (ICE) is an online tool for analyzing Sanger sequencing data, which is cheaper than NGS. ICE is user friendly, open-source, and highly accurate, with results comparable to NGS.

UNDERSTANDING CRISPR: CLUSTERED REGULARLY INTERSPACED SHORT PALINDROMIC REPEATS

CRISPR (short for "clustered regularly interspaced short palindromic repeats") is a technology that research scientists use to selectively modify the DNA of living organisms. CRISPR was adapted for use in the laboratory from naturally occurring genome editing systems found in bacteria.

CRISPR/Cas9

Clustered regularly interspaced short palindromic repeats/Cas9 genome editing system was first discovered as an RNA-guided defense mechanism for bacteria against foreign genetic elements by viruses or phages in order to respond and eliminate their invading genetic elements. There exists three types of CRISPR mechanisms and the type II CRISPR system has been studied the most. The simplicity of the type II CRISPR nuclease system consisting of three components Cas9 protein, the CRISPR RNA (crRNA) and a trans-activating crRNA (tracrRNA) has allowed it to be used as the most favorable genome editing system to date. The type II requires a 20 nucleotide guide sequence as part of the crRNA and tracrRNA partially complementary to the crRNA, and a Cas9 endonuclease protein to cleave the genomic DNA.

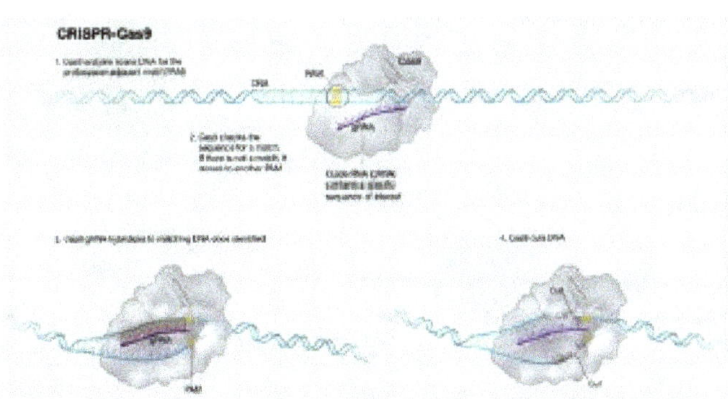

Fig. 4

Recent advances has allowed synthesis of single-guide RNA (sgRNA) consisting of a fusion of crRNA and tracrRNA *in vitro*. In the Cas9 nuclease protein in type II CRISPR/Cas system can cleave target DNA at specific sites producing a double-stranded DNA break.

Following the DNA break, two repair mechanisms exist, non-homologous end joining which is error prone and can lead to mutations by insertion/deletion (indel) or homology-directed repair which is an alternative DNA repair mechanism, in the presence of a repair template, precise and defined modifications are generated at the DNA target sites. Recent improvement in the technology and research has lead to novel ideas such as, CRISPR/Cpf1 genome editing system of the bacterium *Francisella novicida*, CRIPSR Ribonucleoprotein (RNP) and CRISPRa/i (activation/interference). So far, the CRIPSR/Cas9 system has been the only one used in pig genome editing. In 2014, the first use CRISPR/Cas 9 to generate pig's knockout cells, demonstrated its potential in complex genome engineering (Fig. 5).

CRISPR/Cas9 Modified Genes

Modifying the genes which cause the immunologic responses to the porcine grafted organs, namely, HAR, AHXR, immune cell-mediated rejection, and chronic rejection have enhanced survival rate of organ xenotransplantation. Here, after introducing the importance of the CRISPR modified genes viz. cytidine monophosphate-*N*-acetylneuraminic acid hydroxylase (CMAH), B1,4N-acetylgalactosaminyltransferase (B4GALNT2), isoglobotrihexosylceramide synthase (iGb3s), class I MHC, von Willebrand factor (vWF), C3, and porcine endogenous retroviruses (PERVs) in xenotransplantation, we will describe the different approaches which have been used for applying CRISPR/Cas9 to target these genes.

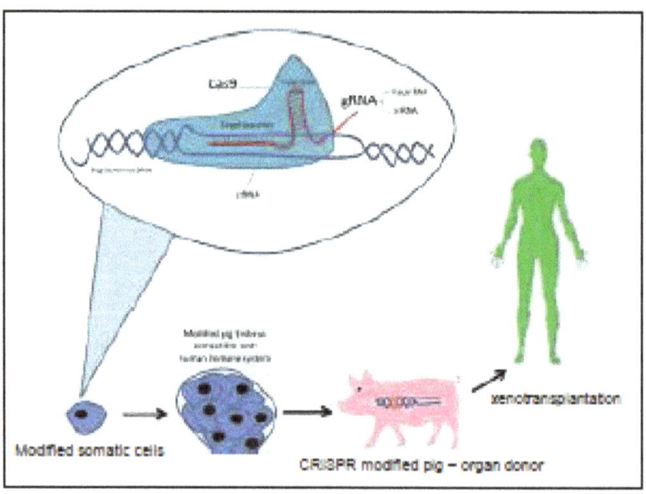

Fig. 5. CRISPR system in xenotransplantation.

RISPR is a powerful tool for editing genomes, meaning it allows researchers to easily alter DNA sequences and modify gene function. It has many potential applications, including correcting genetic defects, treating and preventing the spread of diseases, and improving the growth and resilience of crops. However, despite its promise, the technology also raises ethical concerns.

In popular usage, "CRISPR" (pronounced "crisper") is shorthand for "CRISPR-Cas9." CRISPRs are specialized stretches of DNA, and the protein Cas9 - where Cas stands for "CRISPR-associated" - is an enzyme that acts like a pair of molecular scissors, capable of cutting strands of DNA.

CRISPR technology was adapted from the natural defense mechanisms of bacteria and archaea, a domain of relatively simple single-celled microorganisms. These organisms use CRISPR-derived RNA, a molecular cousin to DNA, and various Cas proteins to foil attacks by viruses. To foil attacks, the organisms chop up the DNA of viruses and then stow bits of that DNA in their own genome, to be used as a weapon against the foreign invaders should those viruses attack again.

When the components of CRISPR are transferred into other, more complex, organisms, those components can then manipulate genes, a process called "gene editing." No one really knew what this process looked like until 2017, when a team of researchers led by Mikihiro Shibata of Kanazawa University in Japan and Hiroshi Nishimasu of the University of Tokyo, showed for the very first time what it looks like when a CRISPR is in action.

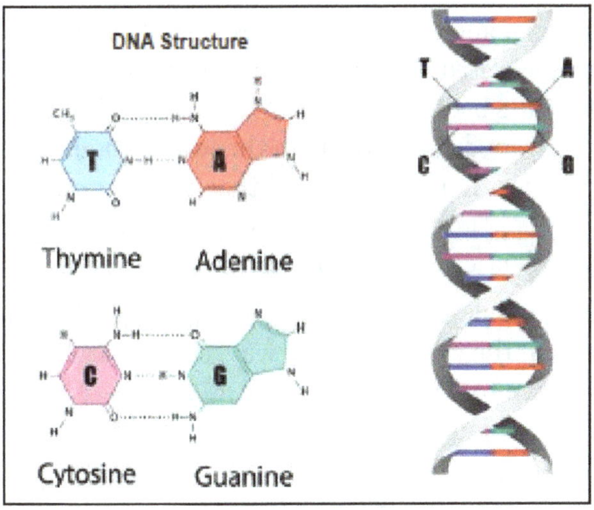

Fig. 6 DNA is a double-stranded molecule whose "rungs" are made up of one of two base pairs: adenine paired with thymine or cytosine paired with guanine.

CRISPRs: The term "CRISPR" stands for "clusters of regularly interspaced short palindromic repeats" and describes a region of DNA made up of short, repeated sequences with so-called "spacers" sandwiched between each repeat.

When we talk about repeats in the genetic code, we're talking about the ordering of rungs within the spiral ladder of a DNA molecule. Each rung contains two chemical bases bound together: A base called adenine (A) links up to another called thymine (T), and the base guanine (G) pairs with cytosine (C).

In a CRISPR region, these bases appear in the same order several times, and in these repeated segments, they form what's known as "palindromic" sequences, according to the Max Planck Institute. A palindrome, like the word "racecar," reads the same forward as it does backward; similarly, in a palindromic sequence, bases on one side of the DNA ladder match those on the opposing side when you read them in opposite directions.

For example, a super simple palindromic sequence might look like this:

- Side 1 - GATC
- Side 2 - CTAG

Short palindromic repeats appear throughout CRISPR regions of DNA, with each repeat bookended by "spacers." Bacteria swipe such spacers from viruses that have attacked them, meaning they incorporate a bit of viral DNA into their own genome. These spacers serve as a bank of memories, which enables the bacteria to recognize the viruses if they should ever attack again. You can also think of spacers like "Wanted" posters, providing a snapshot of the bad guys so they can be easily spotted and brought to justice.

Rodolphe Barrangou and a team of researchers at Danisco, a food ingredients company, first demonstrated this process experimentally. The researchers used *Streptococcus thermophilus* bacteria, which are commonly found in yogurt and other dairy cultures, as their model, according to the Joint Genome Institute, part of the U.S. Department of Energy. They observed that after a viral attack, the bacteria incorporated new spacers into their CRISPR regions. Moreover, the DNA sequence of these spacers was identical to parts of the virus genome.

The team also manipulated the spacers by removing them and inserting new viral DNA sequences in their place. In this way, the researchers were able to alter the bacteria's resistance to an attack by a specific virus, confirming CRISPRs' role in regulating bacterial immunity.

CRISPR RNA (crRNA): CRISPR regions of DNA act as a kind of bank of viral memories; but for that stored information to be useful elsewhere in the cell, it must be copied, or "transcribed," into a different genetic molecule called RNA. Unlike DNA sequences, which remain lodged inside the DNA molecule, this CRISPR RNA (crRNA) can roam about the cell and team up with proteins — namely the molecular scissors that snip viruses to bits.

RNA also differs from DNA in that it's only one strand, rather than two, meaning it looks like just a half of a ladder. To build an RNA molecule, one part of the CRISPR acts as a template and proteins called polymerases swoop in to construct an RNA molecule that is "complementary" to that template, meaning the two strands' bases would fit together like puzzle pieces. For example, a G in the DNA molecule would get transcribed as a C in the RNA.

Each snippet of CRISPR RNA contains a copy of a repeat and a spacer from a CRISPR region of DNA, according to a 2014 review by Jennifer Doudna and Emmanuelle Charpentier, published in the journal Science. The crRNA interacts with the Cas9 protein and another kind of RNA, called "trans-activating crRNA" or tracrRNA, in order to help bacteria fend off viruses.

Cas9: The Cas9 protein is an enzyme that cuts foreign DNA. The protein binds to crRNA and tracrRNA, which together guide Cas9 to a target site on the virus's DNA strand where the protein will make its cut. The target DNA that the Cas9 will cut through is complementary to a 20-nucleotide stretch of the crRNA, where a "nucleotide" is a building block of DNA that contains one base.

Using two separate regions or "domains" on its structure, Cas9 cuts both strands of the DNA double helix, making what is known as a "double-stranded break," according to the 2014 Science article.

There is a built-in safety mechanism that ensures that Cas9 doesn't just cut just anywhere in a genome. Short DNA sequences known as "protospacer adjacent motifs," or PAMs, serve as tags and sit adjacent to the target DNA sequence. If the Cas9 complex doesn't see a PAM next to its target DNA sequence, it won't cut. This is one possible reason that Cas9 doesn't ever attack the CRISPR region in bacteria, according to a 2014 review published in Nature Biotechnology.

Genomes encode a series of messages and instructions within their DNA sequences, and genome editing involves changing those sequences, thereby changing the messages they contain. This can be done by inserting a cut or break in the DNA and tricking a cell's natural DNA repair mechanisms into introducing the targeted changes. CRISPR-Cas9 provides a means to do so.

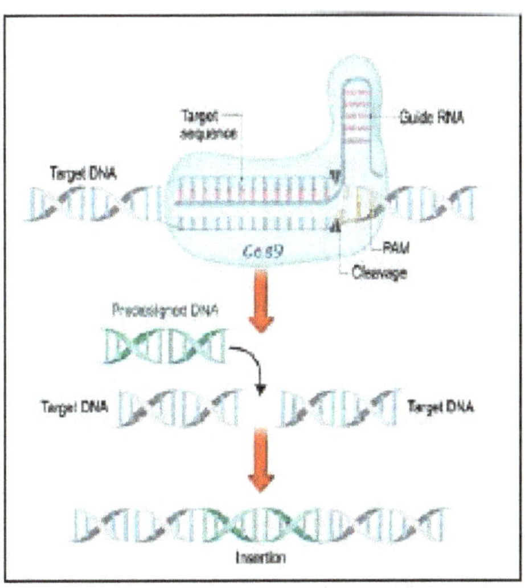

Fig. 7 Here's a breakdown of how Crispr gene-editing works.

In 2012, two pivotal research papers were published in the journals Science and PNAS, describing how the bacterial CRISPR-Cas9 could be used to chop up any DNA, not just that of viruses. In this way, the natural CRISPR system could be transformed into a simple, programmable genome-editing tool.

To direct Cas9 to snip a specific region of DNA, scientists can simply change the sequence of the crRNA, which binds to a complementary sequence in the target DNA, the studies concluded.In the 2012 Science article, Martin Jinek and his colleagues further simplified the system by fusing crRNA and tracrRNA to create a single "guide RNA." Thus, genome editing requires only two components: a guide RNA and the Cas9 protein.

"Operationally, you design a stretch of 20 base pairs that match a gene that you want to edit," and from there, one can figure out what the complementary crRNA sequence would be, George Church, a professor of genetics at Harvard Medical School, told Live Science. Church emphasized the importance of making sure that the nucleotide sequence is found only in the target gene and nowhere else in the genome.

CRISPR-CAS9 SYSTEM: MECHANISM AND COMPONENTS

The CRISPR-Cas system acts in a sequence-specific manner by recognizing and cleaving foreign DNA or RNA. The defence mechanism can be divided into three stages: (i) adaptation or spacer acquisition, (ii) crRNA biogenesis, and (iii) target interference (Fig. 8).

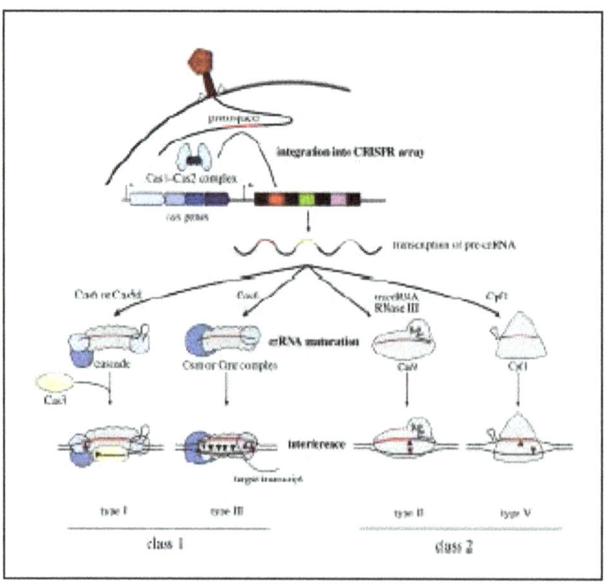

Fig. 8. Simplified model of the immunity mechanisms of class 1 and class 2 CRISPR-Cas systems

Adaptation

In a first phase, a distinct sequence of the invading MGE called a protospacer is incorporated into the CRISPR array yielding a new spacer. This event enables the host organism to memorize the intruder's genetic material and displays the adaptive nature of this immune system. Two proteins, Cas1 and Cas2, seem to be ubiquitously involved in the spacer acquisition process as they can be found in almost all CRISPR-Cas types. Exceptions are the type III-C, III-D and IV CRISPR-Cas systems, which harbour no homologous proteins. Moreover, type V-C shows a minimal composition as it comprises only a putative effector protein termed C2C3 and a Cas1 homologue.

In past years, major advances have been made in revealing the biochemical and genetic principles of CRISPR-Cas immunity. However, the mechanism of spacer acquisition is still not fully understood. The selection of protospacers and their processing before integration remain widely obscure in many CRISPR-Cas types. Recent findings, however, shed light on the biochemistry of the spacer integration process. It has been demonstrated that Cas1 and Cas2 of the type I-E system of *Escherichia coli* form a complex that promotes the integration of new spacers in a manner that is reminiscent of viral integrases and transposases. Although both Cas1 and Cas2 are nucleases, the catalytically active site of Cas2 is dispensable for spacer acquisition. A new spacer is usually incorporated at the leader-repeat boundary of the CRISPR array while the first repeat of the array is duplicated.

The mechanisms of the different CRISPR-Cas types might be conserved only to a certain extent as several studies have shown variations regarding the requirements and targets of the adaptation machinery. While Cas1 and Cas2 are sufficient to promote spacer acquisition in most studied type I CRISPR-Cas systems, type I-B further requires Cas4 for adaptation. The type I-F CRISPR-Cas system of *Pseudomonas aeruginosa* additionally requires the interference machinery to promote the uptake of new spacers. Similarly, type II-A systems require Csn2, Cas9 and tracrRNA (*trans* activating CRISPR RNA—see further details below) for acquisition. Another, so far unique, adaptation mode was revealed for a type III-B Cas1 protein that is fused to a reverse transcriptase. Here, acquisition from both DNA and RNA was reported.

The selection of a target sequence that is integrated into the CRISPR locus is not random. It has been demonstrated that in type I, II and V CRISPR-Cas systems, a short sequence, called the protospacer adjacent motif (PAM), is located directly next to the protospacer and is crucial for acquisition and interference.

In type II-A CRISPR-Cas systems, the PAM-recognizing domain of Cas9 is responsible for protospacer selection. It is believed that after protospacer selection, Cas9 recruits Cas1, Cas2 and possibly Csn2 for integration of the new spacer into the CRISPR array. This feature may be conserved among all class 2 CRISPR-Cas systems although experimental evidence is missing. For type I-E, the Cas1–Cas2 complex is sufficient for spacer selection and integration although it has been reported that the presence of the interference complex increases the frequency of integrated spacers that are adjacent to a proper PAM.

Moreover, in a process called priming, the interference machinery of several type I CRISPR-Cas systems can stimulate the increased uptake of new spacers upon crRNA-guided binding to a protospacer that was selected upon a first infection. This process displays a distinct adaptation mode compared to naive spacer acquisition as it strictly requires a pre-existing spacer matching the target. It usually leads to higher acquisition rates from protospacers that lie in close proximity to the target site. Interestingly, primed spacer acquisition does not depend on target cleavage as it also functions for degenerated target sites that would usually result in impaired interference. The exact mechanism remains obscure but it has been demonstrated that the interference complex can recruit Cas1 and Cas2 during PAM-independent binding to DNA.

Biogenesis

To enable immunity, the CRISPR array is transcribed into a long precursor crRNA (pre-crRNA) that is further processed into mature guide crRNAs containing the memorized sequences of invaders. In type I and III systems, members of the Cas6 family perform the processing step yielding intermediate species of crRNAs that are flanked by a short 52 tag. One exception is given by the type I-C systems, which do not code for Cas6 proteins. Here, the protein Cas5d processes pre-crRNA resulting in intermediate crRNAs with an 11 nt 52 tag.

Further trimming of the 32 end of the intermediate crRNA by an unknown nuclease can occur and yields mature crRNA species composed of a full spacer portion (52 end) and a repeat-portion (32 end), which usually displays a hairpin structure in most type I systems. The maturation of crRNAs in class 2 CRISPR-Cas systems differs significantly. In type II systems, tracrRNA is required for the processing of the pre-crRNA. The anti-repeat sequence of this RNA enables the formation of an RNA duplex with each of the repeats of the pre-crRNA, which is stabilized by Cas9. The duplex is then recognized and processed by the host RNase III yielding an intermediate form of crRNA that undergoes further maturation by a still unknown mechanism to lead to the mature small guide RNA.

An RNase III-independent mechanism was discovered in the type II-C CRISPR-Cas system of *Neisseria meningitidis*. Here, promoter sequences were identified to lie within each repeat and some were able to initiate transcription leading to intermediate crRNA species. Even though RNase III-mediated 32 processing of the crRNA : tracrRNA duplex was observed, it was dispensable for interference. In the type V-A CRISPR-Cas system, it has been shown that Cpf1 has a dual function during CRISPR-Cas immunity. Cpf1 processes premature crRNAs and, following a further maturation event of unknown nature, uses the processed crRNAs that it has generated to cleave target DNA.

Interference

In the last stage of immunity, mature crRNAs are used as guides to specifically interfere with the invading nucleic acids. Class 1 systems employ Cascade (CRISPR-associated complex for antiviral defence)-like complexes to achieve target degradation, while in class 2 systems, a single effector protein is sufficient for target interference. To avoid self-targeting, type I, II and V systems specifically recognize the PAM sequence that is located upstream (types I and V) or downstream (type II) of the protospacer. In type III systems, the discrimination between self and non-self is achieved via the 52 tag of the mature crRNA, which must not base pair with the target to enable degradation by the complex.

In type I systems, Cascade localizes invading DNA in a crRNA-dependent manner and further recruits the nuclease Cas3 for target degradation. Cas3 induces a nick on the foreign DNA and subsequently degrades the target DNA. In type II CRISPR-Cas systems, the tracrRNA:crRNA duplex guides the effector protein Cas9 to introduce a double-strand break in the target DNA. The interference machinery of type III systems comprises Cas10-Csm (types III-A and III-D) and Cas10-Cmr (types III-B and III-C) complexes, which are able to target both DNA and RNA. Intriguingly, it has been shown that interference of type III-A and type III-B systems depends on the transcription of the target DNA. More precisely, the subunit Cas10 cleaves the DNA while Csm3 and Cmr4 cleave the transcribed mRNA in type III-A and type III-B CRISPR-Cas systems, respectively. Interference in type V CRISPR-Cas systems shows similarities to interference in type II. An RNA duplex, consisting of tracrRNA and crRNA, is strictly required for target cleavage in type V-B systems. Type V-A, however, only employ crRNA for target localization and degradation.

Anti-CRISPR mechanisms

Prokaryotes harbour a remarkable arsenal of defence strategies in order to coexist with their viral predators. As a part of the constant arms race between bacteria and their viral counterparts, phages have evolved different strategies to overcome antiviral defence mechanisms. This paragraph summarizes the research on how phages evade the CRISPR-Cas systems.

COMPONENTS

Cas9 endonuclease functions

While native CRISPR/Cas systems have a variety of enzymes responsible for processing foreign DNA as well as the RNA guides required for endonuclease function, when used for genome engineering, the only CRISPR protein required is the Cas9 endonuclease or a variant thereof. This individual protein has all the components necessary to:

Bind to a guide RNA

The guide RNA, which will be described in more detail below, enables Cas9 to cut a specific genomic locus of many possible loci. Without binding to the guide RNA, Cas9 cannot cut.

Bind to target DNA in the presence of a guide RNA provided that target is upstream (5') of a protospacer adjacent motif (PAM)

Cas9 endonuclease binding to the target genomic locus is mediated both by the target sequence contained within the guide RNA and a 3-base pair sequence known as the Protospacer Adjacent Motif or PAM. In order for dsDNA to be cut by Cas9, it must contain a PAM sequence immediately downstream (3') of the site targeted by the guide RNA. In the absence of either the guide RNA or a PAM sequence, Cas9 will neither bind nor cut the target. Cas9 homologs from different organisms or Cas9 mutants developed in a variety of labs have different PAM requirements. These different PAM requirements allow researchers to target many different genomic loci.

Cleave target DNA resulting in a double-strand break (DSB)

Cas9 and its variants have two endonuclease domains: the n-terminal RuvC-like nuclease domain and the HNH-like nuclease domain near the center of the protein. Upon target binding, Cas9 undergoes a conformational change that positions the nuclease domains to cleave opposite strands of the target DNA. Thus, the end result of Cas9-mediated DNA damage is a DSB within the target DNA ~3-4 nucleotides upstream of the PAM sequence.

The synthetic guide RNA or gRNA (sometimes sgRNA)

In the native Type II CRISPR/Cas system, Cas9 is guided to its target sites with the aid of two RNAs: the crRNA which defines the genomic target for Cas9, and the tracrRNA which acts as a scaffold linking the crRNA to Cas9 and facilitates processing of mature crRNAs from pre-crRNAs derived from CRISPR arrays. In most systems used for CRISPR-mediated genome editing, these two small RNAs have been condensed into one RNA sequence known as the guide RNA (gRNA) or single guide RNA (sgRNA). Throughout the remainder of this post, we'll refer to this RNA complex as the "gRNA". The gRNA contains both the 20 nucleotide target sequence to direct Cas9 to a specific genomic locus and the scaffolding sequence necessary for Cas9 binding. When using CRISPR/Cas9 for genome editing, researchers simply need to express a gRNA designed to direct Cas9 to their target sequence of choice (see tips for designing a gRNA) and their prefered Cas9 variant (with the appropriate PAM sequence) to modify the desired genomic locus.

SIGNIFICANCE AND APPLICATIONS

The use of CRISPR-Cas in therapeutic approaches has become increasingly relevant in different fields of medicine. The presence of repetitive sequences interspersed with short spacers, later known as CRISPR, has been exploited for diagnostic purposes and simple typing of *Mycobacterium tuberculosis* strains. This helped to understand ways used by pathogens for their transmission by looking at differences in the spacer content of related strains. This so-called spoligotyping (spacer oligotyping) has also been adapted for *Salmonella enterica*, *Yersinia pestis* and *Corynebacterium diphteriae.*

The use of CRISPR-Cas as a direct antimicrobial tool has been studied recently. Artificial CRISPR arrays have been designed to kill pathogenic bacteria by targeting antibiotic resistance or virulence genes. This elegant way only aims for harmful strains in a bacterial population and allows non-pathogenic strains to overgrow the pathogens. A recent study used lysogenic phages to introduce a CRISPR-Cas system in *E. coli*, which targets antibiotic resistance genes.

The array was designed to additionally target the genomes of lytic phages leading to immunity towards phages only of antibiotic-sensitized bacteria. More precisely, bacteria that are unlikely to acquire antibiotic resistance genes due to their engineered spacer content are also resistant to lytic phages. Thus, in the case of a phage infection, only pathogenic strains would be eradicated from the population.

The medical potential of the CRISPR-Cas systems goes beyond antimicrobial treatment. The introduction of efficient and precise modifications into genes of an organism displays the basis for genome engineering. Programmable nucleases are used that specifically bind genomic regions and cleave the DNA at a desired position. Zinc finger nucleases (ZFN) and transcription activator-like effector nucleases (TALENs) have been widely used to edit DNA. Both genome editing tools rely on the same principle: a sequence-specific DNA binding domain, which provides specificity, is fused to a nuclease. Because of its simplicity, effectiveness and the possibility to target multiple genomic sites simultaneously, use of the CRISPR-Cas9 system is usually favoured over ZFN and TALEN systems. The bacterial defence protein Cas9 is used to target almost any desired DNA sequence with the help of a targeting RNA. This single-guide RNA (sgRNA) is an engineered hybrid of the naturally occurring tracrRNA:crRNA duplex and thus simplifies its application for genome editing purposes.

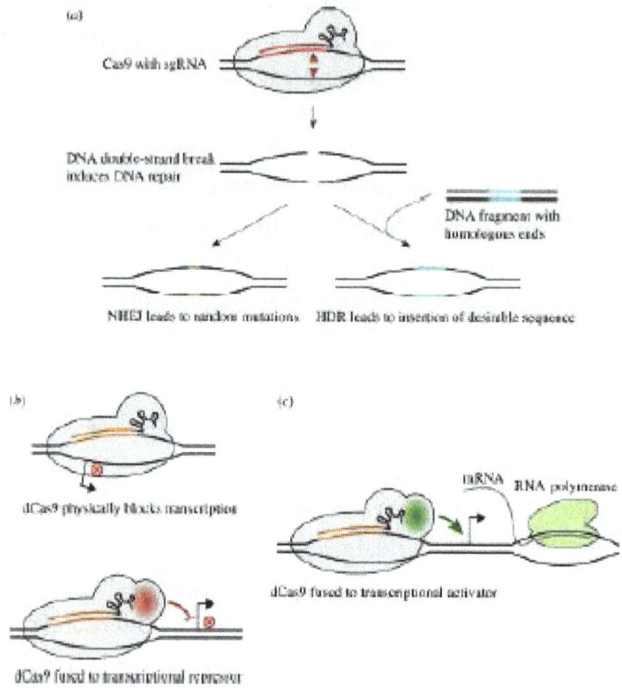

Fig. 9. Applications of the CRISPR-Cas9 technology. (a) Cas9 is guided by a sgRNA to induce a double-strand DNA break at a desired genomic locus. The DNA damage can be repaired by NHEJ yielding short random insertions or deletions at the target site. Alternatively, a DNA sequence that shows partial complementarity to the target site can be inserted during HDR for precise genome editing purposes. (b) Mutations in the catalytical domains of Cas9 yield a dead variant (dCas9) that binds but does not cleave DNA. The approach with dCas9 is used for transcriptional repression by binding to the promoter region of a gene and thus blocking the access for the RNA polymerase. Similarly, dCas9 can be fused to a transcriptional repressor. Red crosses represent inhibition of transcription. (c) The fusion of dCas9 to a transcriptional activator stimulates transcription of an adjacent gene by recruiting the RNA polymerase.

Repurposing the CRISPR-Cas9 system for genome editing exploits the DNA repair mechanisms of eukaryotic cells: after the introduction of a double-strand DNA break, the cell can repair the damage by non-homologous end joining (NHEJ). This process is error-prone and often leads to point mutations, deletions or causes frameshifts that alter the gene product and eventually abolishes its function, which is favoured for genetic knockouts. Precise genome engineering, however, relies on another pathway, termed homology-directed repair (HDR), where a piece of DNA that shows sequence homology to the target site is used to repair the DNA via homologous recombination. This short DNA sequence can harbour any sort of insertion or alteration, allowing the integration of any desirable DNA sequence at the target site (Fig. 9.*a*).

Mutations in the nuclease motifs of Cas9 lead to a 'dead' variant that is unable to cleave DNA and thus can be used to regulate transcription of a desired gene. By targeting the promoter region or the open reading frame of a gene, binding of the RNA polymerase is physically blocked and mRNA elongation is inhibited. Alternatively, dCas9 can be fused to a repressor that controls gene transcription (Fig. 9.*b*). Transcriptional activation can be achieved by fusing dCas9 to a transcription activator that recruits the RNA polymerase and induces gene expression (Fig. 9.*c*). In some cases, gene knock-downs are desired over gene knockouts, e.g. if the targeted gene is essential.

The CRISPR-Cas9 method has also been exploited for epigenome editing that allows the control of gene expression by introducing modifications like DNA methylation or histone acetylation.

One study showed that a fusion protein of dCas9 and the core domain of the human acetyltransferase p300 could activate gene expression at specific sites. Moreover, fusion of dCas9 to the KRAB repressor was able to induce methylation at specific enhancers leading to reduced chromatin accessibility and, thus, silencing of gene expression.

Precise epigenome editing has great potential to reveal site-specific chromatin modification and helps to explore the regulation of gene expression that could lead to new therapeutical strategies. Other approaches—mainly in prokaryotes—exploit the endogenous type I effector complex Cascade for similar experiments if the nuclease Cas3 is absent . In all of the aforementioned genome manipulation strategies, the existence of a PAM adjacent to the target site is a strict requirement. Precise remodelling of the genome can be used to cure gene variants that cause genetic diseases.

Scientists were able to repair mutations that cause cystic fibrosis (CF) by correcting the *cftr* locus in cultured intestinal stem cells of CF patients. Moreover, using the CRISPR-Cas9 technique, a healthy phenotype could be restored in mice suffering from hereditary tyrosinaemia, a genetic disease that causes severe liver damage. Genome editing has been further used to develop antiviral therapeutic approaches. Accordingly, the genome of HIV has been successfully eradicated from latently infected cells.

Indeed, a recent study demonstrated that the generation of NHEJ-induced mutations in the viral genome led to replication defects of the virus. However, it also drove the generation of replication competent mutants that harbour mutations at the target site and thus are no longer targeted by Cas9. Other studies aim to alter a specific surface protein called CCR5 that serves as a co-receptor for the HI-virus to enter a host cell. Mutations in the *ccr5* gene can prevent the virus from infecting a cell leading to a highly resistant but otherwise healthy phenotype. In fact, altering the wild-type *ccr5* gene leads to immunity of monocytes and macrophages against HIV infections.

The CRISPR-Cas9 technique has further simplified genome-scale screens. These screens seek to identify genes that are involved in certain metabolic or pathogenetic processes by abolishing gene function and studying the resulting phenotype. Using this approach, genes that are involved in tumour growth or convey susceptibility towards bacterial toxins could be identified. Previously, RNA interference (RNAi) was used to knock down gene expression in a sequence-specific manner. However, RNAi only decreases the abundance of transcripts, whereas CRISPR-Cas9 enables a full knock-out of candidate genes. Moreover, multiplexing (targeting of several genetic loci at the same time) is crucial for this approach and can be achieved by using a library of different sgRNAs that is usually delivered with Cas9 by a lentiviral vector system.

Some historical notes

CRISPR arrays in bacterial genomes consist of repeated elements separated by unique sequences. When researchers first discovered these arrays, they did not know their functions and simply called the repeated elements "direct repeats" and the unique stretches of DNA between them "spacers".

After years of research, we now know that each direct repeat, combined with its adjacent spacer, ultimately encodes a single crRNA. The direct repeat regions contain sequences required for processing pre-crRNA into mature crRNA and tracrRNA binding. The spacer regions, on the other hand, are the unique, foreign DNA target sequences specific to each individual crRNA.

CHAPTER-5

CRISPR BEYOND GENE EDITING

Originating from the bacterium *Streptococcus pyogenes*, Cas9 was the first Cas protein that scientists repurposed for gene editing. CRISPR-Cas9 technology uses a single guide RNA (sgRNA) to target and cleave DNA. Researchers engineer target-specific sgRNAs by combining two RNA molecules from the bacterial CRISPR system: a sequence that recognizes a specific location in the DNA (crRNA) and a sequence that acts as a binding scaffold for Cas9 (tracrRNA). The modifiable sgRNA sequence allows scientists to program a CRISPR-Cas9 system to target any DNA sequence of interest if it is near a Cas-specific DNA sequence called a protospacer-adjacent motif (PAM). The PAM initiates Cas9-DNA binding, the sgRNA invades the double helix and hybridizes with the target DNA, and Cas9 breaks the unwound double-stranded target DNA immediately in front of the PAM. Repair mechanisms, namely homology-directed repair (HDR) and nonhomologous end joining (NHEJ), repair the break, which can alter the target gene's biological function.

For instance, researchers can use the different repair mechanisms to their advantage to intentionally insert a desired sequence change via template-dependent HDR or introduce random changes through template-independent NHEJ. If scientists provide the cellular repair machinery with a target gene template that contains a mutation, such as a disease-relevant mutation, the HDR process will incorporate the templated mutation into the genome after Cas9 cleavage. In contrast, the more error prone NHEJ pathway repairs the Cas9 cut without a template by introducing random insertions and deletions (indels) that ultimately disrupt gene expression or lead to loss of function. Either of these approaches enable researchers to investigate how silencing or editing a targeted gene affects its downstream pathways.

Brief history of genome-editing efforts

Genomes of eukaryotic organisms are composed of billions of DNA bases. The ability to change these DNA bases at precisely predetermined locations holds tremendous value not only for molecular biology, but also for medicine and biotechnology. Therefore, introducing desired changes into genomes, i.e., "genome editing", has been a long sought-after goal in molecular biology. To this end, the discovery of restriction enzymes that normally protect bacteria against phages in the late 1970s123 was a turning point that fueled the era of recombinant DNA technology. For the first time ever, scientists gained the ability to manipulate DNA in test tubes. Although such efforts drove a number of discoveries in molecular biology and genetics, the ability to precisely alter DNA in living eukaryotic cells came a few decades later. To this end, several key developments were revealed in the mid to late 1980s.

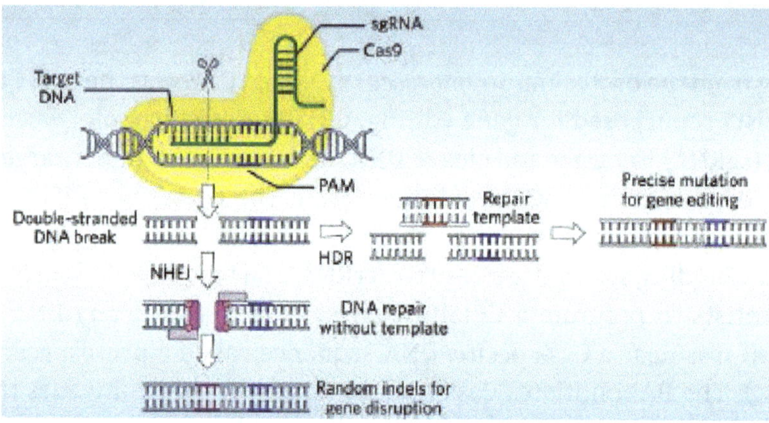

Fig. 1 Scientists program CRISPR-Cas9 to introduce blunt end double-stranded DNA breaks at target genomic loci based on nearby protospacer adjacent motifs (PAMs) and engineered target-specific single guide RNAs (sgRNAs). After Cas9 cleaves the DNA, the target gene becomes a substrate for cellular DNA repair machinery, which catalyzes nonhomologous end joining (NHEJ) or homology-directed repair (HDR) mechanisms that disrupt gene expression or introduce new DNA sequences

Initial targeted gene disruption studies in eukaryotic yeast cells4 followed with breakthrough work by Capecchi and Smithies in mammalian cells. Their studies demonstrated that mammalian cells can incorporate an exogenous copy of DNA into their own genome through a process called homologous recombination. Such targeted gene integration into the genome provided unprecedented power to characterize the functional roles of various genes in model organisms.

However, the feasibility of this approach had several limitations. Firstly, the rate of spontaneous integration of an exogenous DNA copy was extremely low (1 in 10–10 cells). Secondly, the integration rate depended on cell types and cellular states. Finally, and most critically, the approach could result in random integration of the exogenous copy into undesired genomic loci at a frequency similar to or higher than that of the target site.

GENE EDITING TECHNIQUES: TOOLS TO CHANGE THE GENOME

Gene editing might sound simple on paper, but it is far from easy. The history of genome engineering goes back almost 70 years, to the initial discovery of the DNA double helix. Since then, scientists have spent decades trying to uncover ways to edit the genome that balance specificity with time and cost. Here are tsome successful strategies that scientists have used to modify DNA so far

Restriction enzymes: the original genome editors

The ability to edit genes became a reality with the discovery of restriction enzymes in the 1970s. Restriction enzymes recognize specific patterns of nucleotide sequences and cut at that site, presenting an opportunity to insert new DNA material at that location.

Restriction enzymes are not commonly utilized for gene editing these days, since they are limited by the nucleotide patterns they recognize, but they remain widely used today for molecular cloning. Additionally, certain classes of restriction enzymes play key roles in DNA mapping, epigenome mapping, and constructing DNA libraries.

Zinc finger nucleases: increased recognition potential

As time went on, the need for precision in genome editing became more evident. Scientists needed a gene-editing technique that recognized the site they wanted to edit, as off-target effects could be deleterious. The discovery of zinc finger nucleases (ZFN) in the 1980s addressed this issue.

ZFNs are composed of two parts: an engineered nuclease (Fokl) fused to zinc finger DNA-binding domains. The zinc-finger DNA-binding domain recognizes a 3-base pair site on DNA and can be combined to recognize longer sequences. Additionally, the ZFNs function as dimers, increasing the length of the DNA recognition site and consequently increasing specificity. However, while specificity increased with ZFNs, it was not perfect.

One main hurdle with using ZFNs was that the 3-base pair requirement made the design more challenging. Guanine-rich target sites appeared to yield more efficient editing than non-guanine-rich sites. Additionally, since the ZFN interaction with DNA is modular (i.e., each ZF interacts with DNA independently), the editing efficiency was also compromised. Therefore, scientists needed to address these issues if they wanted to have more efficient genome editing.

ZFNs showed real promise in the field of medicine. Notably, scientists used ZFN to disable CCR5 on human T-cells, a major receptor for HIV. Following ZFN-mediated editing, scientists found autologous CD4+ T-cells were safe to use and were an exciting potential for HIV therapy. Additionally, ZFNs have been used to edit tumor-infiltrating lymphocytes as a treatment strategy for metastatic melanoma.

Transcription activator-like effector nucleases: single-nucleotide resolution

In 2011, a new gene-editing technique emerged, which was an improvement over ZFNs. Transcription activator-like effector nucleases (TALENs) are structurally similar to ZFNs. Both methods use the Fokl nuclease to cut DNA and require dimerization to function, however, the DNA binding domains differ. TALENs use transcription activator-like effectors (TALEs), tandem arrays of 33-35 amino acid repeats. The amino acid repeats possess single-nucleotide recognition, thereby increasing targeting capabilities and specificity compared to ZFNs.

Even with the single-nucleotide resolution, using TALENs as a gene-editing tool was still time and cost-intensive and possessed certain design restrictions. The structure of TALENs meant the target site required a 5' thymine and 3' adenine, limiting target customizability. Further, TALENs displayed decreased editing efficiency in heavily methylated regions. Delivery into cells was also challenging, since TALENs are much larger than ZFNs (~6kb vs. ~2kb), increasing the amount of time and money required to create a successful edit. While TALENs showed improvement in genome editing technology, the high labor and monetary cost still hindered its widespread adoption.

Like its predecessor ZFNs, TALENs have been used in the field of medicine, as well as agriculture. Scientists used TALENs to correct COLA7A1 dysfunction in epidermolysis bullosa, a disease characterized by loss of skin integrity leading to potentially fatal skin blisters and increased risk for skin cancer. In agriculture, scientists found a way to create pathogen-resistant rice using TALENs.

CRISPR-Cas9 gene editing: genome editing revolutionized

Scientists were still searching for an easier and faster way to edit genes. Fast forward to 2012, when everything changed. Scientists discovered a new method of genome editing derived from CRISPR-Cas9, a system that has long existed in bacteria to help them fight off invading viruses. The teams led by Jennifer Doudna and Emmanuelle Charpentier rocked the science world with the first biochemical description of CRISPR. In 2013, Feng Zhang described how CRISPR could be used to edit eukaryotic DNA. Since these discoveries, CRISPR as a gene-editing tool has seen unprecedented popularity.

CRISPR is an elegant two-component system consisting of a guide RNA and a Cas9 nuclease. The Cas9 nuclease cuts the DNA within the ~20 nucleotide region defined by the guide RNA. With CRISPR, scientists can customize their guide RNAs, and algorithms have been developed to assess the chances of off-target effects (i.e., does this sequence exist in other places of the genome). However, CRISPR is much more customizable and cost-effective, making it more accessible to scientists that may have budget and time constraints.

The advancements in genome editing techniques have opened up new doors for what genome editing can do to address issues in medicine, agriculture, and beyond. CRISPR has completely revolutionized what genome editing can mean for our future by increasing the speed and breadth of science. We are already feeling the impact of CRISPR in its role in drug discovery, diagnostics, and gene drives, just to name a few. At this rate, don't be surprised if you see more talk about genome editing in the near future.

Base editing: single nucleotide substitutions

Base editing is a relatively new method of genome editing derived from CRISPR-Cas9. Unlike traditional CRISPR systems, base editors (BEs) do not induce double-stranded breaks in the genome. Base editing systems, developed by David Liu's lab at the Broad Institute, use a 'catalytically dead' Cas9 (dCas9), which cannot cleave DNA, fused to bacterial enzymes called DNA deaminases.

Cytidine deaminases, which induce C to T substitutions, are naturally occurring in bacteria, while adenine deaminases, which induce A to G substitutions, were engineered from bacterial enzymes specifically for base editing purposes. Fusing dCas9 to either a cytidine deaminase (CBEs) or an adenine deaminase (ABEs) and providing a sgRNA to direct it to the target sequence, allows researchers to introduce substitutions in DNA.

The ability to induce single nucleotide substitutions was a major step forward for the field, both because a majority of human diseases are caused by single nucleotide polymorphisms and because editing without creating double-stranded breaks avoids many of the potential risks of other forms of editing such as CRISPR-Cas9. However, the current CBE and ABE systems only cover four of 12 possible transition mutations, a fact that led to the development of prime editing systems.

Prime editing: editing without double-stranded breaks

Prime editing systems allow for all possible transition mutations, as well as small insertions of up to 50 nucleotides and deletions of up to 80 nucleotides. The system, also developed by the Liu lab, works using a Cas9 nickase, which induces single-stranded breaks in DNA, fused to a reverse transcriptase enzyme.

Rather than using an sgRNA and a donor template for repair, prime editing uses a single engineered construct known as a prime editing guide RNA (pegRNA), which is made up of the primer binding site (PBS) sequence and a sequence containing the desired edit. After it finds the target, the Cas9 nickase creates a cut in one strand of the DNA and the reverse transcriptase uses the pegRNA as a template for reverse transcription, attaching the corresponding nucleotides to the nicked DNA end.

The original sequence is naturally excised by endonucleases within the cell, however, this leaves a mismatch between the two strands of DNA that must be resolved for the editing to be complete. This is achieved by using a different guide RNA to direct the prime editor to nick the opposing unedited strand. When this occurs, the cell repairs the nick using the new DNA from the first edit as a template for repair, resulting in a complete edit.

The fact that prime editing can generate all possible transition mutations, insertions, and deletions without inducing DSBs makes it much safer to use and therefore highly desirable for therapeutic purposes. Base editors and prime editors are currently being investigated for the treatment of blood disorders such as sickle cell disease, and they have many other potential therapeutic applications.

PASTE: 'Drag-and-Drop' Editing for Large Insertions

Programmable Addition via Site-specific Targeting Elements (PASTE) is the new kid on the block when it comes to genome editing. Developed by Jonathan Gootenberg and Omar Abudayyeh, this system allows for the targeted insertion of large DNA sequences, also without creating double-stranded breaks.

PASTE harnesses the power of serine integrase proteins from bacteriophages to incorporate new genetic information - up to 36 kb - into the genome. For this to work, however, the integrase's attachment site (AttP) must first locate and bind to the correct landing site (AttB), and these are not common in genomes.

To overcome this obstacle, PASTE incorporates prime editing to copy these AttB sites into the genome near the target sequence via reverse transcription.

Conveniently, prime editing systems can insert up to 50 bp of DNA, and the AttB landing sites recognized by integrases are around 46 bp. Once the AttB site has been incorporated near the target sequence via prime editing, it acts as a beacon for the integrase, which will be recruited to the site to perform the insertion of the attached desired DNA sequence.

The development of the PASTE tool is certainly exciting in the field of gene editing because it allows for large-scale gene knock-in without creating DSBs. This indicates its potential in therapeutic applications, such as treating diseases that are caused by multiple pathogenic mutations in large genes. Like base and prime editing systems, PASTE's ability to edit DNA without creating DSBs increases the safety profile of any potential therapeutic application of this technology.

GENOME MODIFICATIONS WITH BASE OR PRIME EDITING

Base editors and prime editors are new generation CRISPR-based technologies that rely on catalytically impaired Cas proteins called nickases, which researchers fuse to other enzymes that modify DNA. Because Cas nickases are less active than the traditional Cas nucleases, they introduce single-stranded DNA breaks (or nicks) instead of double-stranded breaks. These nicks do not typically recruit DNA repair machinery, which allows fusion enzymes to act on the DNA with greater editing accuracy than repair-mediated methods.

Base editing uses modified versions of DNA cutting Cas proteins that retain their DNA binding abilities without introducing double-stranded breaks. These catalytically impaired Cas proteins guide an adenine or cytidine deaminase enzyme to a specific site in the DNA, where they cause C G to T A or A T to G C conversions. In theory, base editing can correct any transition mutation (A to G, or C to T), which represent roughly 30 percent of known disease alleles and are the single largest class of human disease causing mutations.

The prime editing

For prime editing, a nuclease-impaired Cas protein recruits a reverse transcriptase to the desired DNA sequence without introducing double-stranded breaks. A reverse transcriptase converts RNA into DNA. In the context of prime editing, the reverse transcriptase replaces or inserts a DNA sequence encoded by a prime editor guide RNA (pegRNA). These highly versatile prime editors are not limited to the specific pairing of base editors and can catalyze the switch between any base. They can also generate insertions up to 44 base pairs (bps) and deletions of up to 80 bps.

Transcriptome Editing with CRISPR Modulation and dCas

CRISPR modulation changes gene expression without introducing changes to the DNA sequence. Scientist fuse a catalytically inactive Cas nuclease (dead Cas or dCas) with either a repressor domain for CRISPR interference (CRISPRi) or with a transcriptional activation domain for CRISPR activation (CRISPRa). The dCas fusion protein binds to the DNA target sequence without introducing double-stranded breaks, recruits small regulatory RNAs to the locus, and causes either gene expression knockdown (CRISPRi via the repressor domain) or gene expression activation (CRISPRa).

New CRISPR Technology Applications

Base editing has powerful medical potential because of its low risk of creating insertions and deletions. It has been paired with CAR-T cell therapy in clinical trials to treat otherwise untreatable leukemia cases.

Because prime editing can correct most mutations, excluding large duplications, deletions, insertions, or chromosomal rearrangements, researchers are investigating its ability to treat a variety of disorders, ranging from inherited diseases to cancer.

The rise of CRISPR as the genome-editing technology

Although the discovery of artificially designed meganucleases followed by ZFNs and TALENs successively increased the genome-editing efficacy, targeting different sites in the genome required re-design or re-engineering of a new set of proteins. The difficulty in cloning and protein engineering ZFNs and TALENs partially prevented these tools from being broadly adopted by the scientific community. In this respect, CRISPR has revolutionized the field because it is as robust as, if not more so than, the existing tools in terms of editing efficiency. More importantly, it is much simpler and more flexible to use.

The CRISPR gene-editing technology, as we know it today, is composed of an endonuclease protein whose DNA-targeting specificity and cutting activity can be programmed by a short guide RNA. Notably, CRISPR had been simply known as a peculiar prokaryotic DNA repeat element for several decades before it was recognized as the bacterial immune system and subsequently harnessed as a powerful reprog-rammable gene-targeting tool.

CRISPR stands for clustered regularly interspaced short palindromic repeat DNA sequences. Although the name CRISPR was coined much later, these repeat elements were initially noticed in *Escherichia coli* by Dr. Nakata's group. Interestingly, unlike typical tandem repeats in the genome, the CRISPR repeat clusters were separated by non-repeating DNA sequences called spacers. It took more than a decade for researchers to recognize the nature and origin of these spacer sequences. During the human genome project (HGP), the genomes of many other organisms, including many different phages, were also sequenced.

The computational analysis of these genomic sequences led researchers to notice key features of CRISPR repeat and spacer elements. Firstly, the CRISPR sequences are present in more than 40% of sequenced bacteria and 90% of archaea. Secondly, the CRISPR elements are adjacent to multiple well-conserved genes called CRISPR-associated (Cas) genes. Finally and most interestingly, the non-repeating spacer DNA sequences were recognized to belong to viruses and other mobile genetic elements.

These observations sparked the interest of many researchers in studying the functional significance and mechanics of these CRISPR sequences. Although the idea that it could serve as a bacterial immune system started to circulate among researchers, the exact mechanism of action was not known. The key experimental evidence about the potential function of CRISPR systems came from the work of Horvath and colleagues.

They demonstrated that after a viral challenge, *Streptococcus thermophilus* bacteria integrate new spacers derived from the phage genomic sequence into its genome. More importantly, the spacer sequences of CRISPR dictate the targeting specificity of Cas enzymes, which provide defense against the phage. Immediately following this work, other researchers further elucidated the mechanism of action of the CRISPR system. Within a year after this key discovery, it was shown that the activity of Cas enzymes is guided by short CRISPR RNAs (crRNA) transcribed from the spacer sequences and that it can block horizontal DNA transfer from bacterial plasmids. Such exciting publications further stimulated researchers' interest in understanding the molecular mechanism of the CRISPR system. There have been several critical findings that paved the way for CRISPR systems to become the CRISPR genome-editing technology.

One of the key findings was the observation that the acquired spacer sequences are highly similar to each other at regions called protospacer-adjacent motifs (PAMs) and that this sequence is very critical for the CRISPR system to work. Independently, it was revealed that among many Cas proteins, Cas9 was the only one with DNA catalytic activity in *S. thermophilus*. Additionally, the work from the Charpentier group revealed the mechanism of biogenesis of the two short RNAs required for Cas9 action. A final critical discovery was the demonstration that a CRISPR system from one bacterium was transferable to different bacterial strains. Siksnys and colleagues showed that the CRISPR locus from *S. thermophilus* is able to reconstitute the interference in *E. coli*.

These findings were immediately followed by biochemical characterization of the individual components of the CRISPR system. The crucial work, which arguably marked the beginning of CRISPR as a biotechnology tool, has been the demonstration that Cas9 enzymes can be reprogrammed to target a desired DNA sequence in bacteria. Notably, these studies also simplified the CRISPR system by using a single short RNA. The endogenous CRISPR system requires two short RNAs: the mature crRNA and a trans-activating crRNA (tracrRNA). The crRNA is composed of the part that serves as guiding sequence and another part base pairs with the tracrRNA. Both crRNA and tracrRNAs are required to form the Cas9 protein–RNA complex that cleaves DNA with DSBs at target sites.

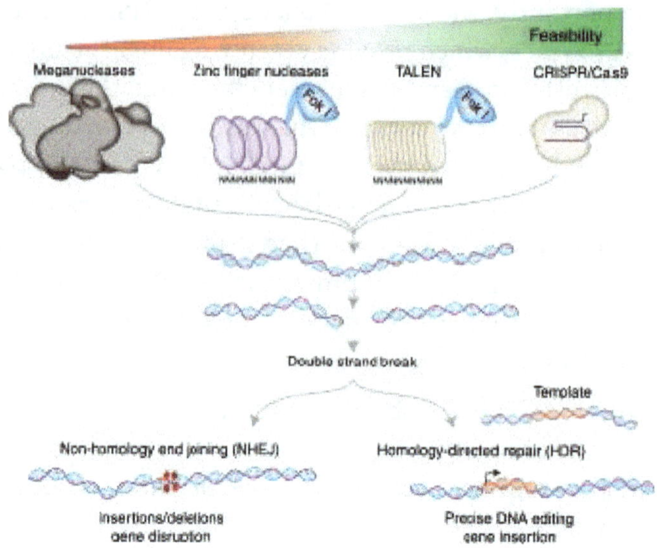

Fig. 2 The basic working principle of major genome-editing technologies. Meganucleases are engineered restriction enzymes that recognize long stretches of DNA sequences. Each zinc finger nuclease recognizes triple DNA code whereas each TALE recognizes an individual base. Unlike protein–DNA recognition in ZFNs and TALENs, simple RNA–DNA base pairing and the PAM sequence determine CRISPR targeting specificity. All these tools result in DNA doublestrand breaks, which are repaired either by error-prone non-homology end joining or homologydirected repair. While NHEJ results in random indels and gene disruption at the target site, HDR can be harnessed to insert a specific DNA template (single stranded or double stranded) at the target site for precise gene editing

Notably, researchers showed that CRISPR-Cas9 can also be guided by a single chimeric RNA formed by the fusion of tracrRNA and crRNA, called single guide RNA (sgRNA). These studies were immediately followed by groundbreaking publications showing that CRISPR can be adapted for *in vivo* genome editing in eukaryotic cells. For the first time ever, researchers had an extremely flexible tool that could be easily guided to target nearly any location in the genome by simply designing a short sgRNA. Due to high editing efficiency and ease of use, researchers from diverse fields quickly adopted CRISPR technology as a method of choice for various genome-targeting purposes. Notably, since its inception as a genome-editing tool in late 2012, more than 9000 research articles have been published about it and the number of publications seems to continue to increase each year.

Different CRISPR systems and their uses in genome editing

The evolutionary arms race between prokaryotes and environmental mobile genetic elements such as phages has been going on for billions of years. This survival struggle yielded various CRISPR-type immune responses as defense mechanisms in bacteria. These CRISPR systems are classified based on the structure of CRISPR-associated (Cas) genes that are typically adjacent to the CRISPR arrays. The classification efforts are yet to be completed as researchers continue to discover new systems and refine the classification system with subclasses, groups, and types based on comparative genomic analyses, structures, and biochemical activities of CRISPR components. Broadly speaking, there are two classes of CRISPR systems, each containing multiple CRISPR types. Class 1 contains type I and type III CRISPR systems that are commonly found in Archaea. Class 2 contains type II, IV, V, and VI CRISPR systems.

Although researchers repurposed many different CRISPR/Cas systems for genome targeting, the most widely used one is the type II CRISPR-Cas9 system from *Streptococcus pyogenes*. Because of the simple NGG PAM sequence requirements, *S. pyogenes'* Cas9 (spCas9) is used in many different applications. However, researchers are still actively exploring other CRISPR systems to identify Cas9-like effector proteins that may have differences in their sizes, PAM requirements, and substrate preferences. In the last few years, more than 10 different CRISPR/Cas proteins have been repurposed for genome editing (Table 1). Among these, some of the recently discovered ones, such as Cpf1 proteins from *Acidaminococcus sp* (AsCpf1) and *Lachnospiraceae bacterium* (LbCpf1), are particularly interesting. In contrast to the native Cas9, which requires two separate short RNAs, Cpf1 naturally requires one sgRNA. Furthermore, it cuts DNA at target sites 32 downstream of the PAM sequence in a staggering fashion, generating a 52 overhang rather than producing blunt ends like Cas9.

CRISPR-BASED DIAGNOSTICS

Since their initial discovery, the number of different CRISPR–Cas systems has expanded rapidly. Currently, CRISPR–Cas systems can be divided, according to evolutionary relationships, into two classes, six types and several subtypes. The classes of CRISPR–Cas system are defined by the nature of the ribonucleoprotein effector complex: class 1 systems are characterized by a complex of multiple effector proteins, and class 2 systems encompass a single crRNA-binding protein. The design of crRNAs for the different effector proteins used in CRISPR diagnostics follow the same principles as those of other CRISPR applications, and are summarized in Table 1. Among the diverse CRISPR systems, class 2 systems have primarily been applied for diagnostics, as these systems are simpler to reconstitute.

Table 1 Design principles for crRNAs used in CRISPR-based diagnostics

Enzyme	Spacer length (bp)	Direct-repeat length (bp)
AsCas12a	20	20
LbCas12a	20	21
LbaCas13a	28	35
LbuCas13a	20	31
LwaCas13a	28	36
CcaCas13b	30	36
PsmCas13b	30	36

The first CRISPR-based diagnostic methods that were developed largely used Cas9 variants, which recognize double-stranded DNA (dsDNA). The main principles of the many different Cas9-based approaches for sensing DNA that have been reported include guide-directed reconstitution of split proteins by catalytically inactive Cas9 partners, Cas9-based destruction of PAM-containing sites and Cas9-induced unwinding of the non-targeted DNA strand as a targeting site for isothermal amplification. The Cas9-based method referred to as NASBACC (for nucleic acid sequence-based amplification (NASBA)-CRISPR cleavage) combines nucleic acid sequence-based amplification for the isothermal preamplification of targets, Cas9 cleavage for PAM-dependent target detection and a toehold sensor for the readout.

In brief, a toehold trigger is attached to a NASBA-amplified RNA fragment through reverse transcription. If a PAM sequence is present in the RNA fragment, Cas9-mediated cleavage leads to a truncated RNA without the trigger sequence. In the absence of the PAM sequence, the trigger containing full-length RNA activates the toehold switch, as indicated by a colour change. By sensing strain-specific PAM sites, the method allows for viral-lineage discrimination, as shown for the detection of the Zika virus (ZIKV) in infected monkey plasma at concentrations in the low femtomolar range (Fig. 3 left).

Reported most recently, a Cas9-based method referred to as LEOPARD (for leveraging engineered tracrRNAs and on-target DNAs for parallel RNA detection) enables the multiplexed detection of different RNA sequences with single-nucleotide specificity. The method is based on the observation that *trans*-activating crRNAs (tracrRNAs)—which in type II systems are necessary for the formation of Cas9–crRNA complexes—hybridize to cellular RNAs, making 'non-canonical' crRNAs. By reprogramming tracrRNAs to bind cellular transcripts of interest while allowing for the formation of Cas9–crRNA complexes, the resulting non-canonical crRNAs enable the guiding of Cas9 to a DNA target.

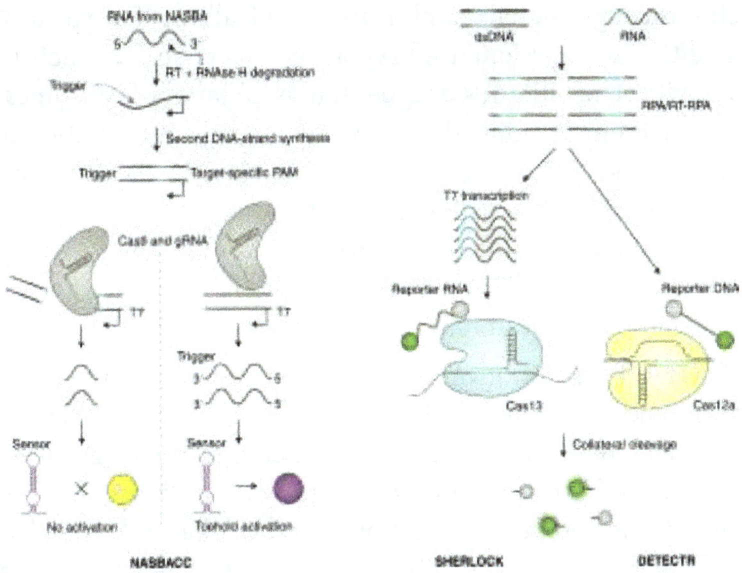

Fig. 3: Two strategies for CRISPR-based diagnostics.

Left: RNA targets are amplified through NASBA, which starts with reverse transcription (RT) to complementary DNA using a sequence-specific primer that appends a trigger sequence (magenta) for the toehold sensor. The RNA from the RNA/DNA hybrid is destroyed by RNase H, enabling a primer that contains a T7 promoter to bind and create a complementary second DNA strand. T7 transcription of the dsDNA template creates the target RNA sequence, which can be used as starting material for a new NASBA cycle or detected by the toehold sensor. If a PAM sequence (blue) is present in the dsDNA amplicon, Cas9-mediated cleavage leads to a truncated template for T7 transcription, which generates a shorter target RNA that cannot activate the toehold sensor. In the absence of the PAM sequence, a full-length target RNA containing the trigger is transcribed, which activates the toehold sensor and produces a visible change in colour. Right: DNA or RNA are amplified by RPA or RT-RPA, respectively. For RNA-targeting CRISPR enzymes (including Cas13a), the amplified RPA product is T7-transcribed into RNA. Binding of the crRNA to the complementary target sequence activates the Cas enzyme and triggers collateral cleavage of quenched fluorescent reporters. Thereby, Cas13a (used in SHERLOCK) or Cas12a (used in DETECTR) indicate the presence of RNA or DNA target sequences, respectively.

The main difference between CRISPR type II (Cas9) systems and type V (Cas12) and type VI (Cas13) systems is the ability of the latter two systems to trigger non-specific collateral cleavage (*trans* cleavage) on target recognition. Collateral activity involves the cleavage of non-targeted single-stranded DNA (ssDNA; Cas12) or single-stranded RNA (ssRNA; Cas13) in solution, which enables the sensing of nucleic acids through signal amplification and allows for various readouts through the addition of functionalized reporter nucleic acids, which are cleaved by collateral activity. The CRISPR–Cas type VI (Cas13) family of enzymes has a size range of ~900–1,300 amino acids, detects ssRNA in *cis* conformation and shows collateral *trans*-cleavage activity against ssRNA *in vitro*. In the Cas13-based assay SHERLOCK (for specific high-sensitivity enzymatic reporter unlocking), DNA or RNA is first isothermally amplified with recombinase polymerase amplification (RPA) or reverse transcription RPA (RT-RPA), respectively, using a forward primer that adds a T7 promoter to the amplicon (Fig. 3, right).

This promoter allows for RNA transcription of the target, which is then recognized and bound by a complex of Cas13a from *Leptotrichia wadeii* (LwaCas13a) and a crRNA that includes a complementary sequence to the target. The activated Cas13 will then cleave both the on-target RNA by *cis* cleavage and, in a target-dependent manner, the ssRNA reporter molecules by collateral *trans* cleavage. The ssRNA reporter molecules consist of a fluorophore and a quencher joined together by a short RNA oligomer, which, once cleaved, allows for the separation of the fluorophore from the quencher, resulting in fluorescence. These processes enable the detection of viral RNA, bacterial DNA, human single-nucleotide polymorphisms (SNPs) and cancer-associated mutations with attomolar (10–18 M) sensitivity.

Version two of the assay (SHERLOCKv2) enabled the quantitative multiplexed sensing of nucleic acids and the target detection at zeptomolar (10–21 M) concentrations, and introduced a lateral-flow readout based on an immunochromatographic assay, where cleaved reporter molecules are detected through antibody-conjugated gold nanoparticles on a paper strip.

Cas12 enzymes that have been used in CRISPR-based diagnostics target dsDNA and ssDNA, require a PAM site in the target region for dsDNA cleavage and collaterally cleave ssDNA. One of the first Cas12-based detection methods was reported in 2018 and referred to as DETECTR (for DNA endonuclease-targeted CRISPR *trans* reporter; Fig. 3, right).

In this method, Cas12a from *Lachnospiraceae bacterium* (LbCas12a) or other organisms is guided to dsDNA targets by a complementary crRNA, triggering collateral cleavage of short ssDNA reporters carrying a fluorophore and a quencher. Similar to SHERLOCK, target recognition and reporter cleavage lead to the separation of the quencher from the fluorophore, which generates a fluorescence signal. DETECTR reached attomolar sensitivity when combined with RPA preamplification. Other Cas12-based techniques include HOLMES (for one-hour low-cost multipurpose highly efficient system), which employs PCR as preamplification together with LbCas12a, and HOLMESv2, which uses loop-mediated isothermal amplification (LAMP) combined with a thermostable Cas12b from *Alicyclobacillus acidoterrestris* (AacCas12b) in a one-pot reaction. Both HOLMES and HOLMESv2 showed a limit of detection (LOD) around 10 aM. Similarly, Cas12f targets dsDNA and ssDNA, but enables better discrimination of SNPs in ssDNA than Cas12a41.

Preamplification-free diagnostics

When using Cas enzymes without upstream preamplification of the target, most CRISPR-based diagnostics have a reported LOD in the picomolar range. This LOD allows for target detection when there is a relatively high concentration of DNA or RNA in the sample. For example, severe acute respiratory syndrome coronavirus 2 (SARS-CoV-2) is apparently shed at high concentration in the early phase of infection (~6.76 × 105 copies per swab), allowing preamplification-free detection. In addition to highly concentrated microbial samples, human genomic DNA (gDNA) and highly expressed human messenger RNAs and microRNAs (miRNAs) have been reported to be amenable to detection without preamplification.

To increase the sensitivity of preamplification-free CRISPR-based diagnostics, LwaCas13a has been combined with the CRISPR type III RNA nuclease Csm6. The nuclease activity of Csm6 is activated by cyclic oligoadenylates (22 ,32 -cyclic phosphate groups). Because the collateral-cleavage activity of both LwaCas13a and PsmCas13b (from *Prevotella* sp. MA2016) generate products with hydroxylated 52 ends and 22 ,32 -cyclic phosphate ends, it was hypothesized that the collateral activity of Cas13 could generate Csm6 activators, thereby enabling amplified signal detection.

In contrast to the cleavage preferences of LwaCas13a for the UU and AU two-base motifs, it was found that Csm6 from *Enterococcus italicus* (EiCsm6), *Lactobacillus salivarius* (LsCsm6) and *Thermus thermophilus* (TtCsm6) had strong cleavage preferences for A-rich and C-rich reporters, allowing for the cleavage activity of LwaCas13a and Csm6 to be independently measured in different channels and for the further design of 'protected' RNA activators. These activators were designed to contain a poly(A) Csm6 activator followed by a poly(U) stretch, on the basis of the rationale that the collateral cleavage of LwaCas13a would only degrade the uridines, thereby generating hexadenylates with 22 ,32 -cyclic phosphate ends and activating Csm6.

Activated Csm6 would in turn cleave additional reporter molecules, freeing up quenched fluorophores and thereby increasing the signal. This design led to a 3.5-fold increase in signal sensitivity when compared with Cas13a alone and when reading Csm6-mediated fluorescence and Cas13a-mediated fluorescence in the same channel. Similarly, the generation of cyclic oligoadenylates from the native Cas10 enzyme in type III systems has been used for detection of SARS-CoV-2 genomes.

More strategies that increase the sensitivity of preamplification-free CRISPR-based diagnostics are needed. This could include the Cas-enzyme-mediated degradation of aptamers or of enzyme-inhibiting nucleic acid linkers (such as calf intestinal alkaline phosphatase or luciferase) that can then be detected via colorimetric or luminescence methods.

CRISPR INTERFERENCE (CRISPRI) AND CRISPR ACTIVATION (CRISPRA)

The advent of CRISPR-Cas9-based genome editing offered a convenient alternative to existing complex techniques such as zinc-finger nucleases. Using CRISPR-Cas9, researchers can edit DNA sequences in the genome with utmost precision. The technique has resulted in easier ways to insert or delete genes, tweak single base pairs in DNA, and has garnered unforeseen adoption in the research community.

But CRISPR technology has the power to go beyond gene editing - a slight tweak in the CRISPR-Cas9 duo turns it into a gene-specific regulator, adding the functionality of a light dimmer to an already formidable on-off switch. The sub-applications of CRISPR that render the brightening and dimming are called CRISPRa and CRISPRi, short for CRISPR activation and interference, respectively.

CRISPRa

CRISPR activation or CRISPRa is a variant of CRISPR in which a catalytically dead (d) Cas9 is fused with a transcriptional effector to modulate target gene expression. Once the guide RNA navigates to the genome locus along with the effector arm, the dCas9 is unable to make a cut, and instead, the effector activates the downstream gene expression.

CRISPRi

CRISPR interference or CRISPRi is also a variant of CRISPR in which a catalytically dead (d) Cas9 is fused with a transcriptional effector to modulate target gene expression. However, in CRISPRi, when guide RNA navigates to the genome locus along with the effector arm, it represses the downstream gene expression instead of activating it.

CRISPRa vs CRISPRi vs CRISPRko

While CRISPRa/i modulate gene expression transiently, gene knockouts disrupt the gene function permanently. While useful to screen weak phenotypes that can be masked by residual low-level gene expression, knockouts are not suited to studying essential genes, because deleting such genes results in cellular toxicity. In addition, gene regulation is useful to study pharmacotherapy because drugs often reduce gene expression, rarely do they wipe out a gene completely. Therefore, knocking down a gene partially can mimic drug action better than a knockout.

Both CRISPRa and CRISPRi work with genes at the transcriptional level. Playing with the levels at which a target gene is expressed can help answer complex questions about gene function in cell signaling cascades. By sliding gene expression up and down, scientists can test the robustness and sensitivity of biological systems to those small changes.

CRISPRa allows scientists the unprecedented ability to increase the expression of genes in their native context. This is in sharp contrast to other gain-of-function techniques such as plasmid and lentiviral-based open reading frames, or ORFs, that can primarily drive exogenous gene expression. CRISPRa also enables the overexpression of large transcripts for which ORF-based methods do not work as well.

CRISPRi is complementary to RNA interference or RNAi, but is advantageous in having fewer sequence-specific off-target effects, and in being able to modulate both coding as well as non-coding genes.

The CRISPR-Cas9 system makes use of an engineered single guide RNA (sgRNA) to home in the Cas9 nuclease to a specific locus in the genome. Cas9 introduces a double-stranded cut at the locus. The DNA damage activates a routine but error-prone process to repair double-stranded breaks, which introduces mutations and results in the complete loss-of-function of the target gene.

In prokaryotes, replacing Cas9 with a catalytically inactive version, dCas9, and targeting it to a promoter region can inhibit downstream gene expression because dCas9 sterically hinders transcriptional machinery. In mammalian cells, dCas9 alone achieves modest repression (60-80%), therefore the CRISPRi system includes dCas9 fused with a repressor domain such as Kruppel associated box, or KRAB, which can help silence genes in an inducible, reversible, and non-toxic manner.

Similarly, CRISPRa uses dCas9 fusions to activate gene expression. dCas9 fused to transcriptional activators such as VP64 and p65 can be targeted to promoter and enhancer regions, which results in higher than usual transcription of the target gene. To enhance gene activation in mammalian cells, dCas9 fusions that recruit multiple activator domains have also been developed, including a protein scaffold called the SunTag system, a sequence containing many copies of the activator recruitment domain of general control protein (GCN4).

Experimental Considerations for CRISPR-based Gene Regulation

A fundamental requirement of CRISPR-based gene regulation is for the sgRNA dCas9 system to be able to find and bind to the promoter region of the target gene.

The first step is to design the sgRNA complementary to the promoter region or transcriptional start site. However, this is a challenge because start sites are not always well annotated in the genome. In addition, sometimes cryptic or alternative promoters located far upstream of the target gene can also come into play, further complicating design considerations. Even if the sgRNA is designed to find the promoter correctly, the promoter site may already be occupied by other protein factors, and therefore be inaccessible for the guide RNA-dCas9 system to bind.

In order to design the perfect sgRNA with the ability to regulate each gene, researchers have employed systematic screens at the genome-scale — by pooling a sgRNA library followed by next-generation sequencing, guide RNA specificity and efficacy can be evaluated for each gene. For example, if the expression of a certain gene went up or down, the specific sgRNA that drove it is considered efficacious. Using these data, researchers have built design algorithms to create lists of top sgRNA sequences for each gene in human and mouse genomes.

Getting from the perfect sgRNA sequence to making it can involve either plasmid-based expression or synthetic production. Cloning using a plasmid vector is one of the earliest methods to produce sgRNA, however, it is also time-consuming, and prone to off-target effects. Instead, synthetic sgRNA production offers a faster and more accurate method to get to high CRISPR editing efficiencies.

In order to further optimize CRISPR-based gene regulation, researchers are also engineering dCas9 to be less toxic by reducing its non-specific binding to NGG sequences (where N can be any nucleotide base).

APPLICATIONS OF CRISPRA AND CRISPRI

CRISPRa and CRISPRi are handy tools to manipulate gene function, adding significantly to the pre-existing arsenal of genetic techniques. These tools lend themselves well to screens for functional genomics — by using a genome-wide library of guide RNAs, researchers can slide gene expression up or down in an unbiased manner, and screen for genes that affect growth and proliferation, or genes that mediate sensitivity to a selective pressure such as a drug or a toxin. Screens using CRISPRa and CRISPRi have been employed to study growth-driving genes in leukemia cells, and genes that mediate resistance to a BRAF inhibitor in melanoma cells. These techniques have found use in both basic and applied life sciences research.

CRISPRa to screen non-coding regions

Resistance to chemotherapy plays a significant role in cancer mortality. Bester et al used CRISPRa to identify genetic units that mediate resistance to chemotherapy, specifically to cytarabine, the cornerstone treatment for acute myeloid leukemia. Using the technique, they screened 14,701 long non-coding RNA genes, and identified novel cell-cycle, survival/apoptosis, and cancer signaling genes of therapeutic relevance.

CRISPRa to drive endogenous gene expression in vivo

Wangensteen et al adapted the CRISPRa system for use *in vivo* to assess whether changing endogenous gene expression levels can lead to functional changes in the native environment of the liver. The authors used CRISPRa screening to identify protein-coding genes that drive hepatocyte proliferation and tumorigenesis in an *in vivo* live mouse model of liver injury and repopulation. The authors reported a significantly enriched proto oncogene and development of hepatocellular carcinoma, both indicative of increased cell division and cancer initiation events.

CRISPRi to probe new cell types and species

In addition to using cancer cell lines, researchers have demonstrated the use of CRISPRi screens in induced pluripotent stem cells (iPSCs). For example, Tian et al used CRISPRi in human iPSC-derived neurons to conduct a growth screen and uncovered genes essential for neurons, but not iPSCs or cancer cells.

Taking CRISPRi to a new species where traditional genetic manipulation is challenging, Walker et al reported adapting CRISPRi to probe gene function in the malaria parasite *Plasmodium yoelii*, which will ultimately help control its transmission.

EPIGENOME EDITING WITH CRISPR

Epigenetics studies the proteins and chemicals that turn genes on and off, without altering the underlying DNA.

For years, scientists and drugmakers have tried to find the right molecular switches that cause a gene to make a protein, and use that information to treat disease. For instance, a chemically linked configuration of carbon and hydrogen atoms - known as a methyl group - can bind to DNA and change how, or whether, certain genes are "read" by specific proteins. Chemical changes to proteins called histones that hold onto DNA can also alter gene expression.

So far, that knowledge has led to limited drug development successes in the form of a few chemical-based cancer medicines. Merck & Co.'s lymphoma drug Zolinza targets a protein that affects the chemical makeup of histones. Epizyme's soft tissue cancer drug Tazverik, meanwhile, blocks an enzyme involved in gene expression.

But in many cases it's been difficult to determine which switches control what genes, or how to get to them without causing other problems. Using new computing tools and advances in genomic research, a number of biotech companies pushed ahead with research, and are now joined by others seeking to use CRISPR-based tools. Their idea is to use CRISPR components to turn genes on or off, or to alter the expression of several at a time without cutting into or changing DNA.

Advantages of epigenetic editing

Epigenetic editing is a bold methodology born from the need to expand the potential of genetic medicine. But what sets it apart from other techniques? The method is unique because it leaves the native DNA intact, has powerful multiplexing capabilities, and thus offers tremendous flexibility in gene output. These benefits expand the potential range of therapeutic targets available to precision medicine beyond those of other therapeutic approaches.

Avoiding DNA damage

As described, epigenetic editing is a 'no-cut' process. This avoids potentially harmful DNA damage such as chromosomal translocations, which are associated with several tumour types. Additionally, because epigenetic editing does not directly alter the genetic source code, on the rare occasion that off-target effects occur, they are generally benign. Put another way, epigenetic editing shares the specificity of nuclease-based genetic editing while minimising the risks associated with directly cutting into the genome.

Multiplexing

Another marked advantage of epigenetic editing is the potential for *multiplexing*, ie, targeting several genes simultaneously. Many disease states are caused by a web of complex interactions rather than single-gene mutations. As a result, the solution to these pathologies requires a multigenic approach. While in theory gene editing can cut several locations simultaneously, in practice, creating multiple double-strand DNA breaks can potentially cause genomic instability. Epigenetic editing eliminates this risk by leaving the DNA intact. This gives epigenetic editing the unique ability to treat complex multigenic diseases and opens the door to enhancing cell and regenerative therapies, which require the manipulation of multiple genes for optimal results.

Nuanced and highly controlled adjustments

Epigenetic editing goes beyond the binary on-off switch state achieved by gene editing and can instead create a dynamic range of expression levels. The field of epigenetic editing includes a growing suite of tools that can deliver precise gene dosages with impressive control. This is particularly critical in cases where over- or under-expression of a gene can be toxic, as in haploinsufficiency syndromes.

Additionally, the epigenetic editing toolkit allows for precise temporal regulation. The innate systems of epigenetic memory allow for the creation of enduring, heritable alterations or brief temporary changes in transcription. This level of control is not achievable with gene editing and is a severe limitation in applications such as T-cell engineering, where precise epigenetic remodelling during differentiation fundamentally shapes the phenotype and function of the cell.

The epigenetic editing toolbox continues to grow increasingly more sophisticated with tools that facilitate greater precision. For example, chemical modifications to the guide RNA – responsible for directing the epigenetic machinery to the region of interest – can fine-tune the intensity and stability of transcriptional changes. Furthermore, scientists are engineering diverse targeting systems such as modified CRISPR-Cas complexes, zinc fingers and transcription activator-like effectors to tackle epigenetic targets with enhanced specificity and durability. As we discover more about the molecular players regulating transcription, new options will emerge, enabling finer control over the dosage and duration of epigenetic changes.

Epigenetic editing in the clinic

With tightly controlled gene regulation and unparalleled multiplexing capabilities, epigenetic editing has profound clinical potential. So, where does epigenetic editing currently stand and how will it transition into the clinic?

Epigenetic editing has a strong foundation in the lab, and ample data has shown safety and effectiveness in non-clinical studies. For example, researchers used it to induce transcriptional silencing in the liver of mice, resulting in an enduring reduction in cholesterol levels. In patient-derived induced pluripotent stem cells with fragile X syndrome, epigenetic editing successfully reversed hypermethylation, effectively restoring healthy expression. Beyond targeted expression changes, epigenetic editing can also perform high-throughput screens to identify functional regulatory elements, which may be critical for effectively engineering stem and immune cells as well as programming tissue regeneration. In each of these examples, as with all epigenetic editing, the base DNA is left completely intact.

The clinic is the natural next step for epigenetic editing. Fortunately, many components of the epigenome editing workflow already exist in the clinical space, making the activation energy required for this transition relatively low. The delivery systems – such as viral vectors and lipid nanoparticles – and the DNA targeting system are already functioning in some clinical capacity. While challenges remain, based on the unique advantages and promising pre-clinical data, epigenetic editing should be ready to enter the clinic within the next few years.

CHAPTER-6

CRISPR IN MEDICINE

Initially discovered in bacteria and archaea as adaptive immune strategies against invading nucleic acids, the clustered regularly interspaced short palindromic repeats-associated protein (CRISPR-Cas) system has now been repurposed as a practical tool for genome editing and other applications. Owing to its flexibility, simplicity, efficiency and precision, the application of CRISPR systems in medicine has attracted significant attention and experienced rapid development. Currently, CRISPR systems have been used to screen novel drug targets, identify pathogenic genes, establish disease research modes, develop therapies against various diseases and so forth.

CRISPR-based genome editing tools provide effective and diverse strategies to prioritize candidate drug targets or validate putative targets. With CRISPR-Cas9 genetic screening in a diverse collection of cancer cell lines, statistical methods were developed to identify core and context-specific fitness genes systematically. Based on the results from CRISPR-Cas9 screens and related genetic biomarkers, further assessments can be made to decide target genes with the most potential for novel anti-cancer drugs. CRISPR systems can be also used to validate putative targets of new drugs. Researchers, for example, used CRISPR-Cas9 mutagenesis to investigate a serious of cancer drug targets in different stages of clinical trials. Upon the silencing of these putative targets, the anti-cancer effectiveness of some of these drugs remained, indicating the existence of unexpected targets.

Besides drug target identification or validation, CRISPR systems can also be used as diagnostic tools, which facilitates accurate detection and thus early intervention for infectious/noninfectious diseases. Researchers developed a platform called SHERLOCKv2 (specific high-sensitivity enzymatic reporter unlocking version 2), which can detect one dsDNA and three ssRNA targets in a single reaction. The multitarget DNA and RNA detection by SHERLOCKv2 can be applied as a portable, rapid and quantitative diagnostic test for Dengue or Zika virus as well as pneumonia pathogens.

Furthermore, the CRISPR-based protective and therapeutic strategies are probably the most promising applications of CRISPR system. CRISPR-Cas9 is revolutionizing the area of gene therapy, and advances have been achieved in Duchenne muscular dystrophy, a1-antitrypsin deficiency, hearing loss and hematopoietic diseases, human immunodeficiency virus type 1 (HIV-1) infection as well as cancers. Several clinical trials are underway to evaluate the feasibility and safety of CRISPR-based treatments (e.g. NCT02793856, NCT03081715, NCT03398967, NCT03655678 and NCT03399448). For example, transplantation of cells with *CCR5* gene artificially disrupted showed a remarkable therapeutic effect against HIV-1 infection.

Researchers established a *CCR5*-ablating system by CRISPR-Cas9 tool in hematopoietic stem and progenitor cells (HSPCs), which conferred HIV-1 resistance *in vivo*. In their recent clinical study, the investigators used CRISPR-Cas9 technology to knock out the *CCR5* gene of the donor stem cells and performed a successful transplantation of the *CCR5*-ablated HSPCs into a patient with HIV-1 infection and acute lymphoblastic leukemia. After 19 months, the acute lymphoblastic leukemia was in complete remission, edited stem cells were engrafted with full donor chimerism and no apparent adverse event was observed. Despite these promising results, further researches remain required to demonstrate the safety and feasibility before clinical practice.

Apart from application in basic and clinical medicine, substantial effort has been paid to enhance the efficiency and precision of CRISPR-based genome editing tools in the past 3 years. Researchers improved the CRISPR system by developing a precise 'search-and-replace' genome editing method (called prime editing), which can induce targeted insertions, deletions and point mutation without DSBs or donor DNA templates. They successfully use prime editing to correct primary mutations that caused sickle cell disease or Tay-Sachs disease in human embryonic kidney cells, with high efficiency and few byproducts. Although this improved editing tool and delivery protocol reduced the off-target activity, concerns remain on the large deletions and complex rearrangements generated by on-target cutting and repair, which may incur potential pathogenic consequences.

Ethical and safety concerns on the application of CRISPR system have received much attention in the past years. Since the first CRISPR-Cas9-based human embryo genome editing in 2015, the speculation over embryo gene editing has never stopped. Recently, the medical genomes editing on twin baby girls by Jiankui He caused fierce controversy in this area. Without considering the long-term adverse consequences, it was irresponsible to inactivate the *CCR5* gene, since this alteration may increase the risk of complications from other viral infections. After in-depth discussions, Lander and colleagues called for a temporary global moratorium on all clinical uses of human germline editing before the establishment of an international framework. Clinical application of germline editing should not be permitted unless the long-term consequences are fully understood.

Creating accurate disease models with CRISPR

To understand and unravel the biological processes that underpin and cause disease, as well as identify and test potential treatment options, a relevant and appropriate model system is crucial. The flexibility and efficiency of CRISPR both *in vitro* and *in vivo* in a myriad of different cell types has allowed for the rapid generation of both cellular and transgenic animal models, allowing for the creation of robust and complex models of disease, including iPSCs, organoids, and precision cancer models. For example, Heckl et al. created a leukemia model by leveraging the ability of CRISPR to be performed in multiplex and knocking out multiple genes simultaneously. CRISPR has also been used to create an accurate model of Alzheimer's Disease, which is inherently difficult to model in animals. Using iPSCs, CRISPR was utilized to create iPSC-neurons displaying an Alzheimer's Disease phenotype alongside an isogenic cell pair where the disease-associated mutations were corrected.

High-throughput screening with CRISPR

As well as revolutionizing gene editing, CRISPR is now a powerful tool for genomic screening, facilitating unbiased interrogation of gene function in a range of cell types, unraveling previously unknown molecular mechanisms. Pooled library screening with CRISPR-Cas9 involves the introduction of various genetic perturbations into a pool of cells—those genes that confer resistance or sensitivity to the biological challenge are then identified by massively parallel sequencing of the pooled cells. Alternatively, arrayed screens can be performed with CRISPR enabling the use of more complex phenotypic analysis. The expanded CRISPR toolkit means that as well as gene knockout, screens can include gene activation and silencing—in addition to high-content readouts such as single-cell RNA sequencing and spatial imaging.

As well as identifying genes, pathways, and molecular mechanisms to understand biological processes that cause disease, CRISPR-based genetic screening enables the identification and prioritization of drug targets. PICKLES v3 is an updated database of pooled *in vitro* CRISPR knockout library essentiality screens that allows visualization of gene essentiality and fitness scores across many different cell lines and tumor types.

CRISPR-based diagnostics

Accurate and timely detection of disease is needed for successful interventions to treat disease, as well as for epidemiological surveillance. Detection of nucleic acids is a sensitive and specific method of diagnosis but comes at a high cost and requirement for trained personnel. An exciting development in molecular sensing is the use of CRISPR. In just a few years, CRISPR-based diagnostics have gone from experimental tool to clinically relevant diagnostic technology that is sensitive, fast, affordable, and portable, enabling testing to be performed at point of care, or even in the home.

The first CRISPR-based diagnostic methods, such as NASBACC (Nucleic Acid Sequence-Based Amplification-CRISPR Cleavage), use Cas9 to detect double-stranded DNA used for detecting Zika virus in the low femtomolar range. Alternatively, SHERLOCK (Specific High-sensitivity Enzymatic Reporter unLOCKing) and the updated SHERLOCKv2 utilize Cas13 variants to detect single DNA or RNA molecules, allowing identification of ssRNA from dengue, Zika virus, and pneumonia pathogens.

The updated SHERLOCKv2 also allows for multiplexed sensing of nucleic acids at the zeptomolar range (10M) where cleaved reporter molecules are detected through antibody-conjugated gold nanoparticles on a paper strip. Chen et al. also used the Cas12a variant to develop DETECTR (DNA Endonuclease-Targeted CRISPR Trans Reporter) to discriminate between HPV16 and HPV18 with high accuracy.

As well as pathogen surveillance, CRISPR-based diagnostics can also be used for disease monitoring, including non-small cell lung carcinoma, colorectal cancer, liquid biopsy samples from adenocarcinoma patients, cervical cancer, and the reproductive system. More recently, researchers has developed an ultrasensitive photoelectrochemical biosensor using the CRISPR-Cas13a system for the specific detection of miRNA-21, a small regulatory RNA involved in many biological and disease processes such as development, inflammation, cardiovascular disease, and cancer.

Fighting cancer with CRISPR

The ability of CRISPR to correct, disrupt, or replace mutated genetic sequences has potential as a therapy for human diseases. There are currently trials underway in a variety of diseases including Duchenne Muscular dystrophy, sickle cell disease, beta thalassemia, hereditary angioedema, Leber Congenital Amaurosis, HIV, diabetes, as well as cancer. In May 2022, a patient with T-cell acute lymphoblastic leukemia was successfully treated with healthy donor-derived T cells that had been base edited to boost its ability to fight the cancer.

Base editing is a method where Cas9 machinery is utilized to deliver a base editor to elicit a single-point mutation—and so could improve patient safety as it does not rely on the introduction of a double-strand break. Researchers is also investigating CRISPR for cancer therapeutics, but is using gene editing to repurpose cancer cells into therapeutic tumor cells (ThTCs). Using CRISPR, tumor cells were made resistant to interferon-b and engineered to release immunomodulatory agents including IFN-b and granuloycte-macrophage colony-stimulating factor (GM-CSF). The ThTCs were then able to eliminate glioblastoma tumors in mice, with increased survival and long-term immunity from cancer.

The therapeutic benefit of CRISPR

CRISPR technology has revolutionized gene editing and has opened up many promising applications for human health. As well as elucidating those biological mechanisms that drive disease, CRISPR is also aiding in the identification and testing of therapeutic agents, and even the detection of viral, bacterial, and human nucleic acids at high sensitivity. CRISPR continues to evolve, with work focusing on improving efficiency and precision of the system, including the development of prime editing, which can induce targeted changes to genetic sequences without a double-strand break or donor DNA template—a system that may provide improved safety for gene therapy in humans. Epigenome editing with CRISPR could also offer novel therapeutic options for a variety of diseases including metabolic disease, cancer, blood disorders, and diseases of the central nervous system.

GENE THERAPY AND CRISPR

Introduction

Gene editing is a technology that precisely modifies the genome sequence to induce insertions, deletions, or base substitutions in the genome. Many diseases are accompanied by changes in gene expression *in vivo*, particularly some genetic diseases caused by mutations in a single gene, and gene-editing technology is expected to control the occurrence of diseases at the genetic level.3 To date, gene-editing technology has undergone three main generations of development: the first generation of gene-editing technology was zinc-finger nucleases (ZFNs); the second generation was transcription activator-like effector nucleases (TALENs); and the most widely used third generation gene-editing technology is clustered regularly interspaced short palindromic repeats (CRISPR)/CRISPR-associated protein 9 (Cas).4 Unlike ZFNs and TALENs, which use proteins to target DNA strands, CRISPR technology directs Cas proteins to a specified location in the genome by changing the base sequence of a small segment of guide RNA, thus improving the efficiency of gene editing and expanding the applicability of gene-editing technology.

CRISPR/ Cas9 is a highly effective gene-editing tool that is widely used in the scientific community. The CRISPR/Cas9 system evolved naturally in bacteria and archaea as a defense mechanism against phage infection and plasmid transfer. Bacteria or archaea acquire a segment of their DNA sequence to insert into the CRISPR spacer region when first infiltrated by an exogenous phage or plasmid. If reinfected with homologous DNA, the bacterium will initiate transcription of the CRISPR region. After a series of processing and maturation processes to generate a single guide RNA (sgRNA), the sgRNA guides Cas9 to shear the DNA strand that disrupts the homologous spacer region. The recognition process of the sgRNA requires the involvement of protospacer-adjacent motifs (PAMs), a short guanine-enriched sequence.

The preferred PAM by *Streptococcus pyogenes* Cas9 (SpCas9) is NGG, which is common in the genomes of most organisms, thereby facilitating the use of CRISPR technology across the fields of plant and animal science, together with biomedicine. By changing the nucleotide sequence of a small segment of guide RNA, CRISPR/Cas9 allows the accurate targeting of almost any desired genomic locus for the purpose of correcting disease-causing mutations or silencing genes associated with disease onset. However, some highly chromatinized regions in the genome may not be accessible to CRISPR/Cas9. Promising applications for this technology include the treatment of cancers, cardiovascular diseases, sickle cell anemia, and neurodegenerative disease.

Wild-type Cas9 only cuts double-stranded DNA to form double-strand breaks (DSBs), which are repaired through DNA repair mechanisms, namely, homology-directed repair (HDR) and nonhomologous end joining (NHEJ). The base sequence of the original gene is damaged, resulting in inactivation, but the inactivation of a single deleterious gene cannot address the complex processes of all disease events.

Therefore, researchers searched for possible ways to modify Cas9 by elucidating the physicochemical structure of Cas9, the mechanism of action by which Cas9 cleaves double chains, and other properties. They endowed Cas9 with new functions by mutating the structural domain of Cas9 and introducing effectors, including transcriptional regulatory tools such as dead Cas9 (dCas9) effectors and single-base substitution tools such as cytosine base editors (CBEs), adenine base editors (ABEs), and prime editors (PEs). Moreover, RNA recognition and cleavage functions can be performed by Cas13a isolated from *Leptotrichia shahii.* These Cas9 variants and derivatives enrich the gene-editing paradigm and can be adapted to additional types of diseases.

Although several experiments have documented the use of gene-editing technology to modify cells in vitro for return to the body to treat some diseases, this approach is not applicable to most disease types. Achieving stable, efficient, and safe delivery *in vivo* is a challenge that must be overcome before CRISPR technology becomes a common treatment modality. CRISPR systems such as plasmid DNA (pDNA), mRNA and ribonucleoproteins (RNPs) are subject to degradation and immune clearance *in vivo* after direct delivery and therefore require the help of delivery vectors.

Adeno-associated virus (AAV) vectors are not suitable for application in most diseases because of the drawbacks of a limited loading capacity, a lack of specific targeting ability and inability to integrate into the host genome. Nonviral vectors have been a hot topic of research in recent years, where lipid nanoparticles have been used in the clinic for the delivery of CRISPR gene drugs. Polymeric nanoparticles, biomimetic nanomaterials, and exosomes have also shown potential for the delivery of CRISPR systems in animal experiments. Further research and development are needed to apply nonviral vectors to a wide range of clinical applications.

Each disease has different characteristics, and our aim is not to develop a universal delivery vehicle but to develop multiple vehicles applicable to different types of diseases. Therefore, studying the pathogenesis of diseases and the pathological characteristics of disease cells and tissues and constructing environment-responsive and ligand-recognizing nanoparticles based on these characteristics will further enrich gene-targeting drugs in diseased tissues. In addition, exosomes and cell membranes from immune cells or diseased organs can effectively avoid immune clearance, and the abundant membrane proteins on the surface enable gene-targeting drugs to be delivered to diseased cells.

In this review, we discuss the development of CRISPR technology and summarize the various types of gene-editing tools that have been developed in recent years. Delivery systems for CRISPR systems in the body are also summarized, with a focus on developing new systems more suitable for different diseases, and finally, the review addresses a collection of problems that may arise when applying CRISPR technology to treat diseases and the corresponding strategies. In conclusion, this approach has positive implications for providing the most effective gene therapy modalities for different diseases.

Discovery and development of CRISPR technology

CRISPR-related gene-editing technology is currently one of the hottest biological tools. Since 2013, explosive growth has been recorded in the study of CRISPR technology, with tens of thousands of CRISPR-related articles published. In October 2020, the Nobel Prize in Chemistry was awarded to French microbiologist Emmanuelle Charpentier and American biologist Jennifer Doudna for "developing a new approach to genome editing". The method had been studied by scientists for nearly three decades before it received widespread attention (Fig. 1).

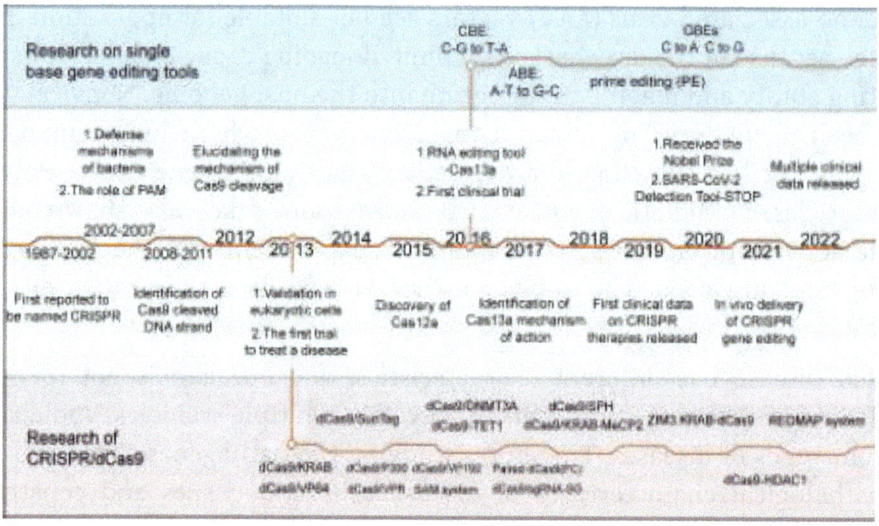

Fig. 1 Timeline of major events in the development of CRISPR/Cas technology and representative Cas9 variants.

Early detection using CRISPR technology

A special sequence of repeated intervals

Like many great discoveries, the discovery of CRISPR technology was born out of an unexpected event. An unusual sequence identified in the 32 end structural domain of the *iap* gene was first reported by Nakata et al. in 1987 while studying the *iap* gene of *E. coli.* The sequence consisted of five highly homologous sequences containing 29 nucleotides separated by 32 nucleotides.

Over the next decade, this particular repeat sequence was detected in a variety of bacteria and archaea. In 2002, Janson et al. provided a generalized summary of the specific repeats that have been identified, naming these repeats as a family and using the acronym CRISPR for clustered regularly interspaced short palindromic repeats.40 In addition, multiple CRISPR-associated proteins (Cas)-Cas1 to Cas4- have been revealed in previous studies.

Bacterial and archaeal defense weapons

In 2005, researchers discovered that the spacer sequences in CRISPR are not unique to each organism.8 Mojica et al. found that most of the spacer sequences were derived from exogenous DNA, with only a small fraction unrelated to the outside world, and they found that viruses were more likely to infect cells without homologous spacer sequences.8 They conjectured that CRISPR is involved in bacterial resistance to infection by external phages and in plasmid transfer. The conjecture was confirmed 2 years later.

When first confronted with phage or plasmid infestation, bacteria containing CRISPR sequences acquire a segment of their DNA sequence, which serves as a spacer region between special repeat sequences. CRISPR RNA (crRNA) then undergoes a series of transcription and maturation processes to produce a single crRNA containing a protospacer sequence of 20 bases that binds to the invading DNA via complementary base pairing. Recognition of the exogenous sequence by crRNA alone does not protect it from the phage; it also must be inactivated by disrupting the exogenous sequence through the cleavage activity of the Cas protein.

The CRISPR/Cas family of proteins is divided into two categories based on genomic and protein structure information, and the best-known protein Cas9 is among the Class II CRISPR/Cas systems. Class I is characterized by a large Cas9 protein complex that shears the DNA strand, while Class II requires only a single shearing protein. Cas9 is characterized by the presence of two ribonuclease structural domains, a RuvC-like nuclease domain near the amino terminus and the HNH nuclease domain in the middle of the protein, both of which have the function of cleaving the DNA strand. Notably, protospacer sequences are not randomly acquired from exogenous sequences but are always accompanied by a guanine-enriched sequence called protospacer-adjacent motifs (PAMs). Subsequent studies have shown that PAM sequences play an important role in the acquisition of the spacer region, where Cas proteins perform cleavage.

CRISPR IN CANCER RESEARCH AND TREATMENT

Cancer is the second most common cause of death globally, and despite years of research, the progress towards cures may seem slow. Until the advent of CRISPR, a tool that has certainly accelerated cancer research. In this blog post, we'll explore why a cure for cancer seems elusive, discuss a variety of cellular immunotherapies for cancer, and how CRISPR has improved their safety and efficacy. We will also cover CRISPR-editing *in vivo* for solid tumors, the current landscape of CRISPR cancer research and trials, and the future of CRISPR cancer therapeutics.

CRISPR-CAS9 SYSTEM?

CRISPR-Cas is a widespread, naturally occurring DNA-cutting system in bacteria and other microorganisms that was initially identified in the 1980s. More recently, in the early 2000s, it was discovered that the CRISPR-Cas system in these organisms serves as a defense mechanism against viruses. In bacteria, invading viral DNA is fragmented into short segments and incorporated into the host genome, organized in arrays known as "clustered regularly interspaced palindromic repeats," or "CRISPR". If the same virus infects the bacteria in the future, RNA molecules produced from the CRISPR array will guide a DNA-cleaving protein called Cas (which stands for "CRISPR-associated") to the corresponding viral DNA. Here, Cas will function as a type of "genetic scissors," cutting the viral DNA into many pieces, thereby leading to its destruction.

Researchers demonstrated that this naturally occurring process could be repurposed into a genome editing tool. Similar to the process in bacteria, CRISPR-Cas9-mediated genome editing relies on an RNA sequence—called a "guide RNA"—to direct the Cas9 protein to the correct DNA site. Once the Cas9 protein is guided to the correct genomic site, it will digest the DNA, thereby deleting the targeted gene.

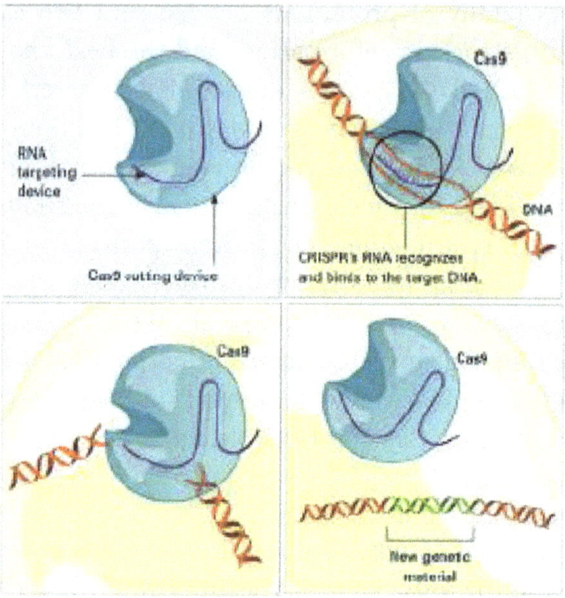

Fig. 2. Overview of the CRISPR-Cas9 genome editing system

In bacteria, the guide RNA forms naturally in response to viral infection, but in the case of genome editing, the guide RNA is carefully designed by the researcher to target a specific region of the genome. In some studies, researchers design their CRISPR-Cas9 assay to simply delete a gene, while in others, they insert a different gene in the place of the one that was deleted. Modifications that inactivate the DNA-cleaving activity of Cas9 are also sometimes used to enable the Cas9 protein to bind DNA, but not digest it. This binding can inactivate the target gene without deleting it.

Compared to other genome editing tools, CRISPR-Cas9 is cheaper, faster, and more accurate. Unlike prior methods, such as zinc finger nucleases and TALENs, the CRISPR-Cas9 method requires only a short stretch of RNA to target a genomic site, making it easier to design and employ. Furthermore, the increased accuracy of CRISPR-Cas9 compared to these other methods reduces the likelihood of off-target effects, making it safer for potential clinical applications.

Use of CRISPR-Cas9 in Cancer Research

Since its inception, the CRISPR-Cas9 technique of genome editing has been used in a myriad of basic cancer research studies. By deleting, inactivating, or otherwise modifying specific genes in cultured cells or in animal models, researchers have been able to study the role of these genes in cancer development and progression, providing insight into how to prevent or treat certain cancers.

In a study, researchers used the technique to understand how inactivation of the BRG1 gene, which is commonly mutated in non-small cell lung cancer, contributes to cancer progression. They utilized CRISPR-Cas9 editing to delete BRG1 in lung cancer cell lines and found that BRG1-deficient cells had increased replication stress and were sensitive to pharmacologic inhibition of the ATR checkpoint protein, revealing a mechanism of cancer progression and a potential therapeutic opportunity for the treatment of lung cancer.

An additional application of CRISPR-Cas9 is in genetic screens, where hundreds of genes are deleted individually to identify key regulatory proteins, potential drug targets, and/or to test responses to experimental drugs in a relatively quick manner. This approach was exemplified by another study, which utilized a CRISPR-Cas9 genetic screen to identify potential therapeutic targets for lung cancer. By individually deleting over 500 different genes that interact with or modify chromatin, researchers determined that deletion of the histone chaperone/ nucleophosmin/ 1 significantly reduced tumor progression in vitro and *in vivo*, thus identifying a potential drug target.

In addition to its widespread use in basic cancer research, the CRISPR-Cas9 system is also being explored as a way to improve cancer treatment. In a keynote address delivered at the AACR Virtual Special Conference: Tumor Heterogeneity, immunologist, oncologist, and Fellow of the AACR Academy Carl June, MD, described the potential utility of CRISPR-Cas9 in adoptive cell therapy, a form of cancer therapy that uses a patient's own immune cells to destroy cancer cells.

In adoptive cell therapy, T cells (a type of immune cell found in blood) are extracted from the patient, engineered to express a T-cell receptor that will recognize the patient's cancer, multiplied, and reintroduced into the patient. Adoptive cell therapy has been successful for the treatment of many blood cancers in both adult and pediatric patients; however, several challenges remain to using this therapy widely, including cancer-mediated activation of immune checkpoints.

Immune checkpoints constitute a normal regulatory process that prevents overactivation of the immune response and subsequent damage to normal tissue. Immune checkpoints are activated by the binding of certain receptors (PD-1 or CTLA-4) on the surface of T cells to their ligands (PD-L1 or B7, respectively) on target cells.

Some cancer cells express high levels of the PD-L1 or B7 ligands even in the absence of immune activation, allowing cancer cells to activate immune checkpoints, turn off the T-cell response, and circumvent the immune system. This presents a problem for adoptive cell therapy: While the engineered T cells can recognize the patient's cancer, they are inactivated before they can mount an antitumor immune response.

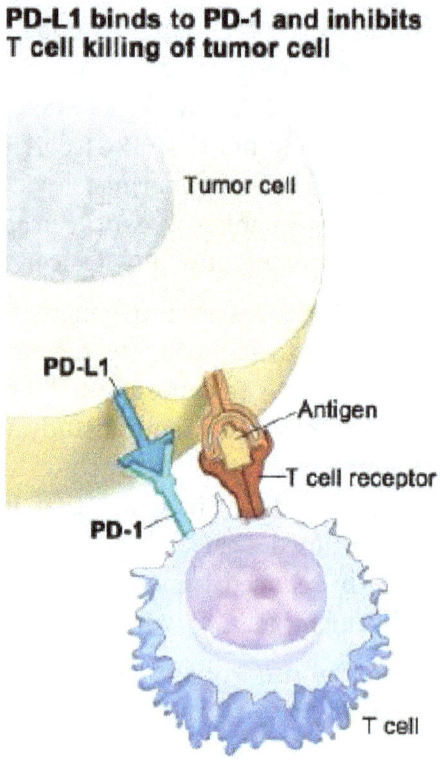

Fig. 3

One approach to addressing this challenge is to utilize CRISPR-Cas9 to delete the gene expressing the PD-1 receptor in the patient's T cells. Researchers hypothesize that the absence of the immune checkpoint would allow the T cells to remain active. Various studies have demonstrated that this strategy improves responses in preclinical models, but the feasibility of this approach in patients remains unclear.

In his presentation, June discussed results from the first-in-human trial of CRISPR-Cas9-modified T-cell therapy, which were published earlier this year. In this study, June and colleagues used CRISPR-Cas9 to delete from T cells the gene expressing PD-1, as well as genes expressing endogenous T-cell receptors to prevent competition with the engineered T-cell receptor. The CRISPR-Cas9-modified T cells were administered to three patients with advanced cancer, including two patients with refractory advanced melanoma and one patient with refractory metastatic sarcoma. Stable levels of the T cells were detected in the blood of all three patients several months after infusion.

None of the patients experienced any serious adverse effects from the treatment, and there was no evidence of autoimmunity or cytokine release syndrome, a potentially lethal side effect that can be associated with adoptive cell therapy. The engineered cells were able to recognize and destroy cancer cells from all three patients, and the treatment led to stable disease in two of the patients. The third patient had a mixed response, which included a 50 percent reduction of one of their tumors but progression of other tumors.

Gene expression analysis of the infused cells in one patient showed that the engineered T cells had increased expression of genes associated with T-cell memory, but not T-cell exhaustion, suggesting that deletion of PD-1 may have prevented the activation of immune checkpoints.

June concluded that editing of T cells using CRISPR-Cas9 technology was safe and feasible for cancer treatment. Further research will be required to determine the potential of this approach to prevent immune checkpoint activation and improve clinical responses.

By discovering a primitive bacterial defense system and transforming it into a powerful genome editing tool, scientists have greatly expanded the possibilities of research in diverse realms of science and medicine. As additional applications of CRISPR-Cas9 are conceived, studied, and applied, the impact of this technology is sure to grow.

Why is Cancer So Difficult to Treat?

Scientists have made huge progress in cancer research in recent years, but a cure remains elusive. This is because cancer is an incredibly complex group of diseases; there are hundreds of different types of cancer, each with a unique signature of genetic mutations. These mutations either switch on oncogenes or switch off tumor-suppression genes, either of which leads to unchecked (cancerous) growth of cells.

The genetic profile is distinct not only between cancer types‘ - for example, between lung and liver cancers‘ - but also within each type, and even within tumors themselves. Among breast cancer patients, for example, it's unlikely that any two tumors will possess the exact same profile of genetic mutations. The complex tumor microenvironment (TME) is immunosuppressive and is comprised of many different types of cells which also have different mutation profiles.

This genetic complexity influences how aggressive cancers are, how they will respond to various treatments, and whether they will re-occur. This means that there can never be a 'one-size-fits-all' cure, and generating accurate models for research is challenging. Additionally, many types of cancer are difficult to treat due to current screening and diagnostic limitations. In some cases, cancer is not diagnosed until it has metastasized and spread to multiple organs.

Once cancer has metastasized, it is more difficult to treat, partly because metastatic tumors are likely to have additional mutations compared to the original tumor and will therefore respond to treatment differently. The accruement of mutations over time can also cause cancers to become drug-resistant. Cancer cells are also very adept at avoiding and/or suppressing the immune system in order to continue to grow. Some cancers are malignancies of the immune cells themselves. For example, lymphoma is a type of cancer that affects lymphocytes, the immune cells that would normally fight cancer and infections.

CRISPR: A Game-Changer in Cancer Therapeutics

Adoptive cell therapies (ACTs) for the treatment of cancer, like TIL, TCR, CAR-T and NK, are desirable over traditional treatments such as chemotherapy and radiotherapy because they are more targeted to cancerous tissue, leaving healthy cells untouched. All cell-based immunotherapies involve expanding (growing) these immune cells ex vivo and re-infusing them into patients so their body has more immune cells available to attack the cancer. These ACTs can be autologous (isolated from the patient) or allogeneic (isolated from a healthy donor).

While all ACTs have encountered various hurdles in their translation into the clinic, in recent years CRISPR has offered new hope for ACT-based cancer treatment. In this section, we'll explore the main types of cellular immunotherapies used to treat cancer, how CRISPR can be used to make them stronger, safer, and more readily available, and how CRISPR can target cancer directly.

TIL and TCR cell therapies

T cells, often called T lymphocytes, encompass several subsets of cells and are a key component of the adaptive immune system. As their name suggests, tumor-infiltrating lymphocytes (TILs) are immune cells that can invade tumors in order to kill cancer cells. Autologous TIL therapy involves extracting TILs from a patient's tumor, expanding them in large numbers ex vivo, and re-infusing them into the patient so they have more TILs available to fight cancer. Despite the early promise of TIL therapy, issues with the isolation and ex vivo expansion of TILs, along with their relatively limited ability to destroy tumors, eventually led to this approach being sidelined in favor of T-cell receptor (TCR) therapy.

Think of TCR therapy as an upgrade to TIL therapy. TCR therapy uses T cells that are genetically modified to express specific TCRs on their surface, making them better at recognizing cancer cells. TCR therapy showed promise, but like TILs, it encountered a few obstacles in its clinical translation. The lack of specificity of TCRs led to problems with the clinical translation of TCR therapy because cancer cells can downregulate the expression of MHC to evade detection by TCR cells.

Expression of endogenous TCRs alongside the transgenic TCR in the cells is also a key problem, causing competition for signaling components, the formation of heterodimers that can cause lethal autoimmunity, and graft-versus-host disease (GvHD). GvHD, a condition where grafts of healthy immune cells recognize the patient's own cells and tissues as foreign and attack them, is a major concern in any form of allogeneic (non-self) cellular immunotherapy and can be lethal.

Before the advent of CRISPR, engineering T cells was time-consuming, difficult, and expensive. Safety concerns associated with the use of retroviral and lentiviral vectors for editing were also a limiting factor. Fortunately, the recent CRISPR revolution has led to a revival of both TIL and TCR cell therapies.

In TILs, CRISPR can be used to knock out a gene that inhibits T cell function called cytokine-induced SH2 (CISH) - disruption of CISH increases TIL aggression towards tumors. In the case of TCR cells, CRISPR can be used for knock-in of the required TCR to specific sites in the T cell genome to increase its expression. It can also be used to knock out the endogenous TCR gene that can cause GvHD and other adverse events. Moreover, CRISPR can generate these edits in very short timeframes and without the use of viral vectors, increasing patient safety outcomes.

CAR-T cell therapy

CAR-T therapy is a gene-edited cell therapy, whereby T cells are edited to express a specific chimeric antigen receptor (CAR) on their surface. The CAR is specific to the type of cancer the patient has, allowing the T cells to recognize the antigens produced by the cancer cells and kill them. Importantly, CAR does not rely on the presence of MHC in order to recognize and kill cancer cells.

CAR-T therapies, both autologous (patient-derived) and allogeneic (healthy donor-derived), currently have some major limitations. For autologous CAR-T, this includes difficulty procuring large numbers of T cells from patients that are lymphopenic due to other treatments and long timeframes for generating a sufficient therapeutic dose. For allogeneic CAR-T, the healthy donor-derived cells can be rejected by the patient immune system, cause toxicity, or induce GvHD.

Unfortunately, some cancers are able to avoid destruction by CAR-T cells by overexpressing certain proteins, like programmed cell death protein 1 (PD-1), on their surface.

CRISPR-Cas9 gene editing has increased the safety and efficacy of CAR-T cell therapies in a variety of ways. Firstly, CRISPR can be used to knock in the CAR in a targeted manner - for example, to a safe harbor site in the genome - to ensure sufficient, long-term expression of the receptor on the cell surface. Secondly,

CRISPR can knock out certain genes in CAR-T cells to increase their cancer-killing activity. Thirdly, it can be used to create edits that minimize CAR-T cell exhaustion and increase their long-term proliferation. Finally, CRISPR edits can be used to generate universal, 'off-the-shelf' CAR-T cells from induced pluripotent stem cells (iPSCs), negating harvesting limitations, long wait times, and toxicity issues.

CAR-T vs. TCR therapy: What's the difference?

The main difference between CAR-T and TCR therapy is the receptors used and their mode of action. Although they sound similar, CAR-T and TCR work very differently. TCRs are naturally-occurring or minimally modified receptors, which rely on major histocompatibility complex (MHC) proteins to work and are therefore non-specific. In contrast, CARs are artificial receptors that are engineered to recognize specific cancer antigens.

Unlike TCRs, which can recognize extracellular and intracellular components, CARs can only recognize extracellular molecules on cancer cells, meaning the range of targets of CAR-T therapy is less than TCR. CAR-T cells also require more antigens to be present in order to be activated compared to TCR cells. Like other cell therapies, CAR-T cells are prone to becoming 'exhausted', a state in which their efficacy is reduced.

NK cell therapy

Natural killer (NK) cells are major players in the immune system, just like T cells. One of the key differences between these cell types is that NK cells are part of the innate immune system, while T cells are a component of the adaptive immune system. NK cells perform immune surveillance, patrolling the bloodstream for abnormal cells such as cancer cells, bacteria, or virus-infected cells. Unlike T cells, NK cells are not antigen-specific and they simply recognize 'non-self' cells, regulated by both activating and inhibitory receptors expressed on their surface. When activated, NK cells release cytotoxic molecules that lyse the non-self cell and kill it.

NK cell therapy is excellent at killing cancer cells and does not cause GvHD, which is a common side effect of other allogeneic immune cell transplants like CAR-T. However, there are several key problems that have prevented their widespread use. This includes issues with the ex vivo expansion of NK cells for clinical use, their relatively short lifespan and low persistence *in vivo*, and a limited ability to infiltrate solid tumors. Some types of cancer also manage to develop mutations that help them evade NK cells.

Similar to the recent progress in CAR-T cell therapies, CRISPR can be used to enhance NK cell therapies. For example, like T cells, NKs can also be edited to express a CAR (CAR-NK). Along with their innate non-self recognition, the addition of a CAR means they can work both specifically and non-specifically, reducing the likelihood that cancer cells will evade treatment. CRISPR can also be used to create edits that increase the lifespan of these typically short-lived cells or to enhance their cancer-targeting ability.

CRISPR gene editing *in vivo* in tumors

In addition to gene-edited cell therapies, like CAR-T, NK, TCR, and TIL, CRISPR-Cas9 can also be used to target tumors directly *in vivo*. Genetic mutations that switch on oncogenes (gain-of-function mutations) stimulate carcinogenesis, and the expression of these oncogenes is specific to cancer cells. Knocking out these genes via CRISPR-Cas9 is an appealing therapeutic target because it will prevent cancer growth. Recent examples include a lipid nanoparticle (LNP)-based delivery approach to disrupt the overexpressed PLK1 gene, and a lentivirus delivery strategy to target multiple cancer-specific indels. In some cases, this can be a bit tricky - for example, when attempting to disrupt fusion oncogenes (FOs), which are chimeric oncogenes resulting from chromosomal rearrangements. However, recent studies have developed strategies to target two introns of the genes involved in these rearrangements using CRISPR-Cas9. In cancers caused by FOs, this targeted disruption causes apoptosis of the malignant cells and eliminates the possibility of further growth. For more information on how CRISPR is being used in cancer research and treatment, you can listen to our recent CRISPR Cuts podcast with Dr. Jesse Boehm, who speaks about his research on rare cancers and his new foundation, Break Through Cancer.

The Current Landscape of CRISPR Cancer Therapeutics

The field of CRISPR cancer therapy is moving at a rapid pace, with lots of exciting proof-of-concept and pre-clinical studies being published regularly, the results of the first clinical trials starting to trickle in, and many new trials beginning. In this section, we'll take a look at the current state of CRISPR cancer research and therapy, including the types of cancer that can be treated, pre-clinical studies and clinical trials, and therapies that have been approved by the FDA.

Hematological cancers versus solid tumors

Hematological (blood) cancers, like leukemia, lymphoma, and multiple myeloma, present very different treatment challenges compared to solid tumors in organs or soft tissue, like breast, lung, or brain cancers. In general, gene-edited cell therapies for hematological malignancies are currently more advanced than those for solid tumors.

While tumors can often be treated surgically, blood cancers cannot be removed surgically because they involve malignant cells circulating freely in the body. However, many tumors cannot be surgically removed, or can only be partially removed. In these cases, treatment options become more difficult because of the complex TME and its immunosuppressive effects.

CAR-T therapies were originally developed to treat blood cancers because it is easier for T cells to attack freely circulating cancer cells rather than tumors, while TILs can be used to target solid tumors. In all cases, CRISPR is being used to enhance these therapies, increasing their safety and efficacy, and reducing their costs and production times.

Current preclinical research and clinical trials

Current CRISPR cancer research includes engineering TILs to be resistant to the molecules that cancer cells express in order to switch lymphocytes off, the creation of universal, off-the-shelf CAR-T cells, and the editing of NK cells for better cancer targeting. For an in-depth look at the different ways CRISPR can be used for cancer immunotherapy, check out this 2021 review.

In a clinical setting, current CRISPR-based trials include treatments for non-small-cell lung cancer (NSCLC), esophageal cancer, cervical cancer, metastatic gastrointestinal cancer, T- and B-cell malignancies, multiple myeloma, melanoma, and acute myeloid leukemia (AML), to name a few. For a full breakdown of the current state of clinical trials for T cell therapies, you can read this 2021 review from the Journal of Experimental & Clinical Cancer Research.

CRISPR Therapeutics is one of the companies leading the charge on gene-edited T cell therapies and is currently developing off-the-shelf, allogeneic CAR-T cells using CRISPR technology. In December 2021, CRISPR Therapeutics also announced a new partnership with NK-specialists Nkarta Therapeutics in order to develop CRISPR-edited NK cell therapy.

Intellia Therapeutics, who were already working on CAR-T therapies, announced a collaboration with ONK Therapeutics to work on several different CRISPR-edited NK therapies of cancer. The FDA has also recently fast-tracked a CRISPR-edited TCR cell therapy for the treatment of AML from Intellia Therapeutics, NTLA-5001.

In a pre-clinical study, Editas Medicine has developed iPSC-derived NK cells (iNK), using CRISPR-Cas12a to knock-in genes that increase NK cell persistence and tumor-killing capacity. Some excellent resources for staying up-to-date with current research and clinical trials using CRISPR technology include the Innovative Genomics Institute website and the CRISPR Medicine News clinical trials page.

For a valuable patient perspective on understanding cancer and getting treatment, you can listen to our interview with Bryce Olson on CRISPR Cuts podcast. Bryce's journey, from being diagnosed with aggressive prostate cancer, to learning about the science of cancer treatment and pursuing genome sequencing and personalized cell therapies, is both interesting and inspiring.

The Future of CRISPR Cancer Therapeutics

Scientists are hopeful that the use of both gene-edited cell therapies and *in vivo* CRISPR gene editing will enable the treatment of many types of cancer. However, we're only just beginning to scratch the surface of what CRISPR is able to do in this area of research. Recent adaptations of CRISPR technology, including newly discovered Cas nucleases, are revealing even more avenues for cancer therapeutics.

TARGETING GENETIC DISORDERS WITH CRISPR

The CRISPR/Cas system was originally discovered as a prokaryotic adaptive immunity system used to recognize and cleave invading nucleic acids. Based on this prokaryotic system, scientists have engineered a series of CRISPR/Cas tools for genome editing in mammalian cells, with the list of CRISPR/Cas systems in use continuing to expand. The most commonly used Cas nuclease comes from *Streptococcus pyogenes* (SpCas9), and belongs to the type II CRISPR system. SpCas9 was the first to be reprogrammed for genome editing in mammalian cells. For specific nucleotide sequence recognition, engineered SpCas9 relies on the guidance of a single-guide RNA (sgRNA). Typically, sgRNA is composed of a scaffold sequence that is bound by the Cas protein, and a custom-designed <“20 nucleotide spacer that defines the genomic target to be modified. Following hybridization of the spacer to a target genomic sequence that is positioned next to a protospacer adjacent motif (PAM), the target DNA is cleaved, leading to a double-strand break (DSB). The Cas-mediated DSB is subsequently repaired by cellular DNA repair machinery via homology- directed repair (HDR) or the non-homologous end joining (NHEJ) pathway. NHEJ can be used to produce insertions and deletions (indels) that disrupt or inactivate the target gene, while HDR can be used for precise nucleotide sequence modifications, such as point mutation correction.

To date, CRISPR/Cas-based techniques have been applied in various cell types and organisms. For therapeutic genome editing to treat monogenic diseases, CRISPR has the potential to be used directly in patients (*in vivo*) or in human cells (*in vitro*). In this review, we focus on CRISPR strategies used to treat human monogenic diseases, and discuss the challenges associated with these approaches.

Recent advances in CRISPR/Cas technology

Shortly after SpCas9 was applied in mammalian cells, other Cas9 proteins have been studied and developed as genome editing tools. For example, smaller Cas9 proteins derived from *Staphylococcus aureus* called SaCas9 and *Neisseria meningitidis* called Nme2Cas9 exhibit gene editing efficiency comparable to that of SpCas9. These smaller Cas9s are more amenable for *in vivo* delivery than the large SpCas9 (~4.3 kb). CRISPR/Cas9 technological advances have also enabled various applications of nuclease-deficient Cas9s, which can bind a specific region of the genome without creating DSBs. For example, catalytically inactive dead Cas9 (dCas9) can be fused with various transcription regulatory domains to create CRISPR activators (CRISPRa) or inhibitors (CRISPRi) that activate or silence the expression of a target gene. dCas9 can also be used as a visualization tool. Chen and colleagues have used dCas9 fused to enhanced green fluorescent protein (EGFP) to visualize repetitive DNA sequences using one sgRNA, or nonrepetitive loci using multiple sgRNAs. In addition, David R. Liu's group has fused D10A Cas9 nickase with either cytidine or adenine deaminase to generate cytidine base editors (CBEs) and adenine base editors (ABEs), respectively.

In addition to DNA editing, Feng Zhang's lab has reported that an RNA-targeting CRISPR system based on Cas13 can target and cleave specific strands of RNA, and subsequently developed strategies called REPAIR (RNA Editing for Programmable A to I Replacement) and RESCUE (RNA Editing for Specific C to U Exchange) to edit RNA. Thus, RNA editing with CRISPR can efficiently modulate target genes at the transcript level in a transient and PAM independent manner. This approach could provide a controllable approach for disease treatment.

Applications of CRISPR in genetic diseases

To date, CRISPR/Cas systems have been used to investigate target genes in genome modification, splicing, transcription and epigenetic regulation, and have been applied in a research setting to investigate and treat genetic diseases, infectious diseases, cancers, and immunological diseases. Among the exciting advances, translational use of CRISPR/Cas in monogenic human genetic diseases has the potential to provide long-term therapy after a single treatment. In this section, we summarize the recent applications of the CRISPR/Cas system in the generation of disease models and in the treatment of genetic diseases *in vitro* and *in vivo*.

Disease modeling using CRISPR/Cas

The generation of disease models is necessary for understanding disease mechanisms and developing new therapeutic strategies. CRISPR/Cas has been widely used for creating disease-related cellular models, such as DMD, aniridia-related keratopathy (ARK), brittle bone, X-linked adrenoleukodystrophy (X-ALD), and Alzheimer's disease. Moreover, researchers have created a series of mouse models using CRISPR/Cas that recapitulate DMD, atherosclerosis, obesity and diabetes, RTHa, and Alzheimer's disease. One example is the development of a mouse model for ryanodine receptor type I-related myopathies (RYR1 RM), which harbors a patient- relevant point mutation (T4706M) engineered into one allele, and a 16-base pair frameshift deletion engineered into the second allele of the RYR1 gene.

Subsequent experiments demonstrated that this mouse model of RYR1 RM is a powerful tool for understanding the pathogenesis of recessive RYR1 RM, and for preclinical testing of therapeutic efficacy. CRISPR/Cas has also been used to generate disease models in large animals, including sheep, rabbit, pig, and monkey. For example, a monkey model was developed to study Parkinson's disease by introducing a *PINK1* deletion and revealed a requirement for functional PINK1 in the developing primate brain. CRISPR/Cas technology offers a flexible and user-friendly means of developing disease models to explore the genetic causes of diseases and evaluate therapeutic strategies.

CRISPR/Cas9 and Gene Editing

CRISPR (clustered regularly interspaced short palindromic repeats)/ Cas9 system, is an RNA-guided DNA nuclease (RGN) are derived from an adaptive immune system that evolved in bacteria to defend against invading plasmids and viruses. Of course, upon invading, this system transcribed and processed CRISPR RNAs (crRNAs) together with a trans-activating crRNAs (tracrRNAs) and forms a complex with CRISPR-associated (Cas) proteins to cleave nucleotide sequence by Cas nucleases at specific location 158-161.

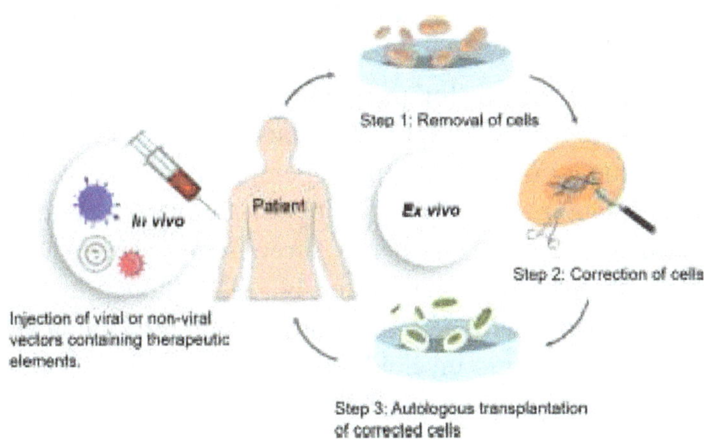

Fig. 4 Ex vivo and in vivo genome editing for clinical therapy. Right: For in ex vivo editing therapy, cells are isolated from a patient to be treated, edited and then re-engrafted back to the patient. To achieve therapeutic success, the target cells must be able to survive in vitro and return to the target tissue after transplantation. Left: For *in vivo* editing therapy, engineered nucleases are delivered by viral or nonviral approaches and directly injected into the patient for systemic or targeted tissue (such as the eye, brain, or muscle) effect.

Ie cleaved double stranded DNA is repaired by either NHEJ (non-homologous end joining) or HDR (homology directed repair). In NHEJ, DNA ends directly join with their counterparts in a highly eŠcient but error-prone manner, which causes small insertions and/or deletions (indels) at the cleave site, which can disrupt the translational reading frame of a coding sequence. In other hand, HDR requires a donor DNA homologous template to repair the cleavage site.

GENOME EDITING FOR DISEASE MODELING AND GENE THERAPY

Targeted gene modification via chimeric genome editing tools (e.g., ZFNs, TALENs, and CRISPR/Cas9) is a powerful method to assess gene function and precisely manipulate cellular behavior and function. These genome editing tools have enabled investigators to use genetically engineered animals to understand the etiology behind various diseases and to clarify molecular mechanisms that can be exploited for better therapeutic strategies .

Cancer research

Oncogenes and mutant tumor suppressor genes provide outstanding opportunities for the use of genome modulating approaches.62 Genome editing technology has accomplished crucial targeted cleavage events in various fundamental studies, from its initial proofs of efficient gene editing in eukaryotes to its recent applications in the engineering of hematopoietic stem cells (HSCs) and tumor-targeted T cells; this technology has established novel concepts of gene modification and has extended to a border field of cancer research.

As an archetypal platform for programmable DNA cleavage, ZFN-mediated targeting has been successfully applied to modify many genes in human cells and a number of model organisms, thus opening the door to the development and application of genome editing technologies. ZFN-driven gene disruption was primarily demonstrated in 1994 when a three-finger protein was constructed to specifically block the expression of the BCR-ABL human oncogene that was transformed into a mouse cell line. After that, a study used a human lymphoblast cell line derived from chronic myeloid leukemia (CML) patients, and a custom-designed ZFN was applied to this cell line to deliver site-specific DSBs to the telomeric portion of the mixed lineage leukemia (MLL) gene breakpoint cluster region as well as to analyze chromosomal rearrangements associated with MLL leukemogenesis via DSB error repair.64 Successful targeted modulation was also achieved using designed ZFNs, which promoted the disruption of endogenous T cell receptor (TCR) b- and a-chain genes.

Lymphocytes treated with ZFNs lacked the surface expression of CD3-TCR and expanded with an increase in interleukin-7 (IL-7) and IL-15. By targeting the promoter function of long terminal repeat (LTR) from human T cell leukemia virus type 1 (HTLV-1), a novel therapeutic ZFN specifically killed HTLV-1-infected cells in an *in vivo* model of adult T cell leukemia (ATL). In addition, it was reported that effective cleavage of the BCR-ABL fusion gene by highly specific ZFNs terminated the translation of the BCR-ABL protein and induced apoptosis in imatinib-resistant CML cells.

Furthermore, cancer-relevant translocations in human Ewing sarcoma and anaplastic large cell lymphoma (ALCL) cells induced by ZFNs demonstrated that precise genomic rearrangements can be achieved in relevant cell types by custom nucleases.

Furthermore, the use of HER2-positive cell-penetrating peptide (CPP) conjugated to mammalian mTOR-specific ZFN made the mTOR locus nonfunctional and inhibited relevant cancer signaling pathways, providing insight into the design of novel molecular targeted therapeutics for breast cancer (in particular) and other types of cancers.

Moreover, as the tumor suppressor gene p53 plays a pivotal role in preventing cancer development, strategies of genome editing to restore wild-type p53 function have been investigated. A yeast-one-hybrid (Y1H) four-finger ZFN was designed to replace mutant p53 with wild-type p53 in several cancer cell lines (from glioblastoma, leukemia and breast cancer) via ZFN-induced HR.70 Although the HR events were not particularly effective in this case, modifications at p53 loci still provided a framework for further investigation. In addition to modifying viral genes associated with tumorigenesis, researchers have applied ZFNs to optimize T cell-mediated antitumor therapy.

For example, by importing a chimeric TCR that comprises an extracellular IL-13 domain (zetakine) and a cytoplasmic CD3 domain into CD8 + T cells, glioblastoma-specific cytolytic T lymphocytes (CTLs) can be generated. To achieve this goal, researchers knocked down the glucocorticoid receptor in the modified CTLs with ZFNs. Consequently, the cytolytic activity of "zetakine" transgenic CTLs against glioblastomas was preserved regardless of the presence of glucocorticoid treatment. This technology has recently been effective in knocking out glucose transport-related genes (MCT4 or BSG) in two glycolytic tumor models: colon adenocarcinoma and glioblastoma.

A milestone of TALENs was achieved when they were primarily applied to efficiently disrupt the endogenous genes NTF3 and CCR5 in human leukemia cells via the introduction of NHEJ- or HDR-induced modification into a coding sequence, demons-trating that TALENs could be designed for selective endogenous gene cleavage. Interestingly, when TALENs and ZFNs were compared abreast at two human loci (CCR5 and IL2RG), TALENs showed a significant reduction in cytotoxicity. Moreover, the CCR5-specific TALEN was able to distinguish between the CCR5 target locus and a highly similar site in CCR2 when compared with ZFN technology. By adopting TALEN gene editing technology, precise disruptions have also been introduced into the T cell receptor á constant (TRAC) gene and the CD52 gene in allogeneic T cells by TALEN-induced HDR. The TALEN used in this study was engineered by a retroviral vector that expressed a chimeric antigen receptor (CAR) targeting CD19+ leukemic B cells, which helped to develop the "universal" CAR T cells (dKO-CART19). Alternatively, a site-specific TALEN was used to disrupt a single allele of the Fms-related tyrosine kinase 3 (FLT3) gene and generate isogenic leukemia cell clones. TALEN-mediated FLT3 haplo-insufficiency impaired cell proliferation and colony formation in vitro. These suppressive effects were maintained *in vivo* and improved the survival rate of NOD/SCID mice transplanted with mutant K562 clones.

The use of engineered TALENs in prostate cancer cells functionally classifies androgen receptor (AR) target gene rearrangements as drivers of resistance. Using TALENs to precisely cut the relevant translocation breakpoints, Piganeau et al. induced cancer-relevant translocations in anaplastic large cell lymphoma (ALCL). Through an analogous strategy, the reversion of ALCL translocation was achieved in a patient cell line, restoring the integrity of the two involved chromosomes. Recent studies have also shown that TALEN gene editing technology used to knock out genes in cancer cells (including cells from prostate cancer, breast cancer, and hepatocellular carcinoma (HCC)) is a powerful and broadly applicable platform to explore gene mutations at the molecular level.

Because of its multiple advantages in genome editing, the CRISPR/Cas9 system has attracted considerable attention, and scientists gradually consider it to be a powerful therapeutic tool for treating diseases associated with genome mutations. The ultimate goal of cancer therapy with CRISPR/Cas9 is to remove malignant mutations and replace them with normal DNA sequences.

In a recent study, the leukemia model was generated by reviving several inactivated oncogenes through the lentiviral delivery of the Cas9-sgRNA system in primary hematopoietic stem and progenitor cells (HSPCs). In this study, the pooled lentiviruses targeted genes, including Tet2, Runx1, Dnmt3a, Nf1, Ezh2, and Smc3. The objective HSPCs were selected via a fluorescent marker; those HSPCs are engaged in the develop-ment of myeloid neoplasia. CRISPR/Cas9 technology has also been adopted to establish organoid tumor models.

For instance, organoid colon cancer models were constructed in vitro with CRISPR technology by introducing mutations of tumor suppressor genes (APC, TP53, SMAD4, etc.) and gene modification of oncogenes (KRAS, PI3K, etc.).

Moreover, guided by colonoscopy, through mucosal injection, Researchers established CRISPR engineered mouse tumor organoids by delivering viral vectors carrying CRISPR/Cas9 components to the distal colon of mice. Such an approach has already been applied in a study modeling tumor progression with an adenoma-carcinoma-metastasis sequence. In the future, the use of CRISPR/Cas9 technology to establish precise cancer models will significantly promote the research of functional cancer genomics and facilitate the advancement of cancer therapies.

Cardiovascular disease

CVD is a serious hazard to human health and is the number one cause of death in many industrialized countries. Many different types of CVD are usually associated with a single genetic mutation or a combination of rare inherited heterozygous mutations.85 In practice, clinical treatments focus on the relief of disease symptoms without addressing potential genetic defects. Currently, the establishment of *in vivo* CVD models with gene editing technology and the in-depth analysis of CVD pathogenic genes as well as their molecular mechanisms have made it possible to test the ability of gene therapy to control specific gene expression and improve gene functions. With the help of genome editing technologies, various research models of cardiovascular conditions have been created.

Researchers used CRISPR/Cas9 to efficiently ablate major histocompatibility complex class II (MHCII) with double gene knockout in normal human endothelial cells. These cells retain the ability to form vascular structures without activating allogeneic CD4+ T cells. It is promising to apply such technology in the field of allograft bioengineering, including the refinement of heart transplant. In addition, CRISPR/Cas9 technology can accurately remove b2M and CCR5 on CD34+ HSCs while retaining its ability to undergo multi-differentiation, which provides the possibility for the future treatment of ischemic heart conditions with HSCs.

In another study, researchers established a cardiac-specific transgenic mouse model by injecting Cas9-containing plasmids into mouse zygotes; the expression of Cas9 was regulated by the Myh6 promoter. In this transgenic model, high levels of Cas9 were expressed exclusively in heart cardiomyocytes. The investigators then intraperitoneally injected sgRNA targeting Myh6 loaded in an adeno-associated virus (AAV) vector, subsequently inducing cardiac-specific gene modification at the Myh6 locus, finally leading to hypertrophic cardiomyopathy.

It has been demonstrated in the whole-exome sequencing of a nuclear family that three missense variants of a single nucleotide in the MKL2, MYH7, and NKX2-5 genes pass on to three offspring with cardiomyopathy with childhood onset. Researchers adopted CRISPR/Cas9 to establish a mouse model that encodes orthologous variants and showed that the complex of heterozygosity of all three variants reproduced the phenotype of human disease. An analysis of mouse heart and human induced pluripotent stem-cell-derived cardiomyocytes provides histological and molecular evidence for the contribution of the NKX2-5 variant as a genetic modifier.

Porcine models resemble human conditions by physiology, anatomy, and genetics and are often considered ideal models for human cardiovascular structure research. Researchers applied ZFN technology with nuclear transfer in somatic cells to generate endogenous gene knockout pigs, which have a specific mutation in peroxisome proliferator-activated receptor gamma (PARP-g). Marfan syndrome (MFS) is an autosomal dominant disease caused by a mutation of heterozygous fibrillin-1 (FBN1) and presents cardiovascular symptoms and skeletal abnormalities. By the same principle, researchers accomplished the establishment of FBN1 mutant cloned pigs (+Glu433AsnfsX98), which exhibited phenotypes similar to those of humans with MFS, such as scoliosis, funnel chest, delayed epiphysis mineralization, and the destruction of elastic fiber structure in the medial aortic tissue.

Human induced pluripotent stem cells (iPSCs) and CRISPR/Cas9 technology can also be combined to generate a congenital heart disease model associated with GATA4 mutations in vitro to investigate the pathogenesis of this gene mutation. Using Barth syndrome (BTHS) iPSC-derived cardiomyocytes (iPSC-CMs) and genome editing, Wang and colleagues demonstrated that TAZ mutation is associated with myocardial metabolism and structural and functional abnormalities. These findings indicate the value of genetically edited animals as models for research on the pathogenesis of CVD and provide new insights into treatment strategies.

By genome editing techniques, potential therapeutic methods of repairing disease-causing mutations or of knocking out specific genes as CVD prevention approaches have also received widespread attention. For example, long QT syndrome (LQTS) is an autosomal dominant congenital heart disease. Hybrid mutations in multiple genes may lead to LQTS, some of which have relatively clear mutation sites with known molecular functions, such as hERG gene mutations in the pore-forming subunit alpha protein that encodes the potassium voltage-gated channel. The hERG gene mainly expresses and functions in cells of the myocardium and smooth muscle, and its mutation can cause fatal ventricular arrhythmia. Repairing hERG gene mutations in cardiomyocytes using CRISPR technology may be an effective strategy to treat such LQTS.

Previous studies have noted that nonsense mutation carriers of the proprotein convertase subtilisin/kexin type 9 (PCSK9) gene have significantly decreased levels of low-density lipoprotein cholesterol (LDL-C) in their blood compared with normal subjects (an allelic mutation corresponds to a 30 to 40% reduction). The blood level of triglyceride (TG) in subjects with nonsense mutations in the apolipoprotein C3 (APOC3) gene was significantly lower than that in unaffected people (an allelic mutation corresponds to a 40% decrease).

The incidence of heart disease in both carriers was lower than that in unaffected subjects by more than 80%, suggesting that the inhibition of PCSK9 and APOC3 gene expression can be used as a potential treatment for cardiovascular disease. Since these two genes are mainly expressed in liver cells, one idea is to directly introduce nonsense mutations to APOC3 or PCSK9 genes in liver cells through genome editing technology, thus fundamentally inhibiting protein synthesis and achieving long-term stable therapeutic effects.

PRKAG2 cardiac syndrome is an autosomal dominant disease induced by a mutation in the PRKAG2 gene encoding the AMP-activated protein kinase g2 regulatory subunit. A recent study suggests that the selective destruction of pathogenic mutations through CRISPR/Cas9 technology *in vivo* is a competent strategy to treat PRKAG2 heart syndrome and other dominant hereditary heart conditions.

Metabolic diseases

Metabolic diseases refer to the pathological state in which the body's protein, fat, carbohydrates, etc. are metabolically disordered. Metabolic diseases include a group of syndromes that are caused by both genetic factors and the environment. Gene editing technology can be applied in functional gene screening, gene therapy and the construction of metabolic disease models, such as obesity, diabetes, and hyperlipidemia. Leptin (Lep) is a hormone secreted by white fat cells that acts on the metabolic regulation center of the hypothalamus through the leptin receptor (LepR).

It has diverse functions, including appetite suppression, energy intake reduction, and fat synthesis inhibition, and can regulate blood sugar concentration, neuroendocrine, etc. A number of animal models have been developed to illustrate the important role of Lep/LepR in glycolipid metabolism, and the most widely used are ob/ob mice against Lep and db/db mice against LepR. Chen and colleagues injected TALEN components into rat zygotes to specifically knockout LepR, thus obtaining three lines of rats with LepR mutations. Phenotypes in these strains manifested as obesity and other metabolic disorders; additionally, the authors established a LepR mutant obese rat model, exhibiting efficient germline transmission. Bao et al.104 successfully established LepR knockout mice using CRISPR/Cas9 technology.

Homozygous LepR-deficient mice are characterized by obesity, hyperphagia, hyperglycemia, insulin resistance, and lipid metabolism disorders, together with some complications of diabetes. The same principle has been used to generate the cytochrome P450 (CYP) 2E1 knockout rat model with CRISPR/Cas9 technology to explore the role of the CYP2E1 gene in biochemical metabolism, toxicology, and diseases (e.g., diabetes and alcoholic cirrhosis).

The FTO allele is associated with obesity, which inhibits the mitochondrial thermogenic effects in adipose precursor cells. FTO gene mutations inhibit the conversion of white fat to brown fat. The FTO gene-regulated thermogenic pathway involves ARID5B, rs1421085, IRX3, and IRX5 factors. rs1421085 can be edited using the CRISPR/Cas9 platform to repair the pattern structure of ARID5B, thereby suppressing the expression of IRX3 and IRX5 and achieving the effect of weight loss.

As an important "experimental tool", the animal model of diabetes can be used for pathological observation, preclinical experiments and drug screening. In a study based on CRISPR/Cas9 technology, pX330 (containing gRNA and Cas9 sequences together with the donor DNA plasmid) was injected into the oocyte to generate new Cre tool mice and achieve the genetic manipulation of pancreatic â cells. The Ins1 (insulin gene) promoter and stop codon sequences served as targets for recombinase Cre insertion. Progeny F1 mice were histologically labeled as Cre-loxP recombination, which was observed in all islets expressing insulin-positive cells and negatively expressed in other tissues. There was no significant difference in glucose tolerance between these genetically edited mice and wild-type mice. Applying CRISPR/Cas9 technology in human iPSCs to target diabetes-related genes has become a promising approach to explore the molecular mechanisms of diabetes.

For example, human iPSCs are isolated from single-gene diabetic MODY patients, and possible mutations in genes such as HNF4A, GCK, PDX-1, and INS are edited by CRISPR; the edited iPSCs then differentiate into pancreatic progenitor cells and are later transplanted into patients.108 In addition, gene editing tools can also structurally modify proteins that promote chromatin structural variation, such as methylase, demethylase, acetylase or deacetylase, to treat diabetes epigenetically.

Gene editing technology is also critically involved in the study of lipid metabolism.110 cAMP responsive element binding protein 3-like 3 (CREB3L3), a transcription factor expressed in the liver and small intestine, controls the energy metabolic equilibrium in fasting response. Researchers used the one-step CRISPR/Cas9 system to establish the CREB3L3-floxed murine model for the first time and subsequently obtained mice that were knocked out of the CREB3L3 gene in the small intestine and liver, respectively. The evidence above provides a new understanding of the role of CREB3L3 in plasma triglyceride metabolism and its contribution to liver and intestinal cholesterol metabolism. Familial hypercholesterolemia is an autosomal single-gene dominant disease correlated with a defect in the low-density lipoprotein receptor (LDLR) gene, which causes a disorder of the body's lipid metabolism. In 2012, researchers used TALEN technology to target LDLR in porcine fetal fibroblasts and obtained miniature swine containing mono- and biallelic mutations in LDLR, thus generating models of familial hypercholesterolemia, which came with critical biomedical significance in simulating lipid metabolic syndrome. Recent genome-wide association studies have identified tribble homolog 1 (TRIB1) to be associated with lipoprotein metabolism in human hepatocytes. Hepatic-specific overexpression of Trib1 reduced plasma TG and cholesterol levels by reducing the production of VLDL; in contrast, Trib1-knockout mice showed elevated plasma TG and cholesterol levels due to the increased production of VLDL. To further explore its regulation of lipid metabolism, Researchers induced the destruction of the chromosome at the TRIB1 locus by delivering the CRISPR/Cas9 system into mouse liver via a nonpathogenic AAV, which increased the transcription of PCKS9 and the secretion of PCKS9 protein; these responses ultimately reduced the level of liver LDL receptors and increased the level of LDL-C in the blood.

Neurodegenerative diseases

Neurodegenerative diseases (NDs), at least including Huntington's disease (HD), Alzheimer's disease (AD), and Parkinson's disease (PD), are a group of conditions that have attracted the most concern because there have been no specific diagnostic approaches or established treatments for them. There are a few potential pathogenic mechanisms behind NDs, including the accumulation of proteins with abnormal structures, impaired ubiquitin-proteasome and/or autophagic lysosomal pathways, oxidative stress and circuit alternations, etc. These mechanisms indicate that NDs are induced by complicated interactions of multiple genetic factors; either alone or in combination, the interactions lead to clinical features. The emergence of gene editing platforms provides a convenient approach to study gene functions related to NDs.

In HD, in vitro investigations demonstrated that via ZFNs, chromosomal expression of the mutant huntingtin (HTT) gene was significantly reduced at both the protein and mRNA levels; *in vivo* studies revealed that via striatal AAV delivery into the HD R6/2 mice, ZFNs extensively suppressed cerebral expression of the HTT gene and ameliorated HD-related symptoms. Additionally, the HTT exon 1 in human iPSCs derived from fibroblasts of HD patients (HD-iPSCs) can be corrected by TALENs. To better understand the pathogenesis of HD, Researchers adopted CRISPR/Cas9 to establish a genome-edited porcine model of HD in 2018, which internally expressed full-length mutant HTT. As a promising breakthrough in the field of NDs, the development of HTT gene knock-in pigs would be of great significance for pathogenesis research and therapy exploration in Huntington disease.

Mutations in the gene encoding amyloid precursor protein (APP) cause familial AD with nearly complete penetrance. Mouse fibroblast cells overexpress APP by receiving electroporated ZFNs designed with a DNA fragment containing the promoter and the protein coding regions of APP. These transgenic cells can be used to elucidate aspects of the molecular mechanisms of AD pathogenesis, particularly those involved in the mutant amyloidogenic pathway affecting the APP coding sequence. The A673V variant near the APP b-secretase cleavage site contributes to AD pathology by increasing Aâ and enhancing its aggregation as well as toxicity; by contrast, the A673T variant, which is adjacent to the aspartyl protease b-site in APP, provides protection against AD progression. When A673V and A673T were induced in normal iPSCs by TALEN technology, these cells differentiated and formed cortical neurons, presenting with different levels of AD-associated biomarkers. In addition, through a gene editing platform based on single-stranded oligonucleotide DNA nucleotides and CRISPR/CAS-blocking mutations, researchers generated human iPSCs with dominant AD-causing mutations in APP and presenilin 1 (PSEN1), both heterozygous and homozygous, leading to early disease onset; thereby, they yielded cortical neurons, which showed genotype-dependent phenotypes associated with AD. Apolipoprotein E4 (APOE4) is a genetic risk factor for late-onset AD, while ApoE2, which differs from APOE4 by only two bases (two C bases in APOE4, corresponding to two U bases in APOE2), is not a risk factor for AD. Researchers introduced APOE4 RNA related to disease risk into cells and successfully changed the APOE4 to APOE2 sequence through the RESCUE (RNA Editing for Specific C to U Exchange) editing system by changing two C bases in APOE4, which is equivalent to converting the disease risk of the AD high-risk population carrying the APOE4 gene to zero.

Alpha-synuclein (SNCA) and leucine-rich repeat kinase 2 (LRRK2) are associated with autosomal dominant PD, whereas another group of genes are associated with autosomal-recessive PD, including parkin, phosphatase and tensin homolog-induced kinase 1 (PINK1), DJ-1, and ATPase type 13A2 (ATP13A2). The missense mutation of SNCA and LRRK2 genes can be corrected by ZFNs in vitro. After correction, the mtDNA damage disappeared in differentiated neural progenitor and neural cells derived from iPSCs. Additionally, researchers combined genome-wide epigenetic information with CRISPR/Cas9 genome editing to generate a genetically precisely controlled experimental system in human iPSCs.

This system has identified PD-associated risk variants in noncoding distal enhancer elements that regulate SNCA expression; it has also confirmed that the transcriptional disorder of SNCA is related to sequence-dependent binding of the brain-specific transcription factors EMX2 and NKX6-1. These results suggest that gene editing techniques can generate specific ND animal models for further exploration into human diseases, and they are potentially capable of offering a robust therapeutic approach against multiple human genetic defects that have been considered incurable.

Viral diseases

Gene editing platforms have emerged recently as antiviral therapeutics for treating infectious diseases, either by altering the host genes required by the virus or by targeting the viral genes necessary for replication. To date, genome editing-based HIV therapy has involved modifying infection-related genes to produce HIV-resistant CD4+ T cells and subsequently reinfusing the edited cells into patients. In 2008, the anti-HIV efficacy of the ZFN system was first presented in preclinical studies by adopting primary human CD4+ T cells.

Approximately 50% of the CCR5 alleles were disrupted with ZFN, which was delivered by the chimeric Ad5/F35 adenoviral vector. HIV-infected mice transfused with ZFN-modified CD4+ T cells also better preserved their original CD4+ T cells and had lower viral loads than nontransfused mice. In 2009, a patient was functionally cured of HIV infection by transplanting allogeneic stem cells from a donor with a homozygous CCR5 d32 allele, suggesting that it is feasible to obtain resistance to HIV by mimicking natural homozygous CCR5 d32 mutations with genome editing technologies. In addition, engineering CD34+ HSPCs instead of CD4+ T cells with the CCR5 ZFN pair provides a durable source of modified cells and protects the CD4+ myeloid cells that are susceptible to HIV-1 as well. Further *in vivo* experiments showed that mice transplanted with ZFN-modified HSPCs experienced rapid selection for CCR5(-/-) cells, which had obviously lower levels of HIV-1 than the control group and maintained human cells throughout their tissues. The disruption of C-X-C chemokine receptor 4 (CXCR4) is also under exploration as a strategy for patients who harbor CXCR4-tropic HIV-1. Simultaneous genetic inactivation of both CCR5 and CXCR4 in human CD4+ T cells by ZFNs confers protection against viruses that exclusively use the targeted coreceptor.

Nuclease platforms based on TALEN143 and CRISPR/Cas are also being applied to disrupt CCR5 in T cells and HSPCs. Laboratory results from showed that CRISPR/Cas9 not only could specifically eradicate latent HIV infection but also could prevent new HIV infection. Similarly, researchers recently demonstrated that the codelivery of chemically modified CCR5 sgRNA with Cas9 mRNA/protein enhanced the genome editing efficiency of human primary CD4+ T cells and CD34+ HSPCs, with no DNA delivery-associated toxicity.

The sustained expression of high-risk human papillomavirus (HPV) oncogenes E6 and E7 is implicated in malignant transformation and is strongly associated with cervical cancer. The targeted mutagenesis of those high-risk HPV genes by gene editing tools may be a potential genetic therapy and may reverse cervical cancer in situ. Researchers constructed a ZFN that could specifically recognize and cleave HPV16/18 E7 DNA. In their study, ZFN-mediated HPV16/18 E7 DNA disruption directly decreased the expression of E7, which resulted in efficient growth inhibition and type-specific apoptosis in HPV16/18-positive cervical cancer cells *in vitro*. When different plasmid-encoded zinc-finger modules were introduced *in vivo*, the therapeutic effects of ZFNs were further confirmed, inhibiting tumor growth in mice bearing cervical cancer cells.

Similar results in another study showed that using ZFNs to target HPV E7 induced specific shear of the E7 gene and attenuated its malignant biological effect. Researchers computationally generated paired zinc-finger arrays (pZFAs) to target and cleave the genomic DNA of HPV-type 16 and 18, respectively. The authors highlighted the therapeutic effect of ZFN-mediated gene disruption in HPV 16/18, which was achieved when HPV-derived viral plasmids or vectors were introduced into precancerous lesions to realize targeted mutagenesis and gene-therapeutic reversal of cervical neoplasia.

Additionally, the combined treatment of ZFNs with two chemotherapeutic drugs (cisplatin and trichostatin A) increased the apoptotic rate by approximately two times more than that of ZFNs used alone in HPV16/18-positive cervical cancer cells. Both chemotherapeutic drugs coordinated with ZFNs to downregulate HPV16/18 E7 expression while elevating retinoblastoma 1 (RB1) expression. TALEN-mediated targeting of HPV oncogenes E6 and E7 within host DNA resulted in restoration of the host tumor suppressors p53 and RB1, which not only reduced tumorigenicity in HPV-positive cell lines but also ameliorated HPV-related cervical malignancy in transgenic mouse models. Furthermore, CRISPR Cas9/HPV16 E6/E7 sensitized cervical cancer cells to cisplatin, indicating the potential of application in cervical cancer therapy.

Hepatitis B virus (HBV) is the most important pathogen of liver disease. Cotransfection of engineered ZFN pairs with a target plasmid containing the HBV genome results in specific cleavage.Researchers designed and screened an efficient gRNA targeting the HBV genomic locus and transmitted the sgRNA/Cas9 system by lentiviral vector to HepG2 cells that were integrated with HBV. Finally, the amount of covalently closed circular DNA (cccDNA) gradually decreased, dropping by 92% on the 36th day; HBV gene expression and replication were also inhibited. One study also attempted to knock out Epstein–Barr virus (EBV)-related genes using CRISPR/cas9 technology to treat latent infections caused by EBV. They used a plasmid containing CRISPR/cas9 to treat Raji cells isolated from Burkitt's lymphoma with EBV latent infection; then, they found that cell proliferation was significantly inhibited and intracellular EBV load was significantly reduced.

Genomic editing technology allows us to gain a deeper understanding of the mechanisms underlying variant diseases associated with viral infection and demonstrates tremendous potential in the development of therapeutic approaches against viral infections, which represent some of the most intractable diseases.

Hereditary eye diseases

In recent years, with the advancement of gene sequencing technology, it is more explicit to make the genetic diagnosis of a variety of hereditary eye diseases, such as congenital cataract, congenital glaucoma, retinitis pigmentosa (RP), congenital corneal dystrophy, Leber congenital amaurosis (LCA), retinoblastoma (RB), and Usher syndrome.

CRISPR/Cas9 has already been used to generate animal models of RP. Receptor expression enhancer protein 6 (REEP6), a member of the REEP/Yop1 family of proteins, influences the structure of the endoplasmic reticulum. Researchers reported that biallelic mutations in REEP6 cause autosomal-recessive retinitis pigmentosa. They identified variants in REEP6 in patients with RP from unrelated families. Moreover, they created a knock-in mouse model of Reep6 p.Leu135Pro via CRISPR/Cas9.

The clinical phenotypes of RP were replicated in the Reep6L135P/L135P homozygous knock-in mice, such as developing photoreceptor degeneration and dysfunction of the rod photoreceptors, which provides a better animal model for future studies of RP. The rodless (rd1) mouse, the most vastly used preclinical model of RP, has been aggressively debated for nearly a century after its occurrence because the cause of the blinding RP phenotype remains undetermined. The rd1 mouse has two homozygous variants in the Pde6b locus of chromosome 5: a nonsense mutation (Y347X) and a murine leukemia virus (Xmv-28) insertion in the reverse orientation in intron 1. Researchers repaired the nonsense point mutation via CRISPR/Cas9 to rescue and ameliorate the disease, demonstrating that the Y347X mutation in rd1 mice is pathogenic. Another animal model of RP, the transgenic S334ter-3 rat, possesses the mutation RhoS334, which shows similar phenotypes to human class I RIIO mistracking mutations, leading to a continual degeneration of photoreceptors and vision decline.

The protospacer adjacent motif (PAM) sequence in RhoS334 (52 -TGG-32) diverges from the PAM in RhoWT (52 -TGC-32) by only one nucleotide. Benjamin et al. reported that an allele-specific disruption of RhoS334 via a single subretinal injection of CRISPR/Cas9 and gRNA by electroporation prevented retinal degeneration and increased visual acuity. Additionally, Latella et al. successfully edited the human rhodopsin (RHO) gene by the electroporation of plasmid-based CRISPR/Cas9 in a P23H transgenic mouse model for autosomal dominant RP and confirmed its efficacy as a genetic engineering tool in photoreceptor cells, which strongly demonstrates that the CRISPR/Cas9 system is an efficient and promising therapeutic tool for retinal degeneration, such as RP. Researchers also determined a CRISPR/Cas9-mediated homology-independent targeted integration (HITI) strategy and demonstrated its efficacy in ameliorating visual function in a rat model of RP. HITI is a targeted integration mediated by NHEJ, and this study is the first time that HITI could play a role in nonmitotic cells. The advantage of HITI technology is that it can be applied to any targeted genome engineering system, not just CRISPR/Cas9.

The combination of CRISPR/Cas9 technology and other methods provides new avenues for the treatment of related eye diseases, such as treatment with AAV and iPSCs. Bassuk et al. first reported that CRISPR/Cas9 precisely repairs retinitis pigmentosa GTPase regulator (RPGF) point mutations, which cause X-linked RP in patient-specific iPSCs; this supports that combining gene editing with autologous iPSCs could be a personalized iPSC transplantation strategy for therapies of various retinal degenerations. Similarly, researchers found that iPSC-derived retinal organoids from three RP patients with different frameshift mutations in the RPGR gene have significant defects in photoreceptors, including defects in their morphology, localization, and electrophysiological activity. The correction of an RPGR mutation via CRISPR/Cas9 reverses ciliopathy and rescues photoreceptor loss, which indicates that CRISPR/Cas9 can serve as an adopted mutation repair strategy.

LCA is a congenital retinal dystrophy that causes significant vision loss at an early age. To verify that mutation in human KCNJ13 causes LCA, Zhong et al. employed CRISPR/Cas9 to create Kcnj13 mutant mice by zygote injection with sgRNA and spCas9 mRNA. Kcnj13 mutant mice showed a declined response to light, a loss of photoreceptors and rhodopsin mislocalization, revealing that the loss of Kcnj13 function could mimic human LCA phenotypes in mice.171 As demonstrated by Zhong et al., CRISPR/Cas9 could accelerate the study of candidate gene function in biology and disease. The centrosomal protein 290 kDa (CEP290) gene, the most frequent mutation in LCA, causes the most common subtype of LCA, which is referred to as LCA10. However, the large size of CEP290 exceeding the capacity of AAV delivery prevents the use of this delivery platform. To overcome this capacity limitation, Ruan et al. used dual recombinant AAV vectors to induce the CRISPR/Cas9-mediated deletion of a specific intronic fragment of the Cep290 gene in mouse photoreceptors.

Additionally, using a smaller *S. aureus* CRISPR/Cas9 system enables a single AAV vector to deliver the Cas9 gene and two gRNAs, which performs a dual-cut excision of the CEP290 mutation-containing region in primary fibroblasts from LCA10 patients.164 Recently, researchers developed a candidate genome editing therapy named EDIT-101 to restore vision loss in LCA10. They delivered the *Staphylococcus aureus* Cas9 and CEP290 gRNA to the photoreceptor via an AAV5 vector. Humanized CEP290 mice showed rapid and continuing CEP290 gene editing after subretinal delivery of EDIT-101. These extraordinary studies provide a roadmap for the preclinical advance of gene therapy for LCA10.

RB is the most common pediatric eye tumor of the developing retina. Approximately one-third of RB cases are caused by biallelic RB1 mutation or deletion. Researchers reported that using TALEN gene editing to inactivate somatic rb1 in adult zebrafish induced tumorigenesis at high frequency. A highly penetrant and rapid RB preclinical model was reported by Naert et al., utilizing the CRISPR/Cas9 system to induce the knockout of rb1 and retinoblastoma-like 1 (rbl1) in *Xenopus tropicalis*. The animal model showed rapid development of RB, and it will be a good model for early stage drug discovery and rapid therapeutic target identification. Researchers generated a pluripotent H1 human embryonic stem cell line with RB1 heterozygous knockout by CRISPR/Cas9 nickase, which provides a valuable cell resource for the study of hereditary retinoblastoma. Glaucoma is the second leading cause of blindness worldwide and is characterized by elevated intraocular pressure (IOP).

Gain-of-function mutations in myocilin (MYOC) have been reported to commonly cause primary open-angle glaucoma (POAG). The accumulation of mutated myocilin inside cells leads to the activation of the unfolded protein response (UPR) cascade and endoplasmic reticulum (ER) stress in the trabecular meshwork (TM). TM cells are sensitive to chronic ER stress and finally die, resulting in increased IOP and glaucoma. Researchers knocked down the expression of mutant MYOC in a mouse model of POAG by CRISPR/Cas9, resulting in the reduction of ER stress, lower IOP, and the preventability of further glaucomatous damage in mouse eyes.

Importantly, they also demonstrated the feasibility of utilizing CRISPR/Cas9 in human eyes with glaucoma. A dominant-negative mutation in KRT12, which causes Meesmann epithelial corneal dystrophy (MECD), results in the occurrence of a novel *Streptococcus pyogenes* PAM. Researchers designed a sgRNA complementary to the sequence adjacent to this PAM and found that this sgRNA has a large effect on the decrease in mRNA and protein of KRT12 *in vitro*. The injection of combined Cas9/sgRNA into the corneal stroma of a humanized MECE mouse model showed frame-shifting deletions of the mutated KRT12 allele. This study is the first to demonstrate the *in vivo* allele-specific CRISPR/Cas9 gene editing of a novel PAM created by a heterozygous disease-causing SNP.

Hematological diseases

Nearly half of hemophilia A cases are caused by the inaccurate expression of factor VIII (F VIII) due to inversion of the chromosome. In one study, iPSCs were derived from somatic cells of hemophilia A patients induced by chromosome inversion, and the F VIII gene of iPSCs was modified by CRISPR/Cas9 technology. The modified iPSCs were induced to differentiate into mature endothelial cells capable of expressing factor VIII and then transplanted into hemophilia mice lacking factor VIII. The results showed that the transplanted mice began to produce factor

VIII, which effectively inhibited bleeding symptoms. Hemophilia B is caused by a deficiency in factor IX (F IX). Coagulation activity can be restored by increasing FIX in plasma.

Researchers found that the F9 gene carries a new mutation, Y371D, in a family of hemophilia B patients, which leads to a more severe hemophilia B phenotype than the previously discovered Y371S mutation. They used naked DNA constructs and adenoviral vectors to deliver Cas9 to adult F9 Y371D mutant mice. After treatment, it was found that when adenovirus was used as a vector to deliver cas9, although the mutation gene was highly efficiently repaired, hepatotoxicity was severe. However, Cas9 with a naked DNA structure successfully repaired more than 0.56% of F9 alleles in hepatocytes in hemophilia B mice, enough to restore hemostasis.

CRISPR technology also provides a quick path to build hemophilia models. Researchers from the Institute of Zoology in the Chinese Academy of Sciences injected the CRISPR/Cas9 system targeting vwF (vascular hemophilia mutant gene) into the fertilized eggs of miniature pigs and obtained the double allele mutant mini-pig quickly and efficiently. These miniature pigs have severe coagulopathy, indicating the successful construction of a miniature pig model of von Willebrand disease by CRISPR technology.

Sickle anemia is the first genetic disease with a clearly understood pathogenesis. A single nucleotide mutation from A to T in the first exon of human â-globin results in a lesion. In 2016, a Stanford University team reported on the use of CRISPR/Cas9 technology to repair â-globin gene (HBB) mutations in patient-derived HSCs *in vitro*. After the modified iPSCs differentiated into red blood cells, normal HBB mRNA could be detected.

This preclinical experiment provided theoretical support for gene editing technology in the treatment of sickle anemia. The CRISPR/Cas9 system has also been used to correct b thalassemia-causing mutations in the HBB. Using CRISPR/Cas9 technology to direct the calibrated DNA sequence to the HBB mutation site, it was possible to correct two different â-thalassemia mutations in the HBB gene of patient iPSCs by HR.

Other hereditary diseases

Duchenne muscular dystrophy (DMD) is the most common form of muscular dystrophy caused by mutations of the DMD gene.196 Current X-linked muscular dystrophy (mdx) mice can only partially mimic human disease conditions. Their small size, limited chronic muscle damage and muscle weakness also impose limitations on disease research and analysis. Therefore, larger animals such as rats, rabbits or pigs are more valuable for preclinical studies. Researchers generated Dmd rats by targeting exon 23 of DMD with TALENs.

These edited rats showed a significant reduction in muscle strength and decreased spontaneous motor activity. Sui's team generated DMD knockout rabbits by coinjecting Cas9 mRNA and sgRNA into rabbit zygotes targeting exon 51 of DMD. These rabbits harbored the typical phenotypes of DMD, and the pathological features in the diaphragm and heart were similar to those of DMD patients. In addition, the monkey dystrophin gene was targeted using CRISPR/Cas9 to create mutations that cause DMD. The detection of the relative targeting rate showed that CRISPR/Cas9 could result in mosaic mutations in up to 87% of the dystrophin alleles in monkey muscle.

Notably, three groups of researchers have recently described the use of CRISPR/Cas9 to remove mutations in the DMD gene encoding dystrophin, which affects protein expression.

The investigators used the CRISPR/Cas9 system to excise the mutant portion of DMD in the mdx mouse model, thereby synthesizing a shorter version of dystrophin protein in the muscle fibers and restoring partial muscle function. This provided a promising method for correcting disease-causing mutations in the muscle tissue of patients.

Patients with primary immunodeficiencies lack a part of their immune system or have immune system dysfunction, and they can be treated with allogeneic HSC transplantation. This may be a high-risk process when leukocyte antigen-matched donors lack tissue compatibility. Correcting a patient's own HSCs through gene therapy provides an attractive option. HSCs can also be used in situ to correct pathogenic mutations and to develop cell or animal models to study the pathogenic effects of specific genetic defects found in immunodeficient patients.

As the most severe immunodeficiency, severe combined immunodeficiency (SCID) is caused by a mutation in the gene encoding the interleukin 2 receptor gamma (IL2RG), which results in the developmental arrest of T cell production and additional primary or secondary defects in B cells. Several research teams have successfully used ZFN and TALEN techniques to induce HDR at the IL2RG locus in various human cell types, including HSCs and embryonic stem cells (ESCs). Other studies have utilized endonucleases to generate different kinds of immunodeficient animal models that were previously unable to be established due to a lack of effective genetic modification. As a result of engineered nuclease-mediated editing of genomic modifications, other animal disease models have been developed, simulating Rett syndrome, hereditary deafness, Wilson disease, Laron syndrome, Niemann–Pick disease, Netherton syndrome, and so on. Advances in genome editing technologies will further expand the application of animal models in disease mechanism research and treatment development.

CHAPTER-7

CRISPR IN AGRICULTURE AND FOOD SECURITY

INTRODUCTION

The scientific and agricultural communities have long understood that climate variability can increase crop failures, potentially causing food shortages. More people are now recognizing that changing and more extreme weather patterns, including drier conditions in growing regions (which can make crops susceptible to disease and pests), will continue to negatively impact global food security.

To help combat growing concerns over food security, some university researchers and agri-companies are using novel gene-editing technologies to modify staple crops to make them more tolerant to the changing climate and its associated perils.

Recently, scientific attention has been directed to a new molecular technique called CRISPR gene-editing technology for introducing commercially desirable traits in crops. It is believed that CRISPR can have a positive impact on food productivity, quality, and environmental sustainability. This will become more important as the global population continues to grow, and less arable land and less water resources are available to grow crops, in part due to environmental change.

This article will explore the variety of trait modifications being pursued using CRISPR technology, the diversity of the agricultural crops that are the subject of these modifications, and the key players in this space identified through patenting and licensing activity. This article will also summarize the key U.S. and European regulations affecting the labeling of food items developed using CRISPR technology and provide reasons why governmental regulatory bodies worldwide should support and harmonize CRISPR-edited agricultural guidelines.

CRISPR

CRISPR is an acronym for Clustered Regularly Interspaced Short Palindromic Repeat, and is a relatively new genome-editing technology. CRISPR technology is a powerful tool that allows researchers to more easily alter DNA sequences and modify gene function in animals or plants.

The various potential products of CRISPR technology carry the promise to contribute to solving many of the great challenges of the twenty-first century, from medical and health issues to food and agricultural production. This may certainly be one of the reasons why the 2020 Nobel Prize in Chemistry was awarded to Emmanuelle Charpentier and Jennifer Doudna for their discovery and development of one of the most popular gene-editing tools: CRISPR-Cas.

Over the years, the working principle behind CRISPR has been widely analyzed and explained in several journal reviews. In general, modification of plant genomes using CRISPR gene editing technology follows the process shown below in Fig. 1.

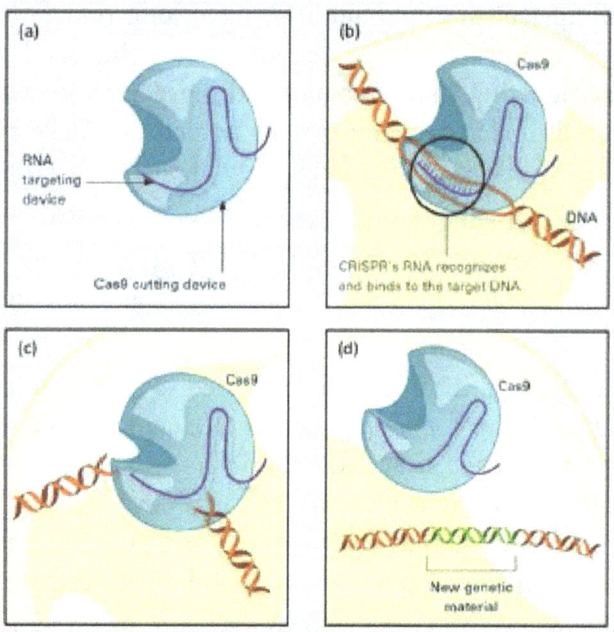

Fig. 1: (a) The CRISPR system has two components joined together: a finely tuned targeting device (a small strand of RNA programmed to bind to a specific DNA sequence) and a strong cutting device (an enzyme called Cas9 that can cut through a double strand of DNA at the binding site); (b) Once inside a cell, the CRISPR system locates the DNA it is programmed to find. The CRISPR seeking device recognizes and binds to the target DNA (circled, black); (c) The Cas9 enzyme cuts both strands of the DNA; (d) Researchers can insert into the cell new sections of DNA. The cell automatically incorporates the new DNA into the gap when it repairs the broken DNA.

The CRISPR process differs from conventional GM (Genetically Modified) methods. In general, CRISPR targets pre-existing nature-derived gene variability within the crop family to bring about targeted variation and a desired output. Specifically, CRISPR is used as a set of molecular scissors to cut and delete (or replace) a gene expressing an economically undesirable trait with a nature-generated "improved" variation of the gene (i.e., existing in the plant family via natural genetic diversity). In other words, CRISPR does not use classical GM crop production techniques, which insert exogenous or foreign DNA from a vector (a vehicle which delivers foreign genetic material) into the plant genome. CRISPR enables desirable crop traits by introducing DNA from nature-generated genetic variations within the crop itself, and not from another reproductively incompatible organism. This eliminates the fear that foreign DNA may be present in the resultant plant and related finished goods.

The CRISPR process' ability not to introduce foreign DNA into the plant genome has enabled the plant products to be considered non-GMO by many scientific researchers and at least some products developed using this technology will not need to be labeled GMO under USDA rules (further discussed below). The regulatory landscape around the use of CRISPR in agriculture will be further explored later in this article.

PATENT LANDSCAPE OF CRISPR IN AGRICULTURE

It should not be surprising that the development of CRISPR's game-changing technology was the subject of a series of patent priority, inventorship, and hence, ownership disputes between high profile research institutions. With the resolution and settlement of these disputes, there has been an increase in commercial applications of CRISPR in economically important plants and agriculture to develop desirable and improved traits.

To maximize the impact of CRISPR for improving agriculture and provide a clear path for obtaining licenses to key portions of the CRISPR intellectual property, in 2017, Pioneer-DuPont (now Corteva AgSciences) and the Broad Institute, U.S.-based holders of the two largest foundational CRISPR patent portfolios, joined forces to license CRISPR (particularly, CRISPR-Cas9) technology. Both agreed to jointly provide non-exclusive licenses for commercial use of the full suite of foundational CRISPR patents in crop agriculture. Just as important, this intellectual property became freely available for use by academic research institutions. Providing licensing opportunities has been one way to encourage scientists in both academia and industry to innovate with CRISPR and explore new ways to improve commercially desirable agricultural traits.

Generally, a good indicator of innovation in a field is the number of patent filings. To obtain an overview of the patent landscape for CRISPR-Cas technology in agriculture, we conducted a series of searches on CRISPR-Cas applications in plants over a series of several months to determine the global hotspots of R&D activity, players, plants (namely commercial crops), and traits modified. We performed a variety of search queries using international public patent databases (such as WIPO [World Intellectual Property Organization], EPO [European Patent Office] and USPTO [United States Patent and Trademark Office]) to obtain search results and arrive at meaningful conclusions. As we performed the searches and evaluated the results, it became clear that the results were complicated by a number of factors. One factor was that the CRISPR tool was a recent technological breakthrough, and use of the tool to enhance traits in plants and agriculture was even more recent. It is apparent from the patent filing data that patent applications relating to use of CRISPR in commercial agriculture began around 2014-15 and have steadily increased annually

Because CRISPR is new, the use of the terminology in patent filings is evolving and applicants have become their own lexicographers in describing CRISPR in applications.

Specifically, we observed that applicants are defining genome editing broadly in the descriptions of patent applications—listing CRISPR-Cas (without any technical support or enabled examples), along with many other gene editing tools. While the apparent intent of the applicants is to generate actual data using the CRISPR tool sometime in the future and claim priority to the earlier filing date, the effect is to create a high "noise to signal" ratio. This is because many patent applications will speculate as to using CRISPR-Cas to alter desired plant traits—without actually having done so. Another point that must be considered is that patent applications are generally published 6 to 24 months after the original filing date, depending on a variety of country-specific factors. For such a young technology, there could be many pending applications on CRISPR plant trait modifications that are not yet publicly available (i.e., unpublished). Therefore, it is expected that the patent application statistics presented here will continue to evolve. With that backdrop, the results were not too surprising given that CRISPR technology was primarily developed through a series of independent and collaborative academic efforts from scientists in research institutions throughout North America and Europe.

Australia and Canada appear to be second and third behind the U.S. in use of CRISPR in commercial agriculture to modify traits. The results indicate that Europe, while it has 38 European Patent Organisation member states, is behind Canada and is ahead of most Asian countries in CRISPR plant modification. However, when analyzing patent data, it is important to consider that the WIPO (international) applications, which make up at least a quarter of the patent applications, are primarily from Chinese universities or governmental research institutions. Interestingly, unlike the U.S., China, which is the most active country in academic genome editing research, has not announced the regulatory status of plant gene editing or of gene-edited foods. The reasons for this are unclear, but it is possible that China wants to have robust patent protection on CRISPR modifications and a commercial product in hand before releasing regulations.

Likewise, there are relatively fewer patent filings in the CRISPR agricultural space coming out of Europe. One reason for this may be that European regulations have inhibited interest and funding for GMO agricultural products, thus impeding the progress of market-oriented agricultural applications.

The patent filings from the leading international players cover all plant types, including rice, tomato, maize, soybean, tobacco, alfalfa, wheat, rapeseed (canola), cotton, and cannabis. The United States appears to have its main patent filings in crops such as rice, tomato, maize, alfalfa, wheat, and cabbage (*Brassicas*).

Among the CRISPR-edited plants, there are a whole host of commercially desirable traits that have been altered by patent applicants. From our patent landscape review, we observed that while some applications aim to satisfy consumer needs, others aim to benefit the farmer. Examples of consumer-focused traits include gluten free, reduction in browning, reduction in allergen levels, improved fruit color, flavor, and shelf life. Farmer-benefiting traits include viral, bacterial, fungal, temperature and drought resistance, stress and drought tolerance, and improved crop yield, among others. Such CRISPR-edited plants will enable new food products to reach the market quickly at affordable prices.

Recent examples of such products include browning-resistant mushrooms, high-amylopectin waxy corn (*Zea mays*), and false flax (*Camelina sativa*) with enhanced omega-3 oil. These were each developed using CRISPR and approved by the U.S. Department of Agriculture (USDA) in record time.

It is not surprising that the first CRISPR-altered crops have been introduced in the United States and not Europe as the EU's GM laws, which CRISPR-altered crops are subject to, are very tough. This is also reflected in the patent landscape—most patent activity is in North America. The relatively permissive regulatory atmosphere for CRISPR technology in the United States is why innovative agricultural companies are proactively licensing CRISPR technology and putting their agri-science expertise to work to make one-of-a-kind fruits and vegetables that can lead the way to better health for people and animals. Therefore, based on the patent landscape, it is anticipated that more CRISPR-edited crops are in the pipelines of various companies and may soon appear on the market, at least in the United States.

Advances in plant and animal breeding

Applications of gene editing

Gene-editing technologies can be used in practical plant and animal breeding to make some of the component processes more efficient and more precise. They are also of considerable use in research to generate information in support of plant and animal breeding. Although many gene-edited organisms have been created or are in the pipeline, very few have been commercialized to date. Gene-edited products that have already entered the market include a gamma-aminobutyric acid enriched tomato, two CRISPR edited fish in Japan, and soybean with improved fatty acid composition in the United States of America.

Disease resistance in plants

Host plant resistance to pathogens is controlled by various genetic mechanisms. One mechanism, involving so-termed S genes, controls resistance to powdery mildew (MLO resistance), a disease that infects a broad range of crops, including wheat and barley. Fifty years ago, Ethiopian barley landraces collected in the 1930s were identified as MLO resistant and the resistance did not conform to the classic gene-for-gene system. Induced mutation was used to develop mlo genes for barley effective against all isolates of the pathogen. Powdery mildew isolates with elevated MLO aggressiveness were produced in barley that were not found in nature.

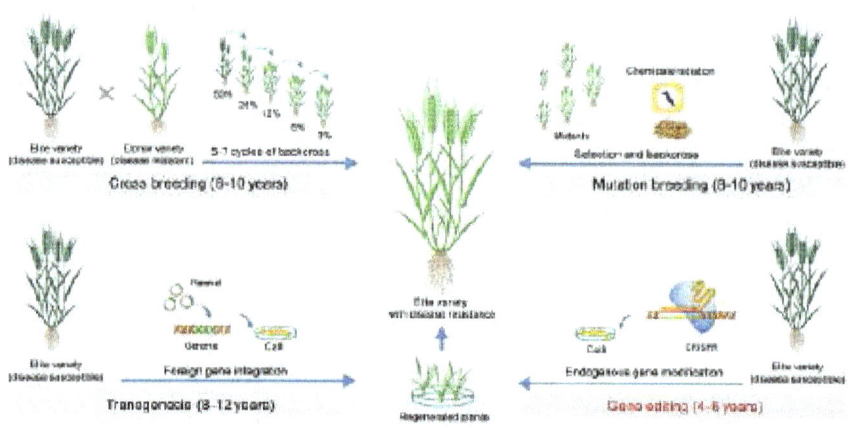

Fig. 2 Developments in plant breeding technologies

Natural mutations that decrease the levels of the MLO protein were identified in barley, which render those barley cultivars resistant to powdery mildew disease. Researchers demonstrated that the use of SDN technologies, which are used for gene editing, can knock out three copies of the MLO gene in wheat to develop resistance. Subsequent studies established that simply knocking out the gene in wheat resulted in decreased crop yields, however. Research subsequently identified a critical additional wheat genome edit that increases wheat yields and rescues any negative pleiotropic growth defects resulting from the MLO knockout.

Rice blast disease reduces rice yields worldwide. Gene editing a critical site in the rice genome enables generation of new rice varieties with durable rice blast resistance. Bacterial blight also reduces rice yields.

A small, targeted deletion edit in the rice genome results in superior improved resistance. These are just three examples of many where gene editing has the potential to improve disease resistance in crop plants to a range of pathogens and pathogen types.

Herbicide tolerance

Herbicide tolerance in crops is an important trait, particularly in intensive agrifood systems. Transgenic breeding for herbicide tolerance relied heavily on the introduction of herbicide tolerance genes from bacteria and various plant species into major crops such as maize and soybean. Geneediting approaches can also be applied to develop new crop varieties with dependable herbicide tolerance. Unlike transgenic breeding, gene editing can be used to edit host plant genomes directly, without introducing extraneous DNA. Precise editing, including base editing and prime editing, can target specific amino acid sequences in genes of interest and thereby confer tolerance. These technologies demonstrate the capacity to develop new plant varieties rapidly by targeting key areas of the host genome.

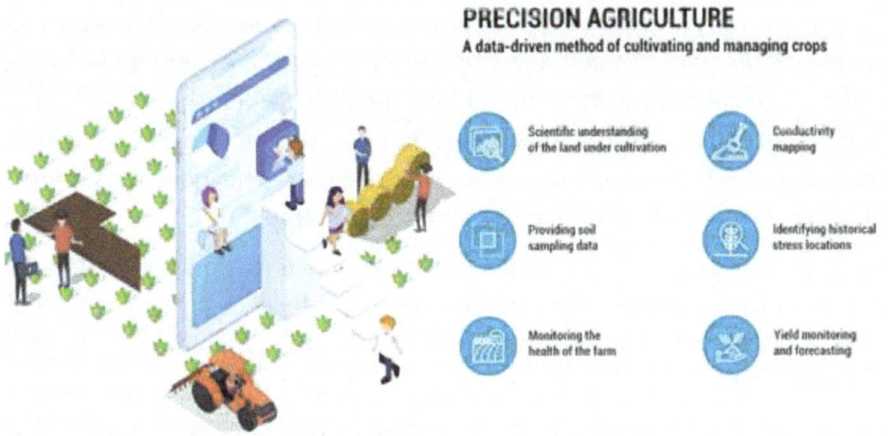

Fig. 3. Precision Agriculture and Crop Improvement

PRECISION AGRICULTURE AND CROP IMPROVEMENT

Precision Agriculture is a farm management concept that revolves around the process of observing, measuring, and responding to various inter-and intra-field variability inputs for modern agriculture.

Popular definitions of Precision Agriculture (PA) or Site-Specific Crop Management (SSCM) describe the term as a technology-enabled approach to farming management that observes, measures, and analyzes the needs of individual fields and crops.

The goal of precision agriculture is to increase efficiency and productivity, reduce input costs, and improve environmental sustainability.

History of Precision Agriculture

Precision agriculture can be traced back to the 1950s and 60s when the first GPS (global positioning system) satellites were launched.

However, it wasn't until the 1980s and 90s that precision ag technologies began to be widely adopted by farmers. This was due to the development of more affordable and user-friendly GPS receivers and mapping software, as well as the availability of satellite imagery and other remote sensing data.

Since then, the field of precision agriculture has continued to evolve and expand, with the introduction of new technologies such as sensors, drones, and autonomous vehicles, as well as advances in data management and analysis tools. Today, precision agriculture is being used by farmers around the world to improve the efficiency, productivity, and sustainability of their operations.

Precision Agriculture and the Farming Revolution

The world is at the edge of the third modern farming revolution and precision farming is an important part of it. The first farming revolution that occurred from 1900 to the 1930s, mechanized agriculture leaving each farmer able to produce enough for 26 people. Long after that, it was the 1990s when the second revolution known as the Green revolution took place. Due to scientific progression genetically modified newer sets of crops that are pest resistant and need less water were introduced, leaving each farmer able to feed 155 people.

The global population is expected to reach 9.6 billion by 2050 and food production must be double the current levels to feed every person. Advanced analytical capabilities and constantly improving IoT capabilities, like precision agriculture drones, will be key elements in the third revolution.

But how is Precision Agriculture different from what farmers and agri-businesses have already been doing?

Precision Agriculture vs Traditional Cultivation

Precision Agriculture and traditional cultivation differ vastly in practice. In traditional agriculture, farmers apply the same amount of pesticides, fertilizers, and irrigation across fields, at prescribed times and frequencies, as per the general recommendations for the region.

However, there are always differences in biological, physical, and chemical parameters even within a single field. Uniform treatment of fields without considering inherent differences results in the overuse of inputs in fertile land and underuse in poor patches. This inefficient use of land, water, fuel, fertilizers, and pesticides increases cost and environmental impact.

Precision agriculture targets the intrinsic differences in farmland and optimizes input by Variable Rate Application (VRA). To implement VRA, detailed spatial data must be collected across fields and locations through geographic information systems (GIS) and crop lifecycles using GPS and remote sensing. The precision farming tools used include precision irrigation systems, yield mapping & monitoring tools, and information management systems. Using advanced descriptive, predictive, and prescriptive analytics engines to analyze the data, precision agriculture arrives at data-driven management decisions to implement cost-effective, environment-friendly sustainable modern farming solutions.

Precision Agriculture and Predictive Farming

According to McKinsey, the invention and adoption of Precision Agriculture is shaped by the following trends:

Big Data and Advanced Analytic Capabilities, and Robotics – aerial imagery, sensors, and sophisticated local weather forecasts. In simple words farming that collects and uses data from plots for managing and optimizing the production of crops is known as predictive farming.

Predictive farming is analogous to taking a pill to target an ailment. The solutions are highly tailored from the type of crop suitable for a plot to the use of pesticides in targeted regions only. Adopting precision agriculture and modern farming reduces production costs and wastage, as the tailored needs of each plot are catered to.

Precision agriculture is practiced by adopting crop management software and technical equipment. Rigorous data collection is done on soil testing, plot measurement, weather pattern analysis, and crop analysis through sensor-equipped devices placed on the fields. The data is calibrated to devise conclusions, and based on those results, a detailed and precise set of practices can be adopted.

Scope of Precision Agriculture

Examples of Precision Farming include the adoption of the same set of practices that use smart farming technologies to cater to the needs of individual plots and crops. Big data analytics software such as Cropin Grow (SmartFarm) or robots such as drones can get detailed information on the plot, soil type, suitable crops, irrigation, and fertilizer needs. The information obtained is used to tailor a very unerring selection of crops, fertilizer quantity, and watering needs. Precision agriculture helps farmers live a debt-free life as production costs and losses are reduced and overall environmental impact is also minimized.

Advantages of Precision Agriculture

By using precision ag technologies and practices, farmers can target their inputs (such as seeds, fertilizers, and pesticides) to specific areas of the field that need them the most, rather than applying them uniformly across the entire field.

This targeted approach can help farmers save on inputs, as well as increase crop yield and quality. In addition, precision ag technologies can help farmers monitor and manage their crops more effectively, allowing them to respond to potential problems (such as pests or diseases) more quickly and effectively.

Farmers can leverage sensors and mapping tools to precisely apply fertilizers at the right rate and location. In addition, precision ag technologies can help farmers optimize their irrigation practices, which can save water and energy.

Since details of areas in a single farm can be traced, precision agriculture benefits farmers in several ways.

Discussed below are some of its key advantages:

- A refined set of cultivation practices and choice of crops based on the suitability of land
- Elimination of volatility and risk
- Waste management
- Reduced production costs
- Minimum environmental impact
- Optimized use of fertilizers
- Water management with optimized irrigation practices
- Improved soil health

Making Precision Agriculture feasible through Farm Digitization

In developing economies, 32% of food losses occur during food production as analyzed by McKinsey on FAO data.

Conventional farming practices are area-centric. There is a general set of crops cultivated throughout an area. All the farmers in that area follow the same procedures with respect to the sowing, nourishing, irrigation, and harvesting period. These practices result in unpredictability, overuse of resources, and uncontrolled waste production.

Before the use of technology in agriculture, a farmer's probability of yielding good produce was as good as tossing a coin and wishing for heads. Since farmers had no information on their farms, there was no way of learning the cause of crop loss. This practice pushed the farmers towards losses and debt. Advancements in big data analytics, IoT, and accessible satellite imagery created optimism for the agriculture sector, giving rise to digital farming, thereby combating the issue of unpredictability.

Precision Agriculture Tools and Technologies

Anyone looking for a precision agriculture solution should be knowledgeable about the tools and technologies available. Computer-based applications are used to create precise farm plans, field maps, crop scouting, yield maps, and to define the exact amount of inputs to be applied to fields. One benefit of this method is the ability to design a friendly agricultural strategy, which helps to lower costs and boost yields. But, because the obtained data cannot be integrated into other supporting systems, these applications offer limited-value data that cannot be used for large-scale precision farming solutions. Let us dive into the tools and technologies needed for precision agriculture.

Global Positioning System (GPS) for Precision ag

The Global Positioning System (GPS) is a satellite-based navigation system that provides location and time information in all weather conditions, anywhere on or near the Earth. In precision agriculture, GPS is used to provide precise location information for mapping and navigation purposes.

One of the main ways GPS is used in precision agriculture is for mapping and field-level data collection. Farmers can use GPS receivers mounted on their vehicles or equipment to collect data on the location, shape, and size of their fields, as well as the location of specific features within the fields such as irrigation systems, drainage ditches, and fence lines. This data can be used to create detailed maps of the fields, which can be used for planning and scheduling field operations, implementation of variable rate technology in agricultural input management, and analyzing field performance.

Farmers can also use GPS-based guidance systems to help them navigate their vehicles and equipment around the fields with a high degree of accuracy, reducing the risk of errors and increasing efficiency.

Geographic Information System (GIS)

A geographical information system (GIS) consists of a computer software database system used to input, store, retrieve, analyze and display, in map-like form, referenced geographical information.

In precision agriculture, GIS is used to create, store, and analyze spatial data related to farming operations, such as field boundaries, topography, soil types, crop types, and other features. Farmers can use GIS to identify the most suitable locations for planting certain crops or to identify areas that are most vulnerable to pests or diseases.

Grid Sampling for targeted farming

Grid sampling is a technique for segmenting fields into around 0.5–5 ha-sized units. Those grids' soil samples will be used to calculate the proper application rates for crop inputs. The Grid Samples are then collected, combined, and delivered to the lab for evaluation.

Grid sampling can be used to collect soil samples from different parts of a field, in order to understand the spatial variability of soil characteristics such as pH, nutrient content, and organic matter. This information can be used to make informed decisions about fertilization, irrigation, etc.

Variable-Rate Technology (VRT)

Variable-rate technology (VRT) consists of farm field equipment with the ability to precisely control the rate of application of crop inputs that can be varied in their application including tillage, insect control, fertilizer, plant population, and irrigation.

VRT is based on the idea that different parts of a field may have different soil types, topography, crop characteristics, and other factors that may affect their requirements for inputs. By using VRT, farmers can optimize their resource use and improve their efficiency and productivity.

The three key ways in which Variable-Rate Technology is used in Precision Agriculture are:

- **Fertilizers:** VRT is used to apply fertilizers at different rates across a field, depending on the soil nutrient levels and crop requirements of each area. This can help farmers save on fertilizers and improve crop yields and quality.
- **Pesticides:** VRT is used to apply pesticides depending on the prevalence and distribution of pests and diseases across the field. This helps farmers control the amount of pesticides and reduce the risk of environmental pollution.
- **Seeding:** VRT is used to sow seeds depending on the soil characteristics and crop requirements of each area. This helps farmers optimize crop yields and quality.

Yield Monitors for Precision Agriculture

A yield monitor is a device that is used in precision agriculture to measure and record crop yields as they are harvested. Yield monitors typically consist of sensors that measure the flow of grain or other crops as they pass through the combine harvester, and a computer or other data recording device that records the yield data.

Yield monitors can be used to create detailed maps of crop yields across a field, showing the spatial variability of yields and identifying areas that are performing poorly or well.

Yield monitors can be used to measure and record the quality of harvested crops, such as moisture content and test weight. This information can be used to optimize storage and handling practices and to improve the overall quality of the crop.

Yield Maps

Data from an adapted combine harvester that has a GPS integrated with a yield tracking system is processed to create yield maps. A yield map is a visual representation of crop yields across a field, showing the spatial variability of yields and identifying areas that are performing poorly or well. In addition to yield monitors, Yield maps can also be created using other data collection tools, such as GPS or remote sensing technologies.

Yield maps can help farmers optimize their resource management practices by identifying areas of the field that have different resource needs. They can also be used to support planning and decision-making by providing a visual and spatial context for data analysis.

Remote Sensors for precise data capture

Remote sensors are generally categorized as aerial or satellite sensors. Changes in soil type, crop development, field boundaries, roads, water, etc., can indicate differences in field color. Remote science in agricultural terms means viewing crops from overhead (from a satellite or low-flying aircraft) without coming into contact, recording what is viewed and displaying the image, and providing the map to pinpoint the field problems earlier and more effectively.

Auto-Guidance Systems for Precision Agriculture

Auto-guidance systems are precision agriculture technologies that use GPS and other sensors to help farmers navigate their vehicles and equipment around the fields with a high degree of accuracy.

These systems typically consist of a GPS receiver, a display unit, and sometimes other sensors (such as cameras or lasers) that provide additional information about the field and the vehicle's position within it.

Auto-guidance systems can be used in a variety of ways in precision agriculture

- **Navigation:** Auto-guidance systems can help farmers navigate their vehicles and equipment around the fields with a high degree of accuracy, reducing the risk of errors.
- **Row Guidance:** Auto-guidance systems can help farmers keep their vehicles and equipment on the correct rows, reducing overlap and skips.
- **Equipment Guidance:** Auto-guidance systems can help farmers control and monitor the position and orientation of agricultural equipment such as planters, sprayers, and harvesters with a high degree of accuracy, improving implement performance and reducing errors.
- **Record keeping and Data Analysis:** Auto-guidance systems can record data on the location and movements of vehicles and equipment, which can be useful for record-keeping and analysis.

Proximate Sensors for precise crop and soil data capture

Proximate sensors can be used to measure soil (N and pH) and crop properties as the tractor pass over the field.

Proximate sensors are sensors that are used in precision agriculture to measure the properties of crops or soil in close proximity to the sensor. The soil sample is scooped and pressed against an electrode, a stabilization period of about 10-15 seconds is allowed, and the reading is taken. These sensors can either be hand-held or mounted on vehicles or other Agri equipment. Proximate sensors are typically used to collect data on a real-time or near-real-time basis, allowing farmers to respond to changing conditions and make adjustments as needed.

Proximate sensors can measure

- **Crop characteristics:** Proximate sensors are used to measure crop characteristics such as height, biomass, and leaf area index. This information can be used to optimize crop management practices and improve crop yields.
- **Soil characteristics:** Proximate sensors are used to measure soil characteristics such as pH, nutrient content, nitrogen, and moisture content. This information can be used to optimize fertilization, irrigation, and other resource management practices.

- **Pests and diseases:** Proximate sensors are used to monitor the prevalence and distribution of pests and diseases in a field, in order to optimize control measures and reduce the risk of outbreaks.

Computer Hardware and Software

Computer support is required to analyze the data gathered by other components of precision farming technology and to make it accessible in formats such as maps, graphs, charts, or reports.

- **Data storage and management systems:** Precision agriculture systems often generate large amounts of data, which needs to be stored and managed in a structured and efficient manner. This can be done using a variety of data storage and management systems, such as databases and cloud-based platforms such as Cropin Cloud.
- **Data analysis and visualization tools:** Precision agriculture systems often require the use of specialized software tools to analyze and visualize data in order to extract insights and make informed decisions.

Sustainable Digitalization of Agriculture with Precision Farming

The recent rapid digitalization has reduced the exhaustive paperwork in banks, hospitals, and most private and public sector organizations seem to diminish as their businesses move online. Digitization has reduced the manual work - which was time-consuming, error-prone, and inefficient - thus saving millions for corporations. Digitization of the economy has broken the barriers and has successfully curtailed the fear of tech dependency, especially among the farming community. Digitalization is slowly also revolutionizing the vast and complex Agriculture sector.

The United Nations projects that by the year 2050 the population of the world will be 9.7 billion. With the relevance of over 60% of the world population relying on agriculture for food, the pressure to increase crop production to meet demands doesn't seem to ease. Coupled with climate change, which is leading to the rise in global temperatures, levels of greenhouse gases, and frequency of extreme weather events like droughts and floods, along with increasing labor costs, high production costs, and unpredictability, poses a major challenge to the future of agriculture. Hence, the goal is to increase productivity in a sustainable way with climate-smart agriculture.

Precision Agriculture and Climate Change

Agri-businesses are highly vulnerable to climate change. Farming is becoming increasingly challenging in many areas due to decreased yields and increased frequency of extreme weather events like drought or flood. Climate change is expected to diminish global agricultural production by 17 percent by 2050, according to a study by the National Academy of Science (NAS).

Precision farming, according to NAS, should play a big role in making farming more sustainable, without compromising production or farmer earnings. With precision agriculture, the World Economic Forum believes world food production could rise 10-15 percent by 2030, and greenhouse gas emissions and water consumption would decrease by 10 to 20 percent, respectively, if 15 to 25 percent of farms adopted the technology.

Site-specific field management and variable-rate application (VRA) technologies are two of the most commonly discussed aspects of precision agriculture. These measures are essential to reducing the environmental effect of agriculture.

PRECISION FARMING AND THE FUTURE OF AGRICULTURE

Precision farming and modern agriculture practices focus on reducing production cost and wastage, as it is tailored to the needs of each plot. It centers on data collection and analysis of farm plots which comprises sensors, drones, and robots for recording the data, and software as a service (SaaS) can be used to adapt to precision farming systems.

Role of Artificial Intelligence and Machine Learning in Precision Agriculture

Although IoT is still at a nascent stage, the governments of agriculture-dominant economies do invest in other cutting-edge technologies like artificial intelligence and machine learning (AI/ML) to make smarter agriculture decisions. In countries like India, the implementation of IoT in agriculture has its own set of unique benefits and challenges.

Firstly, the farmers fear upgrading to Agri tech as they lack knowledge about the applicability of the technology in agriculture. Besides this, the sensors, robots, and drones are expensive, have high maintenance costs, and require technically trained labor to operate. Further, the captured data needs to be analyzed in a lab or using specialized instruments on the farm. Also, a variety of sensors are required for collecting data on different parameters, which need to be analyzed separately, hence driving operational costs higher. Therefore, the digital farming solution must be cost-effective and highly scalable, considering the various sizes of farms.

Precision Agriculture - the Road to Sustainability

Using precision agriculture, farmers can achieve the following:

- Identify the best crops and hybrid seeds suitable for a particular area
- Work only on exactly identified areas to be replanted
- Take specific actions to provide the essential and optimum level of inputs (fertilizers and chemicals)
- Save time and cost and minimize the environmental impact of polluting soil and water
- Create maps for irrigation schedules and use ideal amounts of water to curtail wastage
- Anticipate pest infestations and diseases and take preventative measures before they destroy crops
- Apply weedicides and pesticides without harming biodiversity and killing non-target plants
- Harvest produce when they mature to meet consumer preferences and early enough to extend storage time

The global precision farming market is estimated to reach $12.84 billion by 2026. Both big and small farmers and organizations working with growers benefit by adopting precision and digital farming technologies, which help to optimize agri-input resources without adding costs or workload. With climate change impacting farming in a big way and unpredictable natural disasters such as floods and drought becoming regular events, and the world population growing unabated, precision agriculture is here to stay.

The data obtained from precision agriculture can be used to provide better information for managing all activities on the farm to improve productivity.

GIS Technology In Precision Agriculture

Geographic Information Systems operate with object details and location data to create maps, including digital ones. For dividing the farm into separate zones, remote sensing is needed. All zones have their characteristics, the analysis of which requires GIS and GPS use. Traditionally, these zones are divided according to a soil type, nutrient availability, soil moisture content, pH rate, and pest infestation.

GIS in precision agriculture permits farmers to view records, such as soil survey maps and plant characteristics traditionally grown in the region. Satellite images and aerial photographs provide additional information. Another handy feature of GIS is analyzing multiple farm management options by comparing and manipulating data layers.

UAVs And Dusters in Precision Agriculture

Aerial technologies for crop management are based on using unmanned aerial vehicles (UAVs). Thus, farmers can monitor the yield's condition without scouting all fields in person. At first glance, the easiest option is crop dusters because farmers already use them for watering and sowing. They can attach a hyperspectral camera to any agricultural aircraft and record the needed information. However, it will lead to an additional load on crop dusters, which will be involved in addition to the main tasks for crop management. It will lead to faster wear of the devices and additional environmental pollution. In addition, this method means the employment of a qualified pilot.

Using UAVs or drones in precision agriculture is a more progressive solution. They are controlled remotely and consume less fuel. Moreover, precision agriculture drones can analyze the field thoroughly, conducting complex multispectral, thermal, and hyperspectral soil analyses.

Satellite Remote Sensing In Precision Agriculture

This technology allows growers to observe the yield health using satellite images. They provide up-to-date information on moisture stress, disease, structural anomalies, and nutrient levels. Modern precision agriculture satellite imagery has a high spectral resolution, allowing growers to get the most accurate data.

It is most effective to combine scouting with satellite remote sensing: growers can determine the cause of deviations from the norm. Also, they can use satellite imagery for planning plant treatments and selecting agricultural chemicals.

An essential benefit of this method compared to GPS and UAVs is the absence of additional fuel and labor costs.

Precision Agriculture and its Application Fields

This agriculture system constantly evolves, permitting growers to optimize production and exploitation of resources. Therefore, farmers can improve soil quality and productivity. Here are the most popular application fields.

Future Requirements

Big Data Challenge

Any type of site-specific farming means "information-driven farming." Farming in this context means data generation, data exploration, data modeling, and, finally, data-based operation control. In other words, precision farming approaches the big data challenge, and earlier precision farming models as seen in Figure 10.8 together with the above-defined situation "farmer left alone" approach the model of tomorrow (Fig. 4).

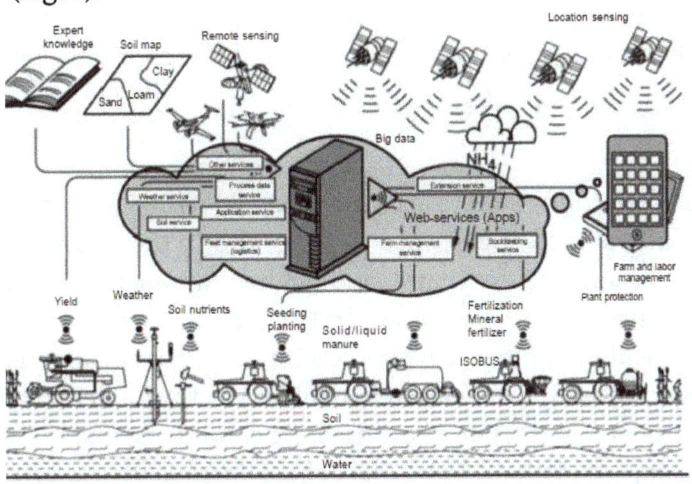

Fig 4 Cloud-based big data in precision cloud farming

The model of tomorrow (Fig. 4). Data of field operations will soon go to the cloud. Web services will then need to ensure that required data explorations or data modeling always use the newest, most advanced, and well-tested scientific algorithms. Farm management itself will evolve to the use of apps with twofold aims:

- Any financial data will still reside at the farm management system remaining as safe and limited-accessibility information.
- Field operation measures derived from big data will go in a timely manner from the cloud to in-field precision farming technology, and in parallel for safe documentation to the farm management system.

But all this will only become a reality in the future if big data in the cloud may be stored and handled in a highly secure way with farm-given access permission and with contractual and financial agreements to any data user.

Models

In today's precision farming practice, existing models are mainly used, evaluated, and approved under homogeneous conditions in conventional farming systems. In the future, there is a big need for well-adapted new models related to the inhomogeneities of different farming types, given farm situations, farm-related rules and laws, and other factors.

All these models may also be more detailed and subdivided into very specific farming systems. Feedback of any information on human overcontrol during operations or other influences or unexpected occasions into the cloud is essential. This allows for learning processes with more and better adjustments to the real world.

Automation

While site-specific farming suffers at small- and mid-sized farms in contrast to the overwhelming acceptance of guidance systems also in these farm types, farmers demonstrate a preference for more and higher automation. In future, auto-steering will be an internal application in tractors and self-propelled machines, and automation will integrate the whole tractor–implement combination. Control will move from the tractor to the implement in tractor implement management (TIM), as implements do the job, know the best conditions, and tell the tractor to react in a welladapted manner. This will become possible if

- Communication between tractor and implement follows well-accepted standards open to any manufacturer, and specifically to small- and medium-sized producers. Safety control is a must; in this case to clearly avoid inappropriate breakdowns and also to allow dedicated maintenance and repair.
- Communication between tractor and implement relays in the response of a so-called full-liner, where both machines come from the same manufacturer. Safety and responsibility in this solution will be clearly defined.

Tractor and implement(s) may then be seen as an integrated self-propelled unit. Adjusted settings given from implements to the tractor will be stop and go, left–right steering, forward speed adoption, PTO-revolution adjustment, hitch position changing, and flow control at auxiliary hydraulic valves. Automation may also spread across field machinery, including cloud settings into the loop either related to a task operation or in real time during job execution. Automation in this case will evolve to the Internet of agricultural utilities.

CRISPR-MODIFIED ORGANISMS FOR SUSTAINABLE AGRICULTURE

Introduction

The term "sustainable agriculture" encloses practices of farming addressed to the production of high-quality and safe agricultural products without compromising natural environments and the social and economic conditions of farmers. According to the Agricultural Sustainability Institute at UC Davis, the main goal of sustainable agriculture remains the preservation of the ability of future generations to meet their own needs, ensuring inclusive economic growth.

In the era of climate change, environmental threats will affect farmers at both the economic and the agronomic level, influencing crop yield and quality and with further negative effects on plant resistance to both abiotic and biotic stress. In order to counteract these environmental threats, and to have the chance to reach a sustainable production of industrial manufacture, new technologies will be used on the basis of the current knowledge in the biotechnology field.

Clustered regularly interspaced palindromic repeats (CRISPR) and CRISPR-associated protein (Cas) represent a new perspective for genetic engineering and the last frontiers of the new breeding techniques (NBTs) and genome editing (GE) tools.

CRISPR/Cas System

CRISPR/Cas systems are part of the adaptative immune system of archaea and bacteria in ensuring protection against viruses. The mechanism of action relies on the recognition and the cleavage of foreign DNA or RNA of invading viruses. One of the main characteristics is the integration of short fragments of the invading DNA (spacers) into the CRISPR locus, conferring the heritable immunity of bacteria. The CRISPR locus consists of an array of unique spacer sequences, derived from foreign invading DNA and interspaced by identical repeat sequences (crisprRNA or crRNA), along with a sequence encoding for the trans-activating crRNA (tracrRNA) and a series of genes encoding CRISPR-associated (Cas) endonucleases, responsible for the cleavage of the genetic material.

The transcription of the CRISPR locus allows for the formation of a single mRNA (pre-crRNA), which is partially complementary to tracrRNA, leading to the formation of an RNA duplex. The RNase III recognizes the RNA duplex and cleaves the double-stranded RNA (dsRNA) to form crRNA–tracrRNA complexes that activate and drive the Cas protein to the target sequence. The Cas protein is capable of introducing a double-strand break (DSB) only in the presence of a short conserved protospacer-adjacent motif (PAM) downstream of the target DNA, representing an essential prerequisite for the recognition of the target sequence.

CRISPR/Cas systems have been divided into two classes, six types and several subgroups based on Cas proteins and the nature of the interference complex. To date, the most common type, used as a genome editing tool, is the Type II Cas9 from *Streptococcus pyogenes* (*Sp*Cas9). The Cas9 protein consists of a bi-lobed architecture, including a large recognition (REC) lobe and a small nuclease (NUC) lobe. The NUC lobe includes a protospacer-adjacent motif (PAM)-interacting domain (PI) and two cleavage domains known as RuvC and HNH domains, each of which cleaves one strand of the target DNA three nucleotides upstream of the PAM sequence. The REC lobe contributes toward activating Cas proteins when combined with the tracrRNA–crRNA complex.

Hanging traits with CRISPR tools

CRISPR genome-editing tools have many applications in plants: researchers use them to learn more about plant biology. Plant breeders use them to create crops with desirable traits. Scientists can even use them to turn plants into leafy manufacturing facilities. Researchers are continually finding new applications of gene editing in plants, many of which go well beyond agriculture. What other applications can you imagine?

CRISPR *tools in plant research*

DNA sequences encode the proteins that give plants their specific traits. Farmers have been selectively breeding plants to get desired traits for thousands of years, but we still have a lot to learn about plant genetics. Scientists often know of a few DNA sequences associated with particular plant traits, but they might not have a precise understanding of what individual DNA sequences do or what other genes or sequences might also be involved. CRISPR tools can help them figure this out.

Improving Growth Under Stressful Conditions

The world population is growing. At the same time, climate change is making it more difficult to grow many important food crops. So, researchers and farmers need to figure out how to adapt agriculture to changing, and sometimes increasingly harsh environmental conditions.

Researchers are actively working on this, and have discovered some DNA variants that increase crop resilience to extreme environments, including drought. They are using CRISPR tools to make these changes in important staple crops like corn, wheat, rice, and sugarcane. Some of these crops are currently being tested in the field, with more to come.

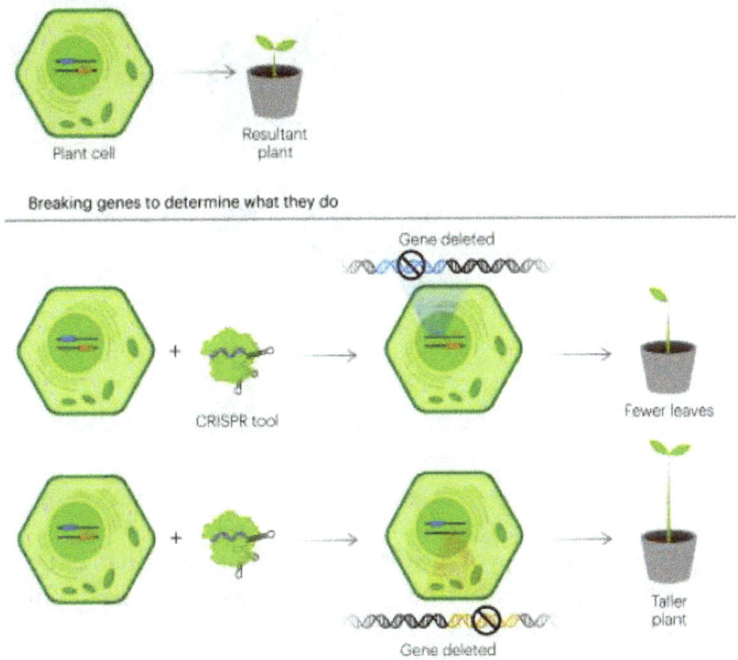

Fig 5. Plant cell with one of its DNA sequences

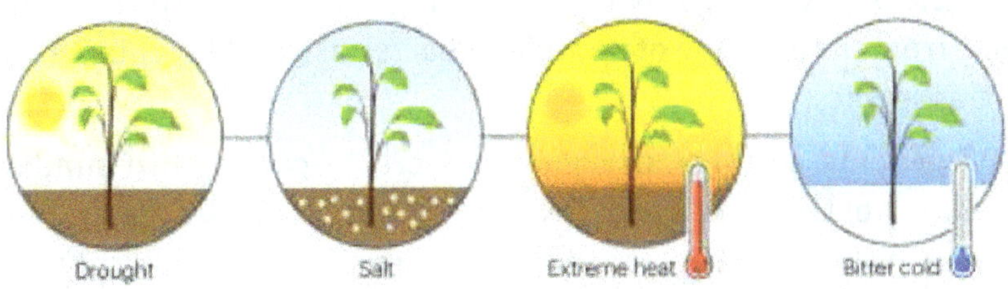

Fig. 6. Climate change threatens crops - Genome editing could be used to make cropsresistant to drought (top left), high salt (top right), extreme heat (bottom left), and bitter cold (bottom right).

Improving pest and disease resistance

Like us, plants get infections. They can also be eaten by pests. Together, these cause huge crop and revenue loss for farmers. They can even wipe out entire plant varieties. Genome-editing technology is one tool that farmers can use to fight pests and plant diseases.

Farmers have been growing insect-resistant genetically-modified crops for many years. Often, these crops produce bacterial proteins that kill or repel insects, preventing the crops from being eaten by insect pests. 'BT crops' are engineered to produce a toxin from the bacterium *Bacillus thuringiensis*, a natural insecticide that protects crops from pests but is not harmful to humans.

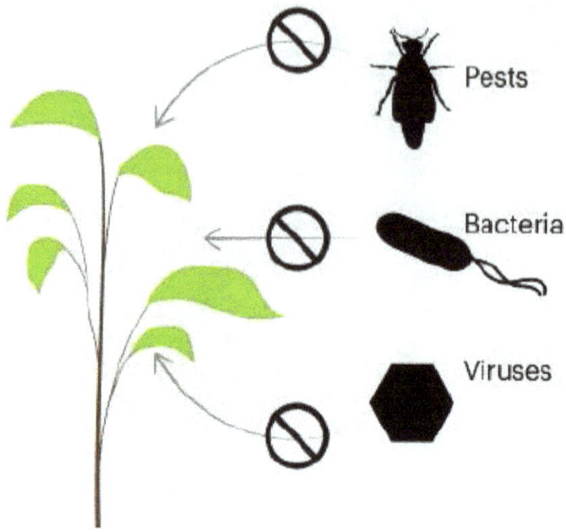

Fig. 7 Genomic strategies for protecting plants - Genome editing can make plants unappealing or toxic to pests who like to eat them, as well as resistant to infection by disease-causing viruses, bacteria, or fungi.

Farmers also grow herbicide-resistant GM crops. These crops produce bacterial proteins that make them resistant to chemicals that kill plants. This allows farmers to use herbicides to kill weeds without harming their crops. The traits in these GM crops come from the addition of non-plant, bacterial DNA to plants. In other words, they are transgenics.

CRISPR genome editing can sometimes create plants with similar traits by targeted editing of the plant's own DNA, rather than by addition of non-plant DNA. For example, many viruses need plant proteins to grow, multiply, and spread. Researchers can use CRISPR tools to break or change the plant proteins a particular virus likes to take advantage of. As a result, the viruses won't be able to spread within the edited plants. Similar techniques can make plants resistant to harmful bacteria, fungi, and insects.

Improving commercial properties

Many of the plants we have today were made with selective breeding and DNA mutagenesis techniques, but these sometimes come with negative side effects. These side effects happen when traits you don't want come along with the ones you do want. On a molecular level, this can happen because DNA for one trait is next to DNA for another, so they get passed down together.

One example of negative side effects is in many of the varieties of big red tomatoes at the supermarket. Farmers bred these tomatoes to be beautiful and transport without turning to mush, but they've lost much of their flavor. More delicate heirloom tomato varieties have sweeter, brighter flavors, but don't transport well.

Researchers may be able to use CRISPR tools to correct these negative side effects, improving crops in ways that make them more appealing to consumers. For example, researchers recently identified DNA sequences responsible for pleasing tomato flavors. With CRISPR tools, they might be able edit or restore these sequences, making more flavorful versions of large supermarket tomatoes.

Some other ways CRISPR tools could change fruits and vegetable traits include:

- Preventing bruising
- Increasing shelf life
- Creating seedless fruits
- Changing color/appearance/flavor
- Improving nutrient profile
- Decaffeinating coffee

Fig. 8. CRISPR genome editing could be used to prevent mushrooms and other produce from turning brown

Beyond these consumer-friendly improvements — or unfriendly, depending on how you like your coffee — researchers can also make plants easier to farm. For example, they can change stalk heights to prevent crops from falling over or prevent plants from scattering their seeds.

Making plants more productive

Plants use photosynthesis to convert sunlight into energy that helps them grow, creating the stems, leaves, seeds, and fruits that make up their *biomass*. To produce food and raw materials for the world's growing population, farmers need ways to grow more plant biomass. And to protect our natural environment, they must do so using fewer resources.

Different kinds of plants have different photosynthetic systems. Some bacteria also have their own photosynthetic systems, and some of these systems are more efficient at converting sunlight into energy and biomass than others. Researchers are using CRISPR tools to edit the proteins involved in photosynthesis to produce plants that are better at converting sunlight into biomass, helping farmers feed the world.

Improving Nutrition

People all over the world suffer from poor nutrition. Researchers can use CRISPR tools to create more nutritious plant varieties. For example, they can make some plants produce more vitamins, like so-called "golden bananas" that have a richer color due to their increased beta carotene. They can make some plants easier for humans and other animals to digest. They can even remove allergens and toxins; for example, researchers are working on removing allergenic proteins from peanuts and cyanide from cassava. If used for these ends, CRISPR tools could make more nutritious, safe foods available to more people.

Turning plants into biological factories

Researchers and biotechnology companies make therapeutics like antibodies, insulin, and vaccines, as well as other organic compounds using cells. They often use bacteria and yeast cells for these processes, but for certain products, researchers use plants instead. For example, the experimental Ebola vaccine, ZMapp, was produced using tobacco plants.

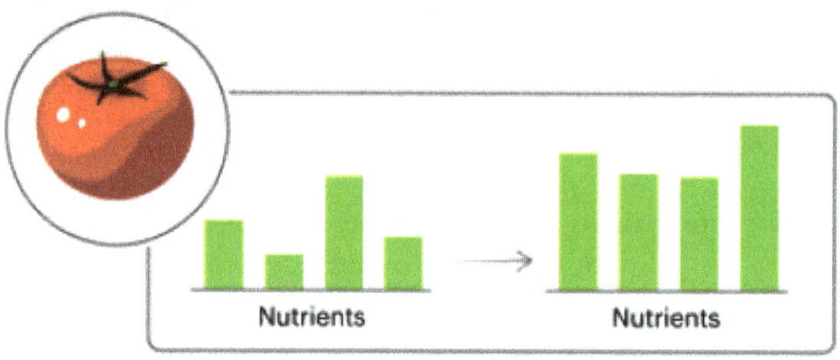

Fig. 9. Genome editing can be used to boost the nutrient content of some common foods.

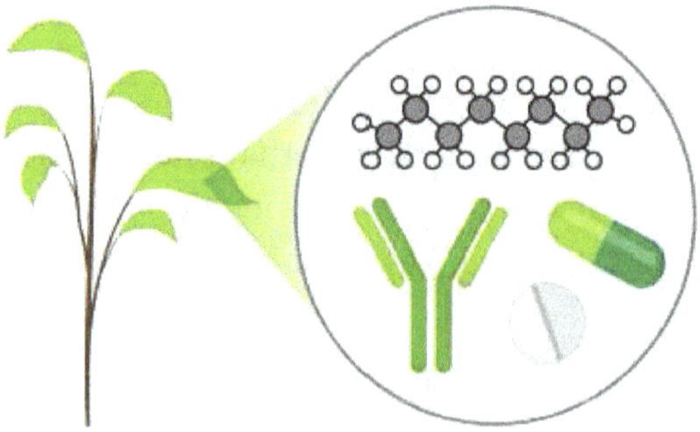

Fig. 10. Genome editing can make plants produce medicinal and industrial compounds.

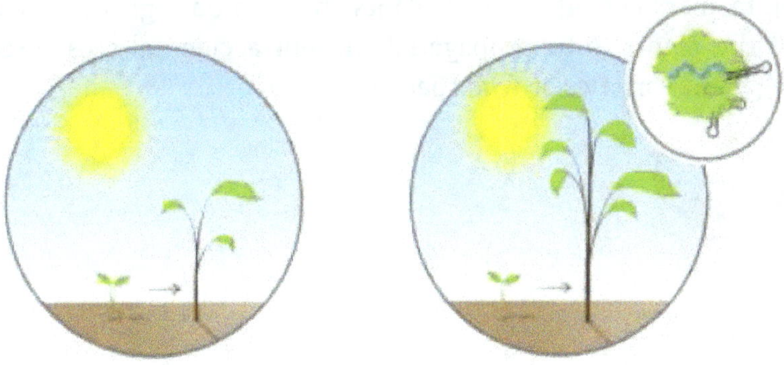

Fig. 11. Genome editing could make plants better at converting the sun's energy into biomass

CRISPR tools can make it much easier to turn plants into therapeutic factories. Using these tools, scientists can quickly design and insert therapeutic-encoding DNA into plants, resulting in effective and efficient therapeutic production.

Researchers can also use plants to create industrial chemicals. For example, they can use CRISPR tools to get plants to make more of their natural oils. They can also alter these oils to make them more useful to humans, including ingredients used in fuels, lubricants, flavors and fragrances. It is also possible to alter the sugar content of plants, making the altered plants easier to ferment into products like ethanol.

Genetic Containment

Some people worry that plants with engineered genetic changes will spread their altered DNA to wild species. This could happen if the altered plants are able to breed with nearby wild plants. Crop plants are often too genetically different from their wild counterparts to be able to breed with them. Many conventional plant breeding practices, including the common use of hybrid crops that are sterile, also prevent crop plants from breeding with nearby wild relatives. This is known as *genetic containment.* Genetic containment methods can be used alongside physical containment methods that put distance or physical structures like walls between altered plants and their wild counterparts.

CRISPR tools could be used to create new methods of genetic containment for genetically altered crops. For example, researchers can create plants that don't producc viable flowers or pollen. For angiosperms, flowers contain plants' reproductive organs, and without them, they can't reproduce sexually. For example, organic farmers in Yolo county, California, grow sunflowers with sterile pollen so they don't cross-pollinate their nearby sunflowers that are grown for seed. Alternatively, farmers can propagate many crops vegetatively. In *vegetative propagation*, farmers cut off parts of plants that can be regrown into full plants. This allows the plants to be propagated without accidental cross-breeding with neighboring plant varieties or wild species.

Fig. 12. Genome editing to stop modified plants from breeding - Genome editing could be used to stop plants from growing sexual organs (in angiosperm plants, flowers). Theseplants would not be able to genetically mix with plants in the surrounding environment

Editing for climate change and conservation

Climate change is the single biggest challenge facing humanity today, and researchers are looking to all possible tools to help society adapt and reverse the effects, including CRISPR genome editing. Researchers are using CRISPR tools to address climate change in three core ways:

- **Helping crops adapt to changing climate conditions** – Researchers are using CRISPR tools to develop crops that can withstand drought, increased temperatures, and increased salinity.
- **Reducing greenhouse gas emissions from agriculture.** Microbes are the largest source of greenhouse gas emissions from agriculture. Researchers are using CRISPR tools to study these microbe communities, or 'microbiomes.' Learn more here:
- Using CRISPR-based genome engineering to enhance the natural ability of plants and soils to capture carbon from the atmosphere and store it for long periods of time. Increasing plant biomass not only helps make agriculture more efficient, it reduces the need for chemical fertilizers and captures more carbon from the atmosphere.

When it comes to conservation, various wild plant species are threatened by pests and infections, and these threats are increasing due to climate change.

Fig. 13. Genome editing could make some endangered plant species more resilient, allowing their wild populations to rebound.

Pros and cons of CRISPR in agriculture

Benefits of CRISPR in agriculture

There are several benefits to utilising CRISPR in genome editing within the realm of agriculture, including:

- Improved farming productivity, which has economic benefits for farmers and society
- Improved food security. Ensuring crops are resistant to adverse conditions and are not struck down by disease represents a significant step towards food security
- Improved population health. CRISPR-based genome editing is an effective way to improve the nutritional value of crops and render crops safe for people with sensitivities or allergies. It can also be used to combat deficiencies, which is particularly important in vulnerable populations
- More sustainable practices. CRISPR can make farming practices more environmentally friendly, which is crucial amid the increasingly urgent climate crisis
- CRISPR is an easy-to-use, versatile, and affordable gene editing technique

Challenges and disadvantages of CRISPR in agriculture

Despite the significant advantages of using CRISPR to genetically modify crops, there are also considerable challenges. These include:

- Precise delivery. Delivering CRISPR/Cas9 components to the correct cellular location remains a significant challenge, particularly *in vivo* in rigid plant cells 7. This may require novel delivery strategies or adaptation of existing ones 8
- Off-target effects. There is significant concern about gene editing techniques, including CRISPR, having off-target effects on non-target genes. Although

the fidelity of CRISPR is relatively high, more work is required to improve this

- **Efficiency.** CRISPR is not efficient in all species of plants or at targeting some traits that are controlled by several different genes. Some crops or gene targets show lower transformation and editing efficiencies and do not obtain the desired genetic changes
- **Loss of genetic diversity.** Many people are concerned that genome editing using CRISPR and other techniques will lead to the narrowing of the genetic pool, making crops more susceptible to pests and disease in the long term and harming biodiversity
- **Legal and regulatory challenges.** Being a relatively new and rapidly advancing area of research, genetic modification of crops using CRISPR is subject to legal and regulatory challenges stemming from the lag in the time taken for laws and regulations to be updated in line with technological advances 9. Moreover, to gain approval, robust evidence must demonstrate safety and efficacy

ADDRESSING GLOBAL FOOD CHALLENGES WITH CRISPR

CRISPR technology - An introduction

The history of agricultural biotechnology can be traced back several decades, with its roots in the broader field of biotechnology. The earliest phase of agricultural biotechnology began with plant tissue culture and plant breeding techniques. In the mid-20th century, scientists began exploring ways to manipulate plant genetics more directly, marking the initial steps toward modern biotechnology.

The discovery of DNA's double helix structure in 1953 by Watson and Crick laid the groundwork for genetic engineering. In the 1970s, recombinant DNA technology emerged, allowing scientists to modify organisms at the molecular level. This led to the development of genetically modified (GM) crops in the 1980s, with the first GM crop, the Flavr Savr tomato, being approved for commercial release in 1994. The evolution of agricultural biotechnology has seen significant advancements, such as the development of Bt crops (plants that produce insecticidal toxins derived from *Bacillus thuringiensis*) and herbicide-tolerant crops, which have become widely adopted globally.

These developments have played a crucial role in increasing crop yields, improving resistance to pests and diseases, and reducing the environmental impact of agriculture. Precision agriculture, also known as site-specific crop management, has evolved significantly over the past few decades. The concept originated in the mid-1980s as a way to optimize field-level management with regard to planting, fertilizing, and harvesting. The introduction of GPS technology in the 1990s was a turning point, allowing farmers to navigate fields with unprecedented accuracy and manage crops with remarkable precision. The progression of precision agriculture is characterized by the integration of various technologies, including Geographic Information Systems (GIS), remote sensing, and various on-the-go sensors, facilitating data-driven decision-making in crop management.

The use of these technologies has enabled farmers to monitor and respond to intra-field variations, enhancing the efficiency of inputs like water, fertilizer, and pesticides. In recent years, the advent of big data analytics, machine learning, and the Internet of Things (IoT) has further revolutionized precision agriculture. These technologies have allowed for more nuanced analysis and utilization of agricultural data, leading to more informed decision-making and optimized agricultural practices.

CRISPR (Clustered Regularly Interspaced Short Palindromic Repeats) technology, a groundbreaking development in the field of genetics, is a genome editing tool that allows scientists to modify an organism's DNA with high precision. The basic principle of CRISPR technology involves the use of a RNA-guided DNA endonuclease, such as Cas9, which is directed to a specific location in the DNA sequence by a guide RNA (gRNA). This system, adapted from a natural defense mechanism found in bacteria, enables targeted cutting of the DNA strand, allowing for the deletion, insertion, or modification of specific genetic sequences. The process begins with the design of a gRNA sequence complementary to the target DNA sequence. Once inside the cell, the Cas9-gRNA complex binds to the target DNA sequence, and the Cas9 enzyme creates a double-strand break at this precise location. The cell's natural repair mechanisms then take over, which can be harnessed to introduce changes to the genetic code. The introduction of CRISPR technology marked a significant advancement over traditional genetic modification techniques.

Traditional methods, such as selective breeding and chemical or radiation-induced mutagenesis, have been used for centuries to enhance crop traits. However, these methods are often time-consuming, less precise, and can result in unintended genetic changes. Genetic engineering techniques developed in the latter part of the 20th century, including transgenic technology, offered more precision by introducing foreign genes into an organism's genome to express desired traits. Despite their effectiveness, these methods faced several challenges, such as the random integration of genes into the genome and the lengthy and costly regulatory processes required for approval.

Food security is an overly broad concept. It is, therefore, difficult to capture the totality of this concept in a few words. Several attempts have been made by global bodies in recent decades before arriving at the current definition. At the World

Food Conference (1974), the concept of 'food security' was introduced, and it was defined as the consistent availability of sufficient global food supplies, including essential nutrients, to enable a steady increase in food consumption and counteract fluctuations in production and prices. The definition was dynamic, as it reflected the prevalent protein–energy deficiency in 1970, which affected more than 25% of the global population.

However, the definition has been consistently revised to broaden its coverage. In 1983, the Food and Agriculture Organisation (FAO) introduced a new concept that emphasized the importance of guaranteeing the universal availability and affordability of basic food requirements for all individuals. Similarly, in 1986, the World Bank incorporated a new notion of ensuring that everyone has access to adequate food to lead a healthy and productive life. These concepts highlight the critical role of equitable and sustainable food systems in promoting human well-being and development.

A broader definition was achieved by the FAO in 1996 , which was revised again in "The State of Food Insecurity in the World 2001" with a social emphasis. Food security was then described as a situation in which all people always have sufficient, affordable, and nutritious food to satisfy their dietary demands for an active and healthy life. However, the definition was deemed inadequate and was revised, adding another component, 'food stability', which emphasizes the ability of food systems to withstand alterations or declines, whether caused by natural or man-induced factors.

Food security is commonly associated with the level of assurance that a population has sufficient access to the required amounts of food from available sources to fulfill their dietary needs at the national or regional level. This is commonly linked to the adequacy of the national food balance.

The degree of food security in a country is measured largely by the minimum per-capita dietary calories required by the lowest nutritional group, assuming that all locations and socioeconomic groups can equally access the available food supply for a prolonged period. To this effect, the world aims to achieve high food security for all people regardless of social status, race, or location.

Biotechnology: A Viable Approach to Curbing Food Insecurity

Biotechnology is an integration of numerous technologies that comprises several fields of science with wide applications. In modern agriculture, biotechnology plays key roles in the advancement of animal husbandry, cropping systems, soil science, soil conservation, plant physiology, seed technology, and crop management, all of which are pivotal to improving the quality and quantity of food production. It has also helped in the production of recombinant proteins, crops that are resistant to pests and pathogens and varieties with high nutrient content, and animals that produce more milk, among other applications.

Although biotechnology is not the only means of crop genetic improvement, it offers a faster and more accurate technology with broader applications to improve crops compared to its alternative, i.e., conventional plant breeding. While genetic modification (GM) targets the introduction of a foreign gene into the plant genome to produce the desired traits, conventional breeding interbreeds plants with the desired characteristics over many generations and then selects individuals with such traits among the progenies. In addition, genetic engineering offers opportunities for improving crops that are sterile or asexually propagated. Although the use of GM is becoming increasingly popular in crop improvement, it is not aimed at completely replacing conventional breeding; both approaches are indispensable, depending on the breeder's goal and selected crop.

Conventional breeding is very affordable, simple, and unrestricted by the government. The main constraint of conventional methods is the sexual procedure, especially the sexual compatibility of the organisms of interest; the desired gene might be in an unrelated species that cannot be effectively cross-pollinated with the recipient or from a non-plant species, such as genes from bacteria or insects. Another limitation is that the conventional approach cannot be used for improving sterile crops. If a valuable gene is identified in a wild relative, the conventional approach to improvement may not be suitable because performing a cross between the high-yield line and its wild relative creates a 'mix' of the parents' genomes, altering the carefully chosen and delicate assemblage of desirable genes present in the high-yield line.

Genetic Modification of Crops: A Silver Bullet to Combat Food Insecurity

As stated in, among the numerous possibilities for crop improvement, genetic modification of crops deserves special attention and could be a 'life jacket' in this ocean of food insecurity. Genetically engineered crops have the potential to reduce food insecurity to a minimum. Genetic modification of crops emerged with considerably lofty expectations and anticipations of enabling farmers to achieve a greater and better agricultural yield, counting on its limitless potential for crop improvement. Although it is still a subject of public debate, "super plants" offer a feasible alternative for addressing world hunger.

Plant genetic modification has occurred in distinct phases throughout history. The earliest crops were modified for resistance or tolerance to pests, diseases, abiotic stress, spoilage, or herbicides. The second generation of modified crops targeted an increase in the yield quality or nutritional content, e.g., fortifying crops with essential vitamins and micro-minerals. The third generation of genetically modified crops will have a wider scope and limitless applications. It will be used for non-food purposes like pharmaceuticals (edible vaccines), biofuel production, and bioremediation, as well as other critical areas. Many of these are already in use today.

Genetically Modified Crops: A Masterpiece

According to researchers, the first instance of a genetically modified crop was reported in 1983. Specifically, a tobacco plant was the first crop to undergo genetic modification with the introduction of an antibiotic-resistance gene into its genome. Generally, tobacco plants dominated early developments in plant genetic modification in the 1980s and early 1990s, with several successes reported using *Agrobacterium*-mediated gene transfer. In 1994, the 'Flavr Savr' tomato became the inaugural genetically modified crop authorized for commercial sale in the United States. It was developed for extended shelf life, and over time, many GM crops have been produced.

The development of plant-genome editing tools has revolutionized the area of plant genetic engineering by making it possible for greater precision and effective modification of plants' genetic compositions. Scientists employ genome editing as a method to explore potential genes and genetic variations that impact favorable characteristics, more effectively utilize genetic variability, and gain insight into the functioning of genes(Fig. 14). The technology allows the precise alteration of plant traits to introduce new characteristics into crops, such as disease resistance, drought tolerance, and improved nutritional content.

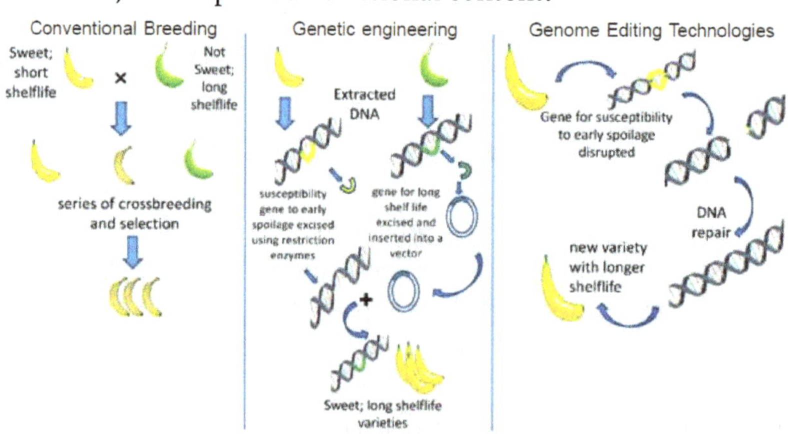

Fig. 14. The concept of conventional breeding, genetic engineering, and genome editing technologies

Plant gene editing uses an array of novel techniques, including CRISPR/Cas9, TALENs, and Zinc-Finger Nucleases. The CRISPR/Cas system is the most widely used gene-editing tool in plants due to its simplicity, efficiency, and versatility. Its ease of use and cost-effectiveness have opened up gene editing to a much wider range of researchers, including those in developing countries. Since the first successful application of CRISPR/Cas9 in plant gene editing in 2013, CRISPR/Cas-based gene editing has been successfully used to generate crops with improved yields, increased resistance to pests and diseases, enhanced nutrient content, and altered plant architecture across a wide range of crops, including soybean, rice, wheat, maize, tomato, cassava, tobacco, and potato, among others.

Additionally, gene editing can be used to reduce the use of pesticides and other harmful chemicals in agriculture by introducing traits that confer natural pest and disease resistance. Another important application of plant gene editing is the reduction or elimination of allergens in crops. Food allergies affect millions of people worldwide and can cause severe and sometimes life-threatening reactions. Gene editing offers a promising solution to this problem by allowing the targeted removal of allergenic proteins from crops without affecting their nutritional value. Researchers have used CRISPR/Cas-based gene editing to create hypoallergenic peanuts, wheat, and soybeans, among other crops.

Despite the potential benefits of plant gene editing, there are also concerns about its safety and ethical implications. Critics argue that the technology could have unintended consequences, such as off-target effects or the creation of new allergens or toxins. With continued research and development, plant gene-editing technologies could help transform the future of agriculture and ensure food security for generations to come.

Developing Crop Varieties with Higher Yield and Extended Shelf Life

Plants genetically modified for enhanced yield have greater productivity and faster growth in the face of limited growth resources and adverse climatic conditions compared to their unmodified relatives. According to a meta-analysis conducted between 1996 and 2013, the yield of genetically modified crops increased by 20%.

Furthermore, in developing nations, yields and producer gains are higher than in developed countries. Researchers discovered that genetically modified maize increased yields by up to 25% and significantly reduced dangerous food contaminants in their critical analysis of over 6000 peer-reviewed publications over 21 years (1996–2016). The findings further favor the global cultivation and consumption of GM maize owing to its improved quality and grain yield, as well as reduced human exposure to mycotoxins. Researchers have estimated the increase in food production attributed to the contributions of GM crops to be about $133 billion. More is achievable if scientists are given the liberty to modify crops in this regard.

Extending the shelf life of agricultural products is another laudable achievement of plant biotechnology, as it can significantly reduce post-harvest losses due to spoilage, which claims a sizeable percentage of agricultural products every year. Since the commencement of this endeavor, researchers have successfully released new varieties of important crops with extended shelf life, many of which are still under development. This has been achieved with tomatoes and bananas. For example, the antisense targeting of genes involved in ethylene generation and sensing in tomato plants has proven to be particularly successful in delaying fruit ripening.

Scientists from Israel's Agricultural Research Organization created transgenic banana plants with increased shelf life by lowering the expression of two transcriptional regulators, *MaMADS1* and *MaMADS2* . Furthermore, the researchers discovered that the flavor and quality of the genetically modified bananas remained unchanged. This technology can be applied to other crops to reduce food shortages caused by post-harvest spoilage, especially in countries without suitable technologies for fruit storage.

Insect Tolerance and Herbicide Resistance

This observation has led to a growing interest in exploring the potential of transferring this *Bt* gene into plants. Sweet corn, cotton, and potatoes are among the most commonly grown commercial crops that possess *Bt* insect resistance. This technology has helped to reduce the devastating impact of the Asian corn borer (Wilson), which causes approximately 40% yield loss in Indonesia's maize plantations. The introduction of new varieties of cotton that are insect-resistant has brought great relief to cotton farmers globally.

For instance, according to, five million farmers in India are growing *Bt* cotton on more than 7.6 million hectares of land. This improved variety has significantly promoted cotton production in India, as well as other parts of the world. The variety protects itself from insects without the use of an external pesticide. The switch to *Bt* cotton has resulted in a 31% increase in output, a 39% decrease in pesticide use, and an increased profit for farmers equivalent to USD 250 per hectare. In 2017, about 100 million hectares of farmland were cultivated with *Bt* crops, producing *Bt* toxins.

To broaden the impact range of GE crops, further pest-tolerance techniques against a wider range of devastating insects, using innovative technologies like RNAi and other non-*Bt* approaches, have been introduced. Plant genetic modification for insect resistance has evolved from directed immunity against a single insect order to a wider immune response with various mechanisms to combat similar or multiple insect species.

Impact on crop quality and sustainability

Crop quality plays an important role in determining the market pricing of crops. CRISPR-Cas in farming is focused on improving different crop qualities which are as follows:

- Crop physical appearance (shape, size, and color modification)
- Crop texture quality
- Prolonging shelf life
- Improving palatability
- Improving eating and cooking quality
- Improving flavor
- Biofortification of nutrient elements or micronutrients (such as carotenoid content)

CRISPR-Cas technology offers tremendous potential for addressing global food security challenges. By revolutionizing crop improvement, it enables the development of climate-resistant crops, enhances crop quality, and promotes sustainable agricultural practices. While there are challenges to overcome, the future of CRISPR-edited crops looks promising in ensuring a more secure and abundant food supply for the growing global population.

Currently, the research emphasis is mainly on the production of biotic/abiotic stress-tolerant crops. Biotic stress includes bacterial, fungal, and viral plant diseases and abiotic stress includes temperature, rain, humidity, etc. Both biotic and abiotic factors directly affect the yield of crops and the crop's shelf-life which are crucial for food security issues. For example, Cacao plantation in West Africa is under threat due to climate change wherein the high temperature supports the growth of mealybug leading to cacao swollen shoot virus disease. CRISPR could enable the Cacao to develop resistance to Mealybug pests thus preserving cacao trees and avoiding their wastage.

The technology simply edits food instead of changing it such as in the case of mushrooms wherein the mushrooms wouldn't turn brown after cutting. So instead of adding a new gene, the scientists just disabled the gene that causes mushrooms to brown.

CRISPR has also increased the yield of crops such as corn kernel and rice along with reducing the environmental impact of agriculture such as minimized water usage for corn, limited methane gas release in rice paddies, drought, and stress-resistant crop varieties, etc.

CRISPR is an economically viable and low-cost technique as compared to GMO and other conventional gene-modification processes thus making it affordable and beneficial for the small-scale agriculture industry.

Along with crops and plants, livestock is also one of the integral parts of the agricultural industry, especially in developing countries wherein cattle are still used for farming activities. Livestock can be considered as the agricultural sub-industry CRISPR technology is actively being employed to improve livestock such as TB-resistant cattle, faster and precise animal breeding, increase milk production, etc.

CASE STUDIES OF CRISPR-MODIFIED CROPS

Disease Resistance

One of the most significant applications of CRISPR in agriculture is in the development of disease-resistant crops. For instance, researchers have successfully used CRISPR to engineer resistance to viral diseases in crops such as tomatoes and cucumbers In one notable study, CRISPR was used to target and modify the eIF4E gene in cucumber plants, which conferred resistance to the cucumber vein yellowing virus, a significant pathogen affecting cucumber yield. Similarly, in wheat, CRISPR has been used to develop resistance against powdery mildew, a common fungal disease, by modifying the mildew resistance locus.

Drought Tolerance

CRISPR technology has also been instrumental in enhancing drought tolerance in crops, a critical trait given the increasing incidence of drought due to climate change. For example, researchers have used CRISPR to edit the ARGOS8 gene in maize, which resulted in improved water use efficiency and yield under drought conditions. This modification did not affect the maize's yield under normal water conditions, demonstrating the precision and effectiveness of CRISPR in improving drought tolerance without compromising crop productivity.

Nutrient Use Efficiency

Improving nutrient use efficiency is another area where CRISPR is making significant strides. By editing genes involved in nutrient uptake and utilization, crops can be made more efficient in using limited resources. For instance, CRISPR has been used to modify the OsNRT1.1B gene in rice, leading to enhanced nitrogen use efficiency. This modification allows rice plants to grow better in conditions with limited nitrogen, a crucial development for sustainable agriculture.

Advantages of using crispr in crop improvement

CRISPR technology stands out in agricultural biotechnology for its precision, speed, and costeffectiveness, marking a significant advancement over traditional breeding and genetic modification methods. Its precision and specificity allow for targeted genome modifications at specific sites, greatly reducing the likelihood of off-target effects and ensuring that only the intended trait is altered, thus maintaining the overall genomic integrity.

This ability to precisely edit genes contrasts sharply with the lengthy timelines associated with traditional breeding methods, which can take years to develop new crop varieties. CRISPR dramatically speeds up this process, enabling the rapid development of crops with desired traits and offering a swift response to emerging agricultural challenges. Moreover, the cost-effectiveness of CRISPR technology cannot be overstated. Unlike traditional genetic engineering, which is often resource-intensive and costly, CRISPR requires fewer resources and is less labor-intensive.

This makes it a more accessible and practical option for researchers and breeders worldwide, including in developing countries where such advancements can have a profound impact on food security.

Precision agriculture trends influenced by CRISPR

The integration of CRISPR technology with precision agriculture marks a transformative shift in modern farming, harmonizing genetic engineering with data-driven agricultural practices. This synergy enables the development of crop varieties finely tuned to specific farm conditions, utilizing CRISPR's capability for precise genetic modifications to cater to local soil, climate, and pest challenges, and complementing these with precision agriculture's targeted management techniques. Further, the advent of data-driven crop breeding, fueled by CRISPR and precision agriculture data, allows for the identification and enhancement of specific genetic traits, leading to the rapid development of crops optimized for particular environmental conditions. This approach has enabled the creation of personalized crop solutions, where CRISPR's efficiency and precision meet the unique environmental and geographical needs of individual farming locales.

However, the integration of CRISPR into precision agriculture is not without ethical and ecological considerations. Ethical concerns arise around the potential impacts of genome editing on biodiversity and ecosystem balance, as well as issues of accessibility and equity in technology deployment.

Ecologically, while CRISPR offers the promise of environmentally friendlier crops, there is a pressing need for rigorous assessment to ensure these innovations do not adversely affect soil health, local flora and fauna, or inadvertently alter gene flow in natural plant populations.

This comprehensive approach, blending CRISPR's genetic precision with the data-centric management of precision agriculture, paves the way for a more sustainable, productive, and ethically considerate agricultural future.

Challenges and limitations

Regulatory Hurdles and Public Perception

One of the primary challenges facing the adoption of CRISPR technology in agriculture is the complex web of regulatory hurdles. Different countries have varied regulatory frameworks governing the use of genetically modified organisms (GMOs), and these regulations are often complex and evolving. In some regions, CRISPR-edited crops are regulated under the same frameworks as traditional GMOs, which involves extensive safety testing and lengthy approval processes. This can delay the introduction of CRISPR-modified crops to the market and increase development costs.

Public perception of genetically modified crops also poses a significant challenge. Despite scientific consensus on the safety of GMOs, public skepticism and opposition remain in many parts of the world.

This skepticism often extends to CRISPR-modified crops, even though they may not involve the introduction of foreign DNA. Misinformation and ethical concerns about 'playing God' with nature contribute to public wariness, which can influence regulatory policies and market acceptance.

Technical Challenges in CRISPR Applications

While CRISPR technology offers unparalleled precision in gene editing, there are still technical challenges that limit its applications. One major issue is the possibility of off-target effects, where CRISPR might inadvertently edit genes other than the intended target, potentially leading to unintended consequences. Although advances in CRISPR design are continually reducing these risks, ensuring specificity remains a critical concern. Another technical challenge is the delivery of CRISPR components into plant cells. Different plant species have varying levels of resistance to genetic transformation, which can make it difficult to introduce CRISPR components effectively into some crop varieties. Additionally, achieving desired edits in all cells of a multicellular organism, which is necessary for stable trait inheritance, can be challenging.

Ethical and Biosafety Concerns

Ethical and biosafety concerns are also significant challenges in the application of CRISPR technology. Ethical concerns primarily revolve around the potential for unintended ecological impacts, such as gene flow to wild relatives and potential effects on non-target species. There are also concerns about the ethical implications of patenting CRISPRmodified organisms and the potential for these patents to limit access to the technology, especially for small-scale farmers in developing countries. Biosafety concerns include the potential for unintended ecological consequences. For instance, the introduction of a CRISPR-modified crop with a new trait could potentially disrupt local ecosystems or reduce biodiversity. The long-term ecological impacts of releasing genetically edited organisms into the environment are not yet fully understood, necessitating cautious and rigorous risk assessment processes.

CHAPTER-8

CRISPR AND ETHICAL CONSIDERATIONS

As with all new technologies that have potential to have such big impacts on humans, it is important to consider the ethics; whether it is morally right to use this technology in a variety of different circumstances.

Gene editing can be done on any cell in the human body. The cells in your body can be categorised into 2 groups: somatic and germline. Germline cells are cells that contain DNA which will be passed onto the next generation: eggs or sperm. Somatic cells are the rest of your body cells that will not be passed on to any of your future children. Using gene editing in somatic cells is generally agreed to be a positive thing, as it can cure a disease for an individual. This is what the licensed gene editing therapies do.

However, if someone has a genetic disease that can be passed on, you could edit their germline cells to prevent them passing the condition on to any future children. See this diagram showing the inheritance of cystic fibrosis. The inheritance pattern is autosomal recessive, meaning that you need one copy of the faulty gene from each parent to have the disease. If you have only one copy, you are a carrier and will not have any symptoms, and likely unaware that you could be a carrier unless someone in your family has cystic fibrosis. Cystic fibrosis is a single gene disorder which affects a protein, causing the airway to fill with thick and sticky mucus which makes it hard to breathe. It also affects many other systems, including the digestive tract. This condition may be cured by gene editing in the future.

Although ensuring a disease is not passed onto future generations sounds beneficial, this technology is new. Until further research can be done, we do not know about the long-lasting effects or if it has any unintended consequences. You would be modifying the genome of not just one person but many future generations. There is also the issue that you cannot obtain consent from someone who is not born yet, as it is their DNA that is being altered. The counter argument however is that it is unethical not to use a technology to cure life limiting diseases such as cystic fibrosis. Eliminating diseases that severely affect the quality of a person's life from the gene pool is a compelling argument to use gene editing on germline cells, once the technology has been rigorously tested and proven.

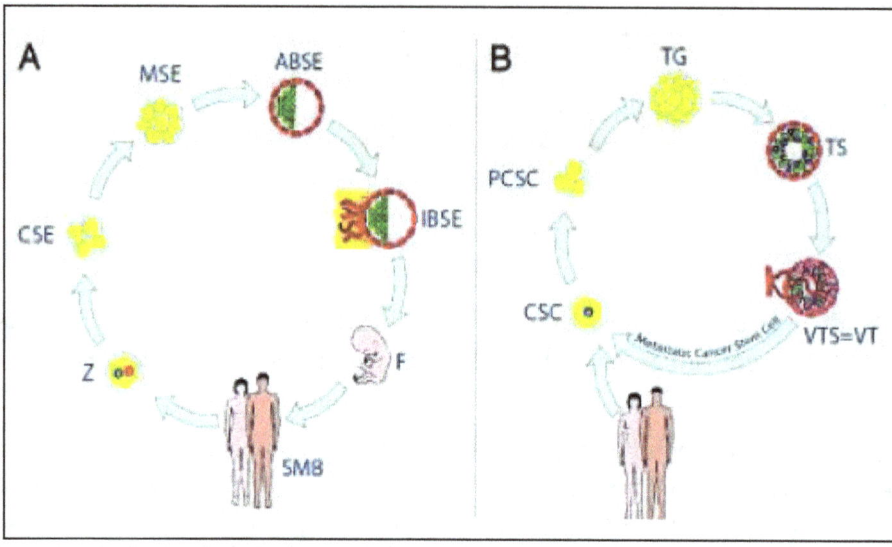

Fig. 1. Stages of the life cycles of germline cells - (A) and oncogerminative cells (B). (A) Z, zygote; CSe, cleavage stage embryo; MSe, morula stage embryo; aBSe, avascular blastocyststage embryo; IBSe, implanted blastocyst-stage embryo; F, fetus. SMB, sexually mature body. (B) CSC, cancer stem cell (i.e., oncogerminative cell); pCSC, parthenogenetic cancer stem cell (a pseudo-cleavage-stage embryo); tG, tumor germ (a morula-stage embryolike structure); tS, tumor spheroid (an imitation avascular blastocyst-stage embryo); VtS/Vt, vascularized tumor spheroid and/or vascularized tumor (an implanted blastocyst-stage embryo-like entity).

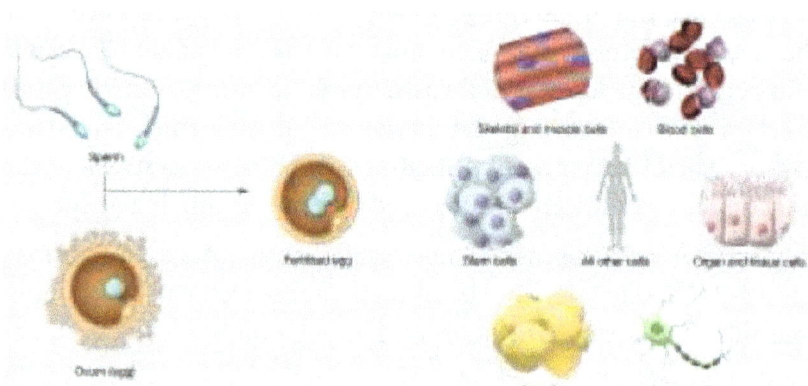

Fig. 2. Somatic cells

CRISPRs are short DNA sequences with unique spacer sequences that, along with CRISPR associated (Cas) proteins, constitute an adaptive immune system in many bacteria and archaea against invading bacteriophages. By using short RNA molecules as a template, Cas makes highly sequence specific cuts in DNA molecules that can be exploited to insert genes or to precisely modify the nucleotide sequence at the cut site. CRISPRs were first identified in the 1980s, but it is only during the past few years that scientists realized their potential to edit the genomes of any organism, from microorganisms to plants to human cells and, most controversially, human embryos. The CRISPR/Cas system is not a breakthrough technology in the sense that it enables genome editing; biologists have been using transcription activator like effector nucleases (TALENs) and zinc finger nucleases (ZFNs) to edit genomes for some time. However, those technologies are expensive, technically challenging, and time consuming, as they require protein engineering to target specific DNA sequences. CRISPR/Cas, in contrast, recognizes its target sequence via guide RNA molecules that can be cheaply and easily synthesized. A standard molecular biology laboratory can now edit genes or whole genomes of many organisms, as CRISPR/Cas does not require sophisticated knowledge or expensive equipment.

This has rekindled the ethical debate about modifying the human germ line. Notwithstanding the talk about "designer babies," CRISPR/Cas offers new possibilities to render humans immune to a range of diseases, or to repair fatal gene defects in a human embryo. Prominent researchers have therefore called for a voluntary moratorium on germ line genome modification in humans until scientists and ethicists have jointly analyzed the implications of doing so. The debate boils down to two sides in a "go/no go" standoff. One group insists that research on human germ line editing should advance in order to reap the scientific and clinical benefits, while the other camp argues that editing the human germ line is too unsafe, or crosses an inviolable ethical line.

However, rather than the use or not of CRISPR to edit human germ cells and embryos, there are more immediate ethical concerns that need to be addressed. CRISPR is already being used to modify insects, animals, plants, and microorganisms and to produce human therapeutics. Since such work has been going on for years—or even decades—the CRISPR technology may not appear to create new ethical problems in these contexts. However, there is a danger that CRISPR's affordability and efficiency could run roughshod over long standing and valid concerns about the generation and release of genetically modified organisms (GMOs). The recent characterization of a new type 2 CRISPR system from *Francisella novicida* demonstrates that the toolbox of genome editing technologies is ever expanding. Consequently, there is an urgent need for effective, global regulations that govern the testing and environmental release of GMOs.

Current national and international regulations provide inadequate guidance and oversight for these applications. As such, they do not foster public trust in the safety of CRISPR edited organisms or the regulatory agencies charged with monitoring them. The concern is that public misunderstanding and mistrust of GMOs will hinder scientific progress and valid uses of CRISPR. Thinking through—and getting right—the regulations and research ethics for these applications of CRISPR might also help to create an ethical framework for human germ line editing.

In the USA, the regulation of genetically modified animals and insects is done by a number of regulatory agencies that comprise the Coordinated Framework for the Regulation of Biotechnology, which was created in 1986 to facilitate inter agency regulation of biotechnology. Its scope and regulatory approach has not been revisited since 1992, but individual agencies within the Coordinated Framework—the Food and Drug Administration (FDA), the US Department of Agriculture (USDA), and the Environmental Protection Agency (EPA)—have issued their own guidelines on particular applications.

FDA guidance issued in 2009 states that the genetic modification of an animal, regardless of the animal's use, meets the criteria for veterinary medicine and is thus regulated by the FDA's Center for Veterinary Medicine (CVM). Genetically modified animals used to study human diseases and drug testing are regulated by the FDA's Center for Biologics Evaluation and Research. The Center for Food Safety and Applied Nutrition (CFSAP) and the USDA are brought in if the effects of a proposed modification will affect processes or products that they oversee—for example, food safety or pest control, respectively. There are potential roles for the EPA, the Department of the Interior and the US Fish and Wildlife Service, on a case by case basis.

The EU has a more centralized regulatory scheme in which the European Food Safety Agency (EFSA) conducts risk assessments, while final approval of a genetically modified animal or plant falls to the European Commission (EC). Analogous to the USA, human therapeutic applications are regulated and approved by the European Medicines Agency (EMA). Other countries with intense biomedical research programs likewise have their own regulatory and oversight schemes. Internationally, there is no unified guidance for the modification of non human organisms other than the Biological and Chemical Weapons Convention, which seeks to prevent research into and development of biological weapons.

Some applications of CRISPR in animals improve current standard practices in the biomedical sciences. For example, some research projects require animal lines that are specifically bred for certain mutations. Using CRISPR to generate these lines produces less genetic variability than standard breeding techniques and helps researchers to introduce mutations that more accurately represent the human genetic defects they study. Though there are standing ethical issues implicated by this practice, such as animal welfare, using CRISPR for this purpose does not challenge existing regulations of laboratory animals.

Other applications in animals, however, pose novel ethical concerns. In particular, CRISPR could be used to replace expensive TALENs, ZFNs, and other methods of genetic modification to improve food for human consumption. For example, CRISPR could be used to increase the muscle mass of animals, render farmed animals less susceptible to disease, enhance nutritional content, or create hornless cattle that are easier to handle. Research groups and private biotech companies are currently assessing whether such genome edits are feasible and safe. So far, no genetically modified animal has ever been approved for human consumption; the approval of genetically modified salmon for human consumption has been pending at the FDA for years. But it is not clear what criteria the FDA—or any other agency involved—uses for assessing the safety of genetically edited animals for human consumption. These regulatory processes must be more transparent and accountable.

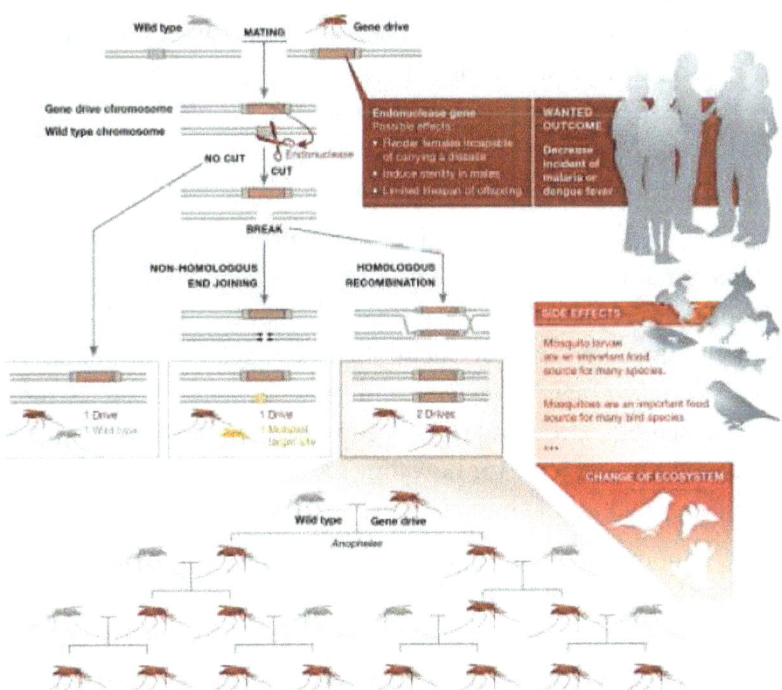

Fig. 3. Gene drives can be used to alter population wide traits

There is another, potentially much more dangerous and controversial, application of CRISPR, namely to potentially eradicate disease by eradicating disease vectors and invasive species. This involves research with the *Aedes aegypti* mosquito, which transmits dengue fever, and certain subspecies of the *Anopheles* mosquito that carry the *Plasmodium* parasite. Researchers at academic centers and private biotech firms are exploring so called gene drives to block disease transmission by editing the female mosquito so as to render it incapable of carrying the disease. Others aim to induce sterility in male mosquitos to prevent reproduction, or limit the lifespan of their offspring. Such methods could effectively destroy an entire species and could have significant environmental consequences.

Gene drive is a powerful tool that makes it more likely that the edited trait will be passed on to offspring through sexual reproduction. When genetically modified organisms are introduced into the environment and mate with wild type organisms, their offspring generally have a 50% chance of inheriting the modified genes (Fig 3).

The introduction of a few edited mosquitos or animals is therefore unlikely to have much of an effect. However, gene drive actively copies a mutation made by CRISPR on one chromosome to its partner chromosome and thereby ensures that all offspring and subsequent generations will inherit the edited genome. Over generations, this would lead to a noticeable effect: for example, in lowering transmission rates of dengue fever or malaria. The use of gene drives, though, also poses a much larger risk to the environment, as they have the potential to decimate an entire species, eliminate a food source for other species, or promote the proliferation of invasive pests.

Scientists have already called for strict biosafety measures and public review when it comes to introducing edited animals and insects into the environment. Yet, many questions remain unanswered: Can off target effects of CRISPR—unanticipated mutations leading to undesirable phenotypes—be controlled? What are the effects on animals or humans who eat genetically edited insects or animals? Will wiping out an entire species—albeit invasive, or disease bearing, such as mosquitos or ticks—upset the ecological balance? Will edited organisms be able to survive in natural environments, and if so, for how long? Addressing these questions requires far more regulatory oversight than currently exists anywhere in the world.

Editing the genomes of crops and trees is not new, and debates over the pros and cons of genetically modified (GM) plants have gone on for decades in the USA and Europe, and, more recently, globally. Agriculturally important plants have been genetically manipulated to make these less susceptible to disease and pests, more productive, and more resilient to changing climates. What makes CRISPR different from other methods of agricultural genetic engineering is that it no longer requires the insertion of foreign DNA into the plant genome using a virus, bacterial plasmid, or other vector system. Various commentators have therefore called for changes in the regulation of GM plants because CRISPR or TALEN edited organisms would no longer classify as transgenic organisms in *sensu strictu*.

In the USA, the Coordinated Framework under the purview of the USDA, the FDA, and the EPA provides guidance on agricultural applications of genome editing, but their regulations only cover "plant pests"—animals, bacteria, fungi, or parasitic plants that can directly or indirectly damage crop plants or parts thereof. This stipulation enters the regulatory process when parts of pest DNA are inserted into a host organism, or when certain viral vectors are used. The plant pest regulations also govern edits to insects that are detrimental to crops, plants, and trees, whereas applications of CRISPR that do not use pests or pest parts to induce genetic edits fall outside current regulations. Since the regulations frame the insertion of DNA as genetic material from a "donor organism," it is also unclear whether the regulations cover copies of pest DNA that are synthesized in the laboratory.

The Animal and Plant Health Inspection Service (APHIS), an arm of the USDA, reviews applications for research on GM crops. APHIS has indicated that products resulting from CRISPR/Cas that only *delete* a gene, in most cases, would not be regulated because no new genetic material is integrated into the recipient genome. Substitutions and insertions of genes would be reviewed on a case by case basis to determine whether the inserted trait counts as a pest. In recent years, APHIS has seen an increase in requests for non regulation status by academic centers and biotech companies asking them to affirm that their products do not fall under current regulations, and so do not warrant review for safety and efficacy by federal agencies. The current trend toward deregulation will promote research into a variety of applications of CRISPR, but the wide implementation of those edits without enforceable oversight could be detrimental to ecosystems, biodiversity, and human health.

In contrast to the USA, the European Union (EU) has much stricter regulatory regime for genetically modified crops in agriculture. It requires an extensive risk assessment by EFSA before the EC decides to grant or withhold approval for use in the EU. EU regulation currently considers all genetically modified crops or animals as transgenic—whether this includes the insertion of foreign DNA or direct genome editing—and therefore subject to regulation and risk assessment. However, there is ongoing debate arguing that CRISPR or TALEN edited plants without any foreign DNA should not be subjected to the same regulatory regime and risk assessment as transgenics. Since the EU is the largest market for agricultural products in the world, other countries are now waiting to see whether the EC will change its definition of transgenic and its regulations before they move on with marketing edited crop plants.

The US Coordinated Framework for the Regulation of Biotechnology was created to facilitate a unified approach to biotech regulation, but it is no longer adequate in the age of CRISPR. Even the EU's stricter regulatory regime is not suitable to address all possible risks—in particular with gene drive—as it is designed to regulate transgenic organisms.

Moreover, given that CRISPR is cheap, easy to use, and does not require sophisticated equipment or expert knowhow, it has become a popular technology worldwide, which will eventually require international standards for testing genetically edited organisms, releasing them into the environment, and assigning liability for damage. Regulations should set clear requirements for testing the safety and efficacy of edited organisms in carefully controlled environments or contained settings that simulate their natural environments. Gene drives in particular should be approved only if the safety and efficacy of desired edits have been rigorously tested. Finally, edited organisms should only be released in typical environments, whether on a farm or in a wild habitat, after public consultation and appropriate consent of potentially affected populations.

Regulations should also require the development of methods to halt the effects of edited insects or animals should they prove harmful to other organisms, the environment, or humans. Such reversal, immunization, and suppression drives would neutralize the effects of already released gene drives by introducing new genes into the population to counter unwanted effects from previous generations. However, these safety mechanisms are limited by the same facts that limit all gene drives.

As the species must reproduce through multiple generations for the desired trait to proliferate, the negative environmental impacts caused by the original gene drive population cannot be immediately halted by a counter gene drive. Furthermore, natural mutations cannot be prevented in the wild and might eliminate an engineered trait—whether the original gene drive edit or the counter edit—anytime after introduction. One approach to address this problem would be so called terminator genes or self limiting genes that limit the lifespan of edited organisms or make engineered organisms more fragile or easy to kill.

In addition, edited insects and animals should also be tagged to be able to assign responsibility and liability for damages. It would also enable researchers to better track the flow of gene edits through a population of insects or animals. These are not merely theoretical scenarios. A private biotech company is developing GE mosquitos in Florida with the aim of lowering the incidence of dengue fever by suppressing the population of *A. aegypti* mosquitos. To date, the FDA has not approved the trial; environmental review and the public comment period are pending.

Some Florida residents strongly oppose the release of the GE mosquitos, citing human safety and environmental concerns. They do have a point, as GE organisms will not always move and behave in predictable ways; GE mosquitos, for instance, even if released on an isolated island, might end up many miles away and have unanticipated effects on the environment such as crossbreeding with related species.

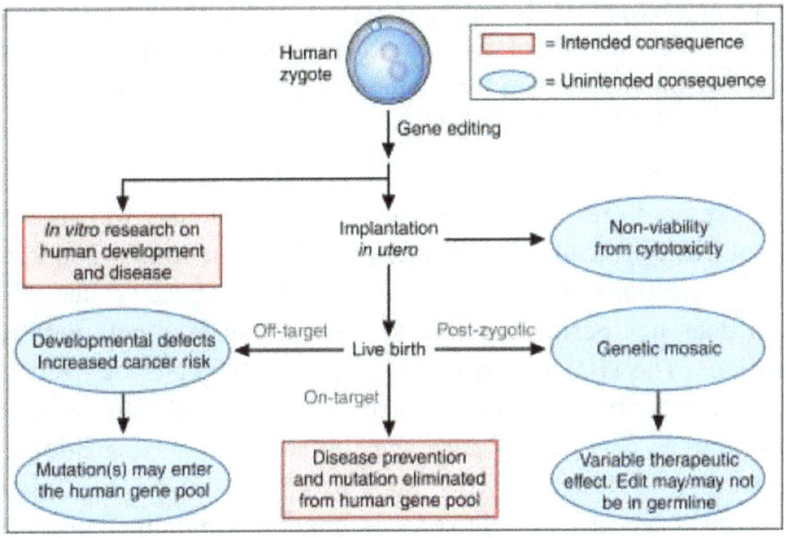

Fig. 4 Consequences of gene editing in human zygotes.

Without clear safety and testing guidelines, and public engagement and discussion, the public's trust in the safety of GE insects and animals will follow the same path as GM food. CRISPR is now being applied in many academic and industry laboratories around the globe. International treaties and policies are therefore required to govern the release of GE organisms into the environment. The WHO's "Guidance framework for testing of genetically modified mosquitos" for instance suggests updating the Cartagena Protocol on Biosafety.

Article 17 of the Protocol obligates parties to notify an International Biosafety Clearinghouse and affected nations of releases that may lead to movements of modified organisms with adverse affects on biological diversity or human health. However, the document does not specify who will enforce the treaty, what prior testing ought to have been conducted, what the limits on organism viability should be, what methods should be used to assess effects, or how to estimate damages or mitigate harms.

The treaty's effectiveness is further limited by voluntary participation. Some significant players in the field of genetic engineering, including the USA and South Korea, are not parties to the Cartagena Protocol. CRISPR is also an enormously powerful tool for synthetic biology to generate microorganisms for a broad range of applications, from the production of pharmaceuticals, biofuels, or chemicals to the remediation of pollution or disease diagnostics and treatment.

Gene editing allows synthetic biologists to design and edit whole genomes of bacteria and viruses with new properties, but it raises the same concerns about accidental or deliberate release of GE microorganisms into the environment. In the USA, the regulation of genetically modified microorganisms is under the purview of various agencies: the FDA, the EPA, and the National Institutes of Health (NIH), but they lack sufficient control and monitoring capacity. The NIH has guidelines for the use of recombinant DNA technology, of which CRISPR is one, that require notification and containment procedures based on the organism's pathogenicity, virulence, communicability, and environmental stability.

However, research not funded by the NIH is not subject to these guidelines. The EPA requires notification of new chemical production, which covers some commercial applications of synthetic biology, but the agency relies on voluntary reports and does not perform proactive audits and does not monitor smaller scale operations. The FDA requires that drugs and biologics be proven safe and effective before entering the market, which covers synthetic biology based human therapeutics, but it does not require specific containment methods to prevent accidental release or design controls such as terminator genes. Only the NIH's guidance was designed specifically to address genetically modified microorganisms, yet it is also the agency with the least regulatory authority.

As CRISPR becomes the primary method of genetic engineering, it would behoove these agencies to require that researchers demonstrate sufficient control mechanisms as a condition of using the CRISPR editing system. There is yet another aspect of the genetic editing of microorganisms to consider, as CRISPR could also be used to synthesize and manipulate pathogens, including smallpox, the Spanish flu virus, avian H5N1 flu virus, and SARS.

It is not unreasonable to think that, in the wrong hands, CRISPR could be used to make dangerous pathogens even more potent. The use of technology to increase the pathogenicity of bacterial or viral disease agents falls under the purview of the Biological and Toxin Weapons Convention (BWC), an international treaty designed to prevent the creation and storage of biological weapons. However, the BWC covers state actors—at least those who have signed it—but it was not designed to address private companies or individuals. Moreover, as the tools needed to design and manipulate pathogenic organisms and the exact genetic sequences and instructions to do so become more readily available, the effectiveness of the BWC to prevent the misuse of biological tools and knowledge is increasingly limited.

One way to achieve some control would be to regulate the tools of synthetic biology, notably DNA synthesis. Many companies that offer DNA primers, molecules, or even whole genome synthesis already monitor orders for specific sequences from pathogenic organisms. While this is an important move by industry to prevent misuse, it does not include all companies; moreover, an increasing number of companies are expanding their customer base beyond academia and industry to private individuals.

One possibility to address this problem is to take the industry's voluntary commitment further and create an international clearinghouse with which genetic sequence producers and sellers must register. It would require all registered companies to monitor their orders and make sure that those who order biological material that could be misused have appropriate credentials, containment facilities, and training. Much of the discussion about the risks of CRISPR technology has focused on using it to edit the human germ line.

Yet, CRISPR has many potential therapeutic applications beyond this specific use, ranging from cancer immunotherapy to treating infectious diseases, to creating stem cell models of disease. These applications constitute genetic editing of human somatic cells and the changes made are therefore not heritable. In cancer immunotherapy, current research focuses on adoptive cell therapies, wherein T cells are harvested from patients, modified *ex vivo* to increase their potential to destroy tumor cells, expanded in number, and infused back into patients.

One particularly promising approach involves chimeric antigen receptor T (CAR T) cells, which are engineered to express receptors with the specificity of monoclonal antibodies on their surface. CAR T therapeutics have proven to be particularly effective in trials against acute lymphoblastic leukemia in both adults and children. As researchers work to elucidate the mechanism by which these therapies achieve a robust response in order to optimize these cells to survive and carry out their effector function *in vivo*, CRISPR is becoming an attractive option to edit the properties of CAR T cells.

Another therapeutic application of CRISPR might help to cure latent infections with HIV or herpes viruses by targeting and "cutting out" viral DNA in infected human cells.

With the rapid application of CRISPR/Cas in clinical research, it is important to consider the ethical implications of such advances. Pertinent issues include accessibility and cost, the need for controlled clinical trials with adequate review, and policies for compassionate use. Many cell based therapies come at a considerable cost, particularly patient specific immunotherapies and stem cell treatments.

Adding customized gene editing on top of that will further push the price of such treatments well out of the reach of those with average means and insurance, to say nothing of those who are uninsured, destitute, or rely on national health services to decide what is to be made available to patients. It also raises the issue of educating patients to secure informed consent for research trials and clinical use.

CRISPR/Cas can be a tricky concept to explain, especially concerning its subtleties and potential for off target genome editing.

As excitement over CRISPR grows, so will demand from patients. Balancing requests from patients desperate for novel treatments with the need for rigorous clinical trials is already a challenge for regulators and will not become easier with the advent of CRISPR. US, European, and corporate policies provide some guidance on when and how to allow compassionate use or expanded access to experimental treatments, but these may have to be adapted to address gene editing. Moreover, and as we have seen with stem cell therapies, there are always those willing to promote misinformation or exaggerate in order to profit from desperate patients and their families.

Ensuring that CRISPR/Cas does not become touted as a panacea for all genetic illness is crucial for proper application and dissemination of the technology. There are specific regulatory challenges and ethical issues pertinent to the various applications of CRISPR technology to edit both somatic and germ line human cells. Far more worrisome, however, is the emerging application of CRISPR to non human organisms. The ability to design first generation organisms with desired characteristics might encourage development without sufficient containment mechanisms, or result in the premature environmental release of those organisms and loss of control over their spread.

In addition, CRISPR could be co opted for nefarious purposes, such as bioterrorism or biowarfare. The ease and efficiency of CRISPR raises the concern that anyone with the appropriate equipment could engineer a vaccine resistant flu virus or invasive species in a crude laboratory. While the new technology has sparked important debate about whether to proceed with human germ line engineering, the risks of the applications described here should serve as a call for discussing domestic and international regulation and guidelines for CRISPR's use.

ETHICAL IMPLICATIONS OF GENOME EDITING

For many years, molecular biologists have sought ways to use cellular repair mechanisms to manipulate DNA through genome editing. In this way, they would have the power to change the genome by correcting a mutation or introducing a new function. For this purpose, genome editing technologies were developed. In recent years, clustered regularly interspaced short palindromic repeats technology (CRISPR-Cas9) has become the most preferred method of gene editing. This technology has advantages such as high accuracy, easy handling, and relatively low cost compared to previous technologies, such as zinc-finger nuclease (ZFN) and transcription activator-like effector nuclease (TALEN). Thanks to these benefits, CRISPR-Cas9 technology can be easily applied in any molecular biology laboratory.

Genome editing technologies are used in the formation of human disease models in experimental animals and for the understanding of basic gene functions. They also have great therapeutic potential for future treatment of untreated diseases such as certain cancers, genetic disorders, and HIV/AIDS. Today, genome editing in somatic cells is one of the promising areas of therapeutic development. However, various bioethical issues have arisen due to the potential impact of these technologies on the safety of food stocks and clinical applications.

Application areas

Animal models

CRISPR-Cas9 can be used to create animal models to mimic human diseases and to understand disease development by mutating or silencing genes. A mouse model has been developed to determine the harmful effects of mutations in cancer by making mutations that cause the loss of function in tumor suppressor genes or give functions to protooncogenes.

Conventional genetically modified (GM) mouse models are produced by gene targeting in embryonic stem cells or transgenesis, which are time-consuming and highly expensive. With CRISPR-Cas9, GM mice can be efficiently produced in a much shorter time. It can be applied to nonhuman primates such as monkeys. Nonhuman primates are more similar to humans in anatomical, physiological, and genetic terms than rodents. Therefore, they are more suitable models than rodents in understanding human biology and disease development. The first successful application of CRISPR-Cas9 in nonhuman primates, from which a knockout monkey was produced, was realized in 2014. However, genome sequences of many nonhuman primates are not yet fully identified. This makes it difficult to design selected single-guide RNAs (sgRNAs). Therefore, the application of CRISPR-Cas9 in nonhuman primates is still at an early stage.

Genome editing in specific tissues

Researchers have been able to modify the genomes of specific tissues such as liver and brain tissues using hydrodynamic injection and adeno-associated virus (AAV). In a study, CRISPR-Cas9 has been successfully and effectively applied to the mammalian nervous system. A mixture of green fluorescent protein (GFP)-labeled AAV-spCas9 and AAV-spGuide plasmids was transferred *in vivo* to the hippocampal toothed brain folds of adult male mice (Swiech et al., 2015). It is thought that the number of such applications will grow in the fields of cancer and neuroscience in the following years.

Multiple gene mutations

CRISPR-Cas9 can be used to generate mutants for target genes. In the first such study by Li et al., six sgRNAs targeting Cas9 mRNA and six different genomic regions encoding the *Tet1*, *Tet2*, and *Tet3* genes were transferred to the cytoplasm of rat embryos. Findings showed that all three Tet genes carried the desired mutations in 59% of newborn rat pups. Successful results were also reported in studies with zebrafish embryos and *Arabidopsis* as well.

Epigenome studies

Epigenome studies can be performed in two different ways: genome or epigenome editing. In genome editing, nuclease is used to modify the DNA sequence, whereas in epigenome editing, an effector domain is used and the DNA sequence is not changed. This function is achieved by catalytic inactivation of the Cas9-associated effector domain by replacing the Cas9 protein.

Altered effector proteins are used to activate or suppress transcription. In epigenome editing, the epigenome can be modified by changing the proteins that maintain and protect the epigenome. Suppression of DNA methylation as a result of degradation of catalytic domains that accelerate the loss of spherical DNA methylation in human cells and lead to cell death is a good example of epigenome studies using CRISPR-Cas9.

Another application is the editing of long nonencoded RNAs (lncRNAs) and enhancer RNAs (eRNAs) that can control gene expression and epigenome processes. In a study, eRNA-expression factors and IncRNA-expression enhancers were suppressed by stimulating deletion mutations in a lymphoma cell line using CRISPR-Cas9. It is predicted that CRISPR-Cas9 may allow different levels of epigenome modification and facilitate further changes to humans.

Treatment of diseases

CRISPR-Cas9 can be applied to cells *in vivo* or ex vivo. In the *in vivo* approach, CRISPR-Cas9 is directly transferred to cells in the body using either viral or nonviral methods. In the ex vivo approach, first the cells are removed from the body; then CRISPR is applied to the cells and they are transferred back to the body. This approach has great potential to develop tissue-based therapies. Using CRISPR-Cas9, the mutation in the dystrophin protein responsible for the most common form of Duchenne muscular dystrophy was successfully removed.

There are studies to prevent and treat AIDS by inhibiting the entry of HIV into the cell or by removing the HIV genome integrated into the host genome using CRISPR-Cas9 (Saayman et al., 2015). Induced pluripotent stem cells (iPSCs) were successfully produced from cystic fibrosis patients with confirmed F508 deletion in the cystic fibrosis transmembrane regulator (*CFTR*) gene by CRISPR-Cas9. There are also studies for cataracts (Wu et al., 2015; Yang et al., 2016) and Parkinson's disease. However, recent studies have shown that CRISPR-Cas9 activates the type 1 interferon (INF) pathway, causing a type 1 INF-mediated immune response. These findings currently limit the use of CRISPR-Cas9 in treatment.

Industrial uses

CRISPR was first used for commercial purposes to make bacterial cultures used in cheese and yogurt production resistant to viral infections. One of the applications in agriculture is to produce GM crops. There are attempts to increase the yield in the livestock industry. It can be used to control invasive pest species to reverse pesticide and herbicide resistance in insects and weeds or to prevent disease spread. Researchers have succeeded in preventing the spread of genes protecting mosquitoes from harmful malaria parasites and making female mosquitoes infertile in the laboratory. Vaccine development is another significant area of interest. The smallpox virus vector (VACV) is used in the eradication and vaccination of smallpox. Using CRISPR-Cas9, the efficiency of marker-free VACV vectors has been increased. Another example is the hepatitis B vaccine. In order to prevent viral gene expression and replication, specific regions of the hepatitis B genome were targeted and cut by CRISPR-Cas9.

RNA editing

Single-stranded RNA (ssRNA) sequences can also be edited by CRISPR-Cas9. In RNA editing, CRISPR-Cas9 consists of a DNA oligonucleotide presenting the PAM (protospacer adjacent motif) region (PAMmer), ssRNA, guide RNA (gRNA), and Cas9 protein. PAMmer acts as a PAM region specifically recognized by Cas9 and directs Cas9 to bind and cut the target ssRNA. 5'-Elongated PAMmers containing bases paired with different ssRNAs and immediately in front of PAM are required for specific binding of target ssRNAs. Since RNA molecules have different functions than DNA, CRISPR-Cas9 can offer a much more flexible application than other genome editing methods.

Military applications

One of the lesser-discussed application areas of CRISPR-Cas9 technology is its use for military purposes. As is known, a substantial portion of genome editing studies are supported by the defense ministries of the countries. These studies are commonly focused on increasing the tolerance of soldiers against biological or chemical warfare. This technology has the potential to influence human performance optimization. Studies are usually concentrated on discovering different genes that can be harnessed from other species and identifying new genes that can be associated with posttraumatic stress disorder, which is frequently experienced by soldiers. In a study by researchers, researchers developed dog embryos with higher muscle mass using CRISPR-Cas9. Another interesting study showed that the *CMG2* gene, known to cause low sensitivity to anthrax toxin when expressed in small amounts, could be silenced by this technology (Arévalo et al., 2014). However, it should be noted that far more research needs to be conducted for using CRISPR technology in humans as a defense tool against biological and chemical weapons.

Ethical considerations in therapeutic gene editing

Although genome editing is a general process, the ethical implications of editing are not simply about the process, but instead are directly related to the purpose for which it is used. In "Germ line" and "Somatic cells," we outline some thoughts on the ethical considerations of using genome editing to modify the germ line and somatic cells.

The ethical issues surrounding the use of gene modification that may affect the human germ line have been extensively discussed previously. The National Academy of Sciences (NAS) and National Academy of Medicine (NAM) in conjunction with the Royal Academy of Sciences and Chinese Academy of Sciences have assembled a "Committee on Human Gene Editing: Scientific, Medical, and Ethical Considerations" chaired by Professors Alta Charo and Richard Hynes to perform a year-long, in-depth study of the issue during which input and guidance from multiple stakeholders from around the world will be solicited before issuing a consensus study report.

The Charo/Hynes committee represents the second step the NAS/NAM has taken on this issue following the highly successful and informative "International Summit on Human Gene Editing" held in December 2015 and chaired by Dr David Baltimore. This review touches on just a few of the issues that the committee will address in their more comprehensive study.

Unintended germ line editing

The ethical issues for unintended modification of germ cells in the course of an *in vivo* gene-editing therapy are not different from those for therapeutic approaches that would use gene addition. It has been a founding tenet of the field of human gene therapy that the human germ line in patients should not be perturbed genetically. Preclinical studies required for early clinical trials using *in vivo* viral-mediated gene delivery (e.g., by adenovirus or AAV vectors) demonstrated that there was not permanent gene transfer to egg or sperm in rodents and large animal models.

Similarly, it would be reasonable to demonstrate that a specific *in vivo* gene-editing approach would also not detectably modify germ cells in appropriate preclinical models as part of preclinical safety assessment to minimize risks of inadvertent modification of the germ line in patient's germ cells. Of course, there are also risks of spontaneous germ line alterations in embryos due to the ex vivo manipulations per se, in addition to effects for the genome-editing reagents.

Intentional germ line editing

The devastating nature of certain genetic diseases morally drives physicians and scientists to find ways of curing or preventing such diseases. One approach to these diseases, currently only theoretic, would be to use genome editing of germ cells or zygotes to correct disease-causing mutations in all cells of the person to curatively prevent these diseases from developing (Fig. 4). Although the goal may be to correct the genetic defect in some or all of the somatic cells of the person, their germ cell–forming tissues (testes or ovaries) would also be gene modified and this modification could be passed in the germ line to successive generations. One of the inadvertent but accepted effects of the successes of modern medicine is that by providing improved survival it has indirectly altered the allele frequency of certain disease-causing mutations in the human gene pool. However, there is a higher ethical concern for the direct intentional alteration of the germ line to eliminate pathogenic mutations that may through childbearing introduce the modified genomes into the human gene pool.

In addition to the issues of direct intentional editing of the germ line, other ethical considerations concerning germ line editing include the goal of the intervention: disease curative or "genetic enhancement" and whether the change will re-create what is naturally found in human genetics or whether it will create a human genome that is not normally found in the human population, such as by using editing to add a gene into a specific genomic location. Given that our understanding of the genome is still rudimentary, editing with intent to "improve" is fraught with many caveats, not the least of which is that we do not know, may never know, or nor even be able to define what an "improved" genome would look like.

Although there are moral arguments for and against germ line editing, there are 3 important practical limitations to germ line genome editing. The first is that there are only a tiny number of instances in which preimplantation genetic diagnosis could not achieve what one might want to achieve with germ line genome editing. Nonetheless, there are examples in which preimplantation genetic diagnosis would not work, most notably when couples who share a homozygous-recessive disease have children or when 1 parent is homozygous for an autosomal-dominant disease.

The second is that current procedures of zygote editing, even in lower rodents, are still inefficient with multiple zygotes having to be manipulated and implanted in order to generate a mosaic genetically edited progeny. The third, and perhaps most problematic, practical limitation is that one cannot assure that a harmful unintended heritable modification does not occur in the context of an intentional germ line editing procedure, when there is no capacity for characterization of a modified germ cell postprocedure. Even with only the best of intentions to remove a disease allele, present technology cannot assure us that unintended modifications created through an editing procedure would not result in a devastating long-term outcome such as cancer or adverse developmental effects if one were to modify a zygote.

These, perhaps insurmountable, safety concerns might be abrogated by editing germ line stem cells followed by deep sequencing to assure no unintended modifications have occurred. The transplantation of human germ line stem cells or the generation of human germ cells from germ line stem cells, even without modification, however, has not been performed.

Although modifying human zygotes and implanting them to generate humans perhaps has insurmountable safety and ethical limitations, the use of editing as a research tool to understand early human development or disease pathogenesis and treatment without implantation is probably ethically permissible based on current guidelines (for examples, see National Institutes of Health [NIH]). The major ethical issue for the research use of editing in human zygotes is that it would require the specific generation of 1-cell zygotes for study as zygotes generated by in vitro fertilization are already at the multicellular blastocyst stage and thus beyond the very earliest stages of zygote development. Moreover, at the multicellular blastocyst stage, any editing approach would likely create a mosaic of edited cells in which different cells acquired different genomic changes, thus confounding any experimental results. In fact, the Human Fertilization and Embryology Authority in the United Kingdom has recently given permission to Kathy Niakan, a developmental biologist working at the Francis Crick Institute in London, to perform in vitro studies using clustered regularly-interspaced short palindromic repeats (CRISPR)/CRISPR-associated protein 9 (Cas9) genome manipulation in human embryos.17

Somatic cells

Modification of somatic cells by gene editing does not raise unique ethical issues above those raised by gene addition. Both approaches have ranges of efficacy and potential adverse effects, and their use is predicated on defining a risk-benefit ratio that favors benefit to the patient. For gene-editing procedures, as with any other medical procedure, this balance will be variably defined depending on the type of cell being edited (e.g., terminally differentiated somatic cells vs stem cells), the nature and extent of the editing procedure (eg, number of loci being edited and their potential oncogenic activities), and the therapeutic application (eg, life-threatening disease vs enhancement/cosmetic). Gene-edited cells have been used in clinical trials for HIV-1, by zinc finger nuclease (ZFN)-mediated disruption of the CCR5 HIV coreceptor, and more recently for the treatment of leukemia through the use of allogeneic T cells armed with a chimeric antigen receptor, with transcription activator-like (TAL) effector nuclease (TALEN)-mediated editing to disrupt the endogenous T-cell receptor (TCR) gene to prevent alloreactivity.

Some have raised the possibility that genome editing of somatic cells for "enhancement" poses a specific ethical issue. The ethical issue of enhancement by genome editing, however, is probably not ethically distinct from enhancement using other gene-therapy strategies, including the use of lentiviral vectors or AAV vectors and thus is a relatively mature ethical issue. We note that "enhancement" ranges from engineering a cell to overexpress a therapeutic protein, as was done using lentiviral modification of HSCs to treat metachromatic leukodystrophy, to using genome editing to knock out CCR5 to "enhance" resistance to HIV infection, to as-yet-unproven genetic modifications that might result in cosmetic changes (intelligence, muscular function, height, etc).

As noted previously, given our still rudimentary knowledge of the complex and interacting genome, the issue of whether we know what an "improved" genome really is remains as the concept of unintended/unanticipated consequences comes to the fore. Nonetheless, any proposed gene-therapy approach aimed at enhancement must still pass the well-established threshold that the potential risk/benefit ratio is acceptable, informed consent is provided under local institutional review board guidance, and all required regulatory approvals are obtained from the relevant oversight entities (eg, the NIH Recombinant DNA Advisory Committee [RAC] and the Food and Drug Administration [FDA] in the United States, and the European Medicines Agency [EMA], etc, in the European Union).

RESPONSIBLE USE OF CRISPR TECHNOLOGY

Over the last few years, the exuberant development of genome editing has revolutionized research on the human genome, which has enabled investigators to better understand the contribution of a single-gene product to a disease in an organism. In the 1970s, the development of genetic engineering (manipulation of DNA or RNA) established a novel frontier in genome editing. Based on engineered or bacterial nucleases, genome editing technologies have been developed at a rapid pace over the past 10 years and have begun to show extraordinary utility in various fields, ranging from basic research to applied biotechnology and biomedical research.

Genome editing can be achieved in vitro or *in vivo* by delivering the editing machinery in situ, which powerfully adds, ablates and "corrects" genes as well as performs other highly targeted genomic modifications. Targeted DNA alterations begin from the generation of nuclease-induced double-stranded breaks (DSBs), which leads to the stimulation of highly efficient recombination mechanisms of cellular DNA in mammalian cells. Nuclease-induced DNA DSBs can be repaired by one of the two major mechanisms that occur in almost all cell types and organisms: homology-directed repair (HDR) and nonhomologous end-joining (NHEJ), resulting in targeted integration or gene disruptions, respectively.

Historically, homologous recombination (HR), in which undamaged homologous DNA fragments are used as templates, has been the approach to realize targeted gene addition, replacement, or inactivation; however, the utility of HR is heavily limited due to its inefficiency in mammalian cells and model organisms. After it was discovered that DSBs could raise the incidence of HDR by multiple orders of magnitude, targeted nucleases have been found as an alternative approach to increase the efficiency of HDR-mediated genetic alteration. Once a targeted DSB has been made, HDR may reconstruct the cleaved DNA using an exogenous DNA template analog to the break site sequence.

This mechanism may be used to introduce precise mutations by delivering an appropriately designed repair template into targeted cells directly, thereby, in a site-specific manner, resulting in mutation correction or new sequence insertion. Alternatively, NHEJ-mediated repair tends to result in errors because it leads to efficient formation of gene insertion or deletion (indels) in diverse lengths at the DSB site, which eventually causes gene inactivation. If indels occur in the coding sequence, there will be frameshift mutations, which will result in mRNA degradation or nonfunctional truncated protein production by nonsense-mediated decay. This approach and its applications are thought to be simpler than HR-based methods because

- There is no need for a repair matrix and
- The cell type has less impact on modification efficacy (contrary to HR, NHEJ may be active all through the cell cycle).

Thus, similar to RNAi, NHEJ may be applied in immortalized cell lines to generate the inactivation of a single gene or multiple genes, but by creating loss-of-function mutations, it may lead to permanent gene inactivation.

In the early development stage of genome editing, to induce the desired DSBs at each particular DNA target site, the engineering of distinct zinc-finger nucleases (ZFNs) or meganucleases has been the research focus. These nuclease systems required specialized competence to generate artificial proteins consisting of customizable sequence-specific DNA-binding domains, each connected to a nonspecific nuclease for target cleavage, providing researchers with unprecedented tools to perform genetic manipulation. Subsequently, a new class of a *Flavobacterium okeanokoites* (FokI) catalytic domain derived from bacterial proteins termed transcription activator-like effectors (TALEs) has shed light on new possibilities for precise genome editing.

TALE-based programmable nucleases can cleave any DNA sequence of interest with relatively high frequency. However, the main challenges for transcription activator-like effector nucleases (TALEN) approaches are the design of a complex molecular cloning for each new DNA target and its low efficiency of genome screening in successfully targeted cells. Clustered regularly interspaced short palindromic repeat (CRISPR)-associated 9 (Cas9) nuclease is a recently discovered, robust gene editing platform derived from a bacterial adaptive immune defense system. This system can be efficiently programmed to modify the genome of eukaryotic cells via an RNA-guided DNA cleavage module and has emerged as a potential alternative to ZFNs and TALENs to induce targeted genetic modifications. Since 2013, when it was first applied in mammalian cells as a tool to edit the genome, the versatile CRISPR/Cas9 technology has been rapidly expanding its use in modulating gene expression, ranging from genomic sequence correction or alteration to epigenetic and transcriptional modifications.

The advent of programmable nucleases has greatly accelerated the proceedings of gene editing from concept to clinical practice and unprecedentedly enabled scientists with a powerful tool to maneuver literally any gene in a wide variety of cell types and species. Current preclinical research on genome editing primarily concentrates on viral infections, cardiovascular diseases (CVDs), metabolic disorders, primary defects of the immune system, hemophilia, muscular dystrophy, and development of T cell-based anticancer immunotherapies. Some of these methods have gone beyond preclinical research and are recently undergoing phase I/II clinical trials. Here, we review recent improvements of the three main genome editing platforms (ZFN, TALENs, and CRISPR/Cas9) and discuss applications of their derivative reagents as gene editing tools in various human diseases and in promising future therapies, focusing on eukaryotic cells and animal models. Finally, we outline the clinical trials applying genome editing platforms for disease treatment and some of the challenges in the implementation of this technology.

Genome editing for disease modeling and gene therapy

Targeted gene modification via chimeric genome editing tools (e.g., ZFNs, TALENs, and CRISPR/Cas9) is a powerful method to assess gene function and precisely manipulate cellular behavior and function. These genome editing tools have enabled investigators to use genetically engineered animals to understand the etiology behind various diseases and to clarify molecular mechanisms that can be exploited for better therapeutic strategies.

Discovery and development

CRISPR-Cas9 technology has revolutionized genome editing due to its simplicity, efficiency, and versatility. The CRISPR-Cas9 system was first harnessed for genome editing by Jennifer Doudna and Emmanuelle Charpentier, who demonstrated its potential in a seminal study published in Science in 2014. They showed that the Cas9 nuclease could be guided by a synthetic single-guide RNA (sgRNA) to introduce targeted DNA cleavage in vitro. This groundbreaking work laid the foundation for the rapid advancement of CRISPR-Cas9 technology.

Since its initial discovery, CRISPR-Cas9 has been widely adopted and extensively optimized for various organisms and cell types. Numerous studies have further refined the system and expanded its applications. For example, Mali et al. demonstrated the successful use of CRISPRCas9 to edit the genomes of human cells, highlighting its potential for biomedical research.

The specificity of CRISPR-Cas9 technology has been a topic of great interest and improvement. Studies have explored strategies to reduce off-target effects and enhance precision. Researchers developed a modified Cas9 variant, termed SpCas9-HF1, which exhibited reduced offtarget activity while maintaining high on-target efficiency. This study demonstrated the feasibility of engineering the Cas9 nuclease to improve its specificity.

The versatility of CRISPR-Cas9 extends beyond gene knockout. Researchers have harnessed the technology for a range of applications, including gene activation and repression. In a study by the researchers, the dCas9 protein (a catalytically inactive form of Cas9) was fused to transcriptional activators or repressors, enabling precise control of gene expression. This approach has facilitated the investigation of gene function and regulatory networks.

CRISPR-Cas9 technology has transformed the field of genome editing. Key studies by Doudna and Charpentier have paved the way for its widespread adoption and optimization. Advances in specificity, versatility, and applications continue to propel the field forward, opening up new possibilities for scientific research, therapeutic interventions, and agricultural advancements.

CRISPR-Cas9 technology application in agriculture

CRISPR-Cas9 technology holds significant promise for revolutionizing agriculture by enabling precise and efficient genetic modifications in crops and livestock (Figure 5). This subsection highlights the applications and advancements of CRISPR-Cas9 in agricultural settings. Crop improvement is a major focus of CRISPR-Cas9 technology in agriculture. The ability to precisely edit plant genomes allows for the enhancement of desirable traits, such as disease resistance, yield, nutritional content, and stress tolerance. For example, in a study by Shan et al., CRISPR-Cas9 was used to successfully target the susceptibility gene in rice (Figure 5, leading to enhanced resistance against bacterial blight.

CRISPR-Cas9 has also shown promise in accelerating the breeding process. Traditional breeding methods often require multiple generations to achieve desired traits. With CRISPRCas9, targeted genetic modifications can be introduced directly into the germline of plants, significantly reducing the breeding timeline. In a study by researchers, CRISPR-Cas9 was utilized to enhance rice grain yield by precisely editing genes associated with grain size regulation. Furthermore, CRISPR-Cas9 technology has the potential to address global food security challenges. By modifying crop genomes, it is possible to develop crops that are more resistant to pests, diseases, and environmental stresses, thereby increasing productivity and reducing yield losses. This technology also offers opportunities for the production of nutritionally fortified crops, which can help combat malnutrition and improve human health.

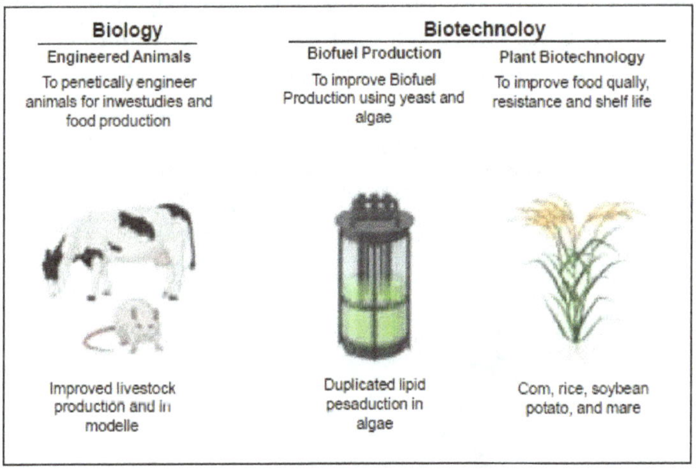

Fig. 5. Applications of CRISPR-Cas9 technology for genome editing in the field of agriculture

The use of CRISPR-Cas9 in livestock breeding is another emerging area in agricultural applications. Researchers are exploring the potential to introduce specific genetic modifications in livestock to improve traits such as meat quality, disease resistance, and animal welfare. While progress in this field is still in its early stages, CRISPR-Cas9 holds promise for revolutionizing livestock production and addressing challenges faced by the livestock industry. Although the application of CRISPR-Cas9 in agriculture offers immense potential, it is not without challenges. Regulatory frameworks and public acceptance remain important considerations, particularly regarding genetically modified organisms (GMOs). Efforts are underway to address these concerns and establish guidelines for the responsible use of CRISPRCas9 in agriculture.

Furthermore, CRISPR-Cas9 technology presents significant opportunities for agricultural advancements. Its precise genome editing capabilities enable targeted modifications in crops and livestock, leading to improved traits, increased productivity, and enhanced food security. As research progresses and regulatory frameworks develop, CRISPR-Cas9 holds the potential to revolutionize agricultural practices and contribute to a sustainable and resilient global food system.

CRISPR-Cas9 technology in medicine

CRISPR-Cas9 technology has emerged as a powerful tool with transformative implications in the field of medicine (Figure 6). This sub-section provides an overview of the applications and advancements of CRISPR-Cas9 in medical research and therapy. One of the key applications of CRISPR-Cas9 in medicine is in the study of gene function and disease mechanisms. By selectively modifying genes in human cells or model organisms, researchers can gain insights into the underlying causes of genetic diseases. For example, in a study by Shalem et al., CRISPR-Cas9 was employed to systematically knock out each gene in the human genome, leading to the identification of genes essential for cancer cell survival.

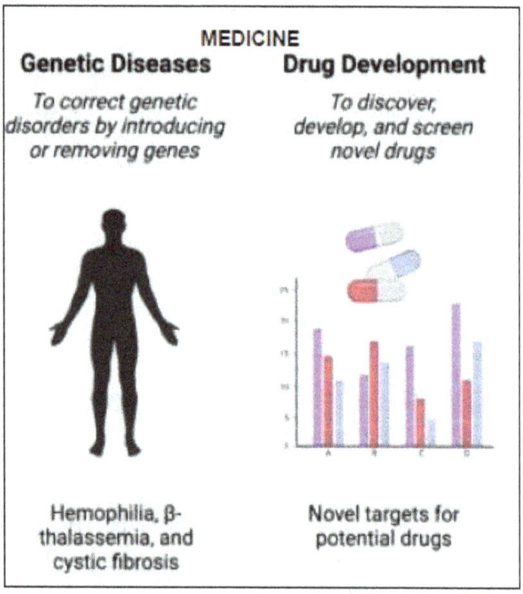

Fig. 6. Applications of CRISPR-Cas9 technology for genome editing in the field of medicine

CRISPR-Cas9 also holds promise for the development of gene therapies. It offers the potential to correct disease-causing mutations directly in the genomes of patient cells. In addition, CRISPR-Cas9 technology is possible to correct a mutation associated with a genetic blood disorder, b-thalassemia, in patient-derived stem cells.

Furthermore, CRISPR-Cas9 technology has opened up new avenues for cancer research and therapy. It allows for the targeted disruption of genes involved in cancer progression or the introduction of specific modifications to sensitize cancer cells to existing treatments. In a study, CRISPR-Cas9 was used to knock out a gene involved in chemotherapy resistance, enhancing the effectiveness of the treatment in cancer cells. The potential of CRISPR-Cas9 extends beyond gene editing. It has been harnessed for diagnostic purposes, such as the detection of specific DNA sequences associated with diseases. CRISPR-based diagnostic tools, such as specific high-sensitivity enzymatic reporter UnLOCKing (SHERLOCK) and DNA endonuclease-targeted CRISPR trans reporter (DETECTR) have been developed to enable rapid and accurate detection of viral infections and genetic mutations.

Moreover, the combination use of CRISPR-innovative technology and stem cell research could enable the creation of disease models for the investigation of therapy. The gene editing by using CRISPR-Cas9 was applied to clarify the mechanism autosomal dominant polycystic kidney disease (ADPKD) by deleting the PKD2 gene. Furthermore, CRISPR also helped to explain the hypoglycemia by insertion of MEN1 gene with artificial specific point mutation into iPSCs. The CRISPR-Cas9 is also subjected for drug screening by inducing deoxyguanosine kinase (DGUOK) knockout to the stem cells which developed as hepatocytes. It generated the mitochondrial dysfunction in hepatocytes as a board for compound screening.

While the potential of CRISPR Cas9 in medicine is immense, there are challenges that need to be addressed. These include off-target effects, delivery methods, and ethical considerations, particularly when it comes to germline editing. Ongoing research aims to improve the specificity and safety of CRISPR-Cas9, and discussions around ethical and regulatory frameworks continue to evolve. Nonetheless, CRISPR-Cas9 technology holds significant promise in medicine. Its applications range from understanding gene function and disease mechanisms to developing precise gene therapies and diagnostic tools. As research progresses and challenges are overcome, CRISPR-Cas9 is poised to revolutionize the field of medicine, offering new possibilities for treating genetic diseases and advancing personalized medicine.

CRISPR-Cas9 technology in clinical sciences

The application of CRISPR-Cas9 technology in clinical sciences holds great promise for advancing therapeutic interventions and precision medicine. This sub-section provides an overview of the use of CRISPR-Cas9 in clinical research and potential clinical applications.

Clinical research utilizing CRISPR-Cas9 has contributed to a better understanding of disease mechanisms and the development of potential treatments.

For instance, CRISPR-Cas9 was used to target the PCSK9 gene, known to be involved in cholesterol regulation. The study demonstrated the potential of CRISPR-Cas9 as a therapeutic approach for lowering cholesterol levels in individuals with familial hypercholesterolemia.

The potential of CRISPR-Cas9 for treating genetic diseases is a major focus of clinical applications. A study explored the use of CRISPR-Cas9 to correct disease-causing mutations in patient cells or tissues. In a landmark study, CRISPR-Cas9 was employed to edit the CCR5 gene in hematopoietic stem cells, aiming to confer resistance to HIV infection.

Another area of interest is the development of CRISPR-based cancer therapies. Researchers are investigating the potential of using CRISPR-Cas9 to target and eliminate cancer cells with precision. A study demonstrated the feasibility of using CRISPR-Cas9 to selectively kill cancer cells by targeting specific fusion genes involved in leukemia. Furthermore, CRISPR-Cas9 technology has been explored for the development of ex vivo gene therapies. In this approach, patient cells are isolated, edited using CRISPR-Cas9, and then reinfused back into the patient.

A notable example is a clinical trial conducted at the University of Pennsylvania, where CRISPR-Cas9 was used to modify T cells from patients with advanced cancer to enhance their ability to recognize and kill cancer cells. It is important to note that while CRISPR-Cas9 technology shows tremendous potential, there are still challenges and ethical considerations that need to be addressed before widespread clinical implementation. Issues such as off-target effects, delivery methods, long-term safety, and ethical guidelines surrounding germline editing require careful evaluation and regulation.

The application of CRISPR-Cas9 in clinical sciences has the potential to revolutionize therapeutic interventions and precision medicine. Ongoing research and clinical trials are exploring the feasibility and safety of CRISPR-based treatments for genetic diseases, cancer, and other conditions. With further advancements, CRISPR-Cas9 holds promise for addressing unmet medical needs and improving patient outcomes.

CRISPR-Cas9 technology in environmental sciences

The application of CRISPR-Cas9 technology in environmental sciences presents exciting possibilities for addressing environmental challenges, conservation efforts, and sustainable practices. This sub-section provides an overview of the use of CRISPR-Cas9 in environmental research and potential applications. One area of interest in environmental sciences is the use of CRISPR-Cas9 for conservation purposes.

Researchers are exploring the potential of CRISPR-Cas9 to mitigate threats to endangered species by editing their genomes to enhance their resilience or adaptability to changing environments. The successful use of CRISPR-Cas9 to introduce genetic resistance to white-nose syndrome in bats, a devastating fungal disease that has caused significant declines in bat populations.

CRISPR-Cas9 technology is also being investigated for controlling or eradicating invasive species. Invasive species pose significant threats to ecosystems and native biodiversity. Researchers are exploring the possibility of using CRISPR-Cas9 to genetically modify invasive species to suppress their populations or make them less harmful to native species. While still in the early stages of research, this approach has the potential to provide effective and environmentally friendly solutions for invasive species management. Additionally, CRISPR-Cas9 has the potential to enhance agricultural practices and promote sustainable farming methods. The use of CRISPR-Cas9 to develop crops with improved nutrient uptake, drought resistance, and reduced pesticide reliance. These genetic modifications can contribute to increased crop productivity, reduced environmental impact, and more sustainable food production.

Furthermore, CRISPR-Cas9 technology offers opportunities for environmental monitoring and assessment. A study stated that the detection and identification of specific organisms, pathogens, or contaminants in environmental samples. This can aid in the early detection of harmful agents, helping to prevent or mitigate environmental pollution or disease outbreaks. CRISPR-based diagnostic tools have shown promise in detecting waterborne pathogens and monitoring environmental health.

While the application of CRISPR-Cas9 in environmental sciences is still in its early stages, it holds tremendous potential for addressing environmental challenges and promoting sustainable practices. However, it is crucial to consider potential ecological and ethical implications, and careful risk assessments and regulatory frameworks should accompany the development and deployment of CRISPR-based environmental interventions. CRISPR-Cas9 technology presents exciting opportunities for environmental sciences. Its applications range from conservation efforts and invasive species management to sustainable agriculture and environmental monitoring. Continued research and responsible implementation of CRISPR-Cas9 in environmental contexts can contribute to the preservation and restoration of ecosystems and support sustainable practices.

CRISPR-Cas9 technology in biotechnology

The innovative CRISPR-Cas9 technology has revolutionized the field of biotechnology, offering powerful tools for genome editing, gene regulation, and synthetic biology.

This subsection provides an overview of the applications and advancements of CRISPR-Cas9 in biotechnology. Genome editing is one of the primary applications of CRISPR-Cas9 in biotechnology. It enables precise modifications in the DNA sequences of various organisms, including bacteria, yeast, plants, and animals. This capability has accelerated the development of genetically modified organisms (GMOs) with improved traits or novel functionalities. For instance, CRISPRCas9 was utilized to engineer a strain of yeast, demonstrating the potential of CRISPR-Cas9 in the production of valuable compounds.

CRISPR-Cas9 technology also plays a crucial role in gene regulation and functional genomics. It allows for the precise manipulation of gene expression, enabling researchers to investigate gene function and unravel the complexities of biological systems. By using CRISPR-Cas9 to target and control gene regulatory elements, such as promoters or enhancers, researchers can gain insights into the roles of specific genes and their interactions.

A study demonstrated the application of CRISPR-Cas9 in activating or repressing specific genes, providing a valuable tool for functional genomics studies. Synthetic biology, a field focused on engineering biological systems with novel functions, has been greatly empowered by CRISPR-Cas9 technology. CRISPR-based tools, such as CRISPR interference (CRISPRi) and CRISPR activation (CRISPRa), allow for precise control of gene expression and modulation of cellular pathways. These tools have been applied in the design of biosensors, metabolic engineering, and the development of new biotechnological applications.

Moreover, the use of CRISPR-Cas9 also could improve the quality of crops. The mutation induction of OsProDH gene in rice could increase the accumulation of proline. It played an essential role for protection from abiotic and biotic stress by reactive oxygen species (ROS). Therefore, the manipulation of OsProDH gene could escalate the thermotolerance of rice plant.

Furthermore, CRISPR-Cas9 technology has enabled the rapid and cost-effective generation of animal models for studying human diseases. It allows for the introduction of specific mutations associated with human genetic disorders, providing valuable tools for understanding disease mechanisms and developing therapeutic interventions. In a study, CRISPR-Cas9 was used to create a pig model of cardiovascular disease, demonstrating the potential of CRISPR-Cas9 in modeling complex human diseases in animals.

As with any powerful technology, responsible use and ethical considerations are important in the field of biotechnology. Ongoing research and discussions continue to address safety, regulatory, and ethical concerns surrounding the use of CRISPR-Cas9 in biotechnological applications. CRISPR-Cas9 technology has revolutionized biotechnology, enabling precise genome editing, gene regulation, and synthetic biology. Its applications range from the development of GMOs with improved traits to the generation of animal models for disease research. As research and advancements continue, CRISPR-Cas9 holds great promise for shaping the future of biotechnology and driving innovation in various sectors.

CRISPR-Cas9 technology in fisheries

The application of CRISPR-Cas9 technology in fisheries holds potential for improving aquaculture practices, enhancing fish health, and addressing challenges in fisheries management. This sub-section provides an overview of the use of CRISPR-Cas9 in fisheries research and potential applications. CRISPR-Cas9 has been explored as a tool for genetic improvement in aquaculture species. Liu et al. demonstrating the use of CRISPR-Cas9 to introduce beneficial traits, such as disease resistance, improved growth rates, and enhanced stress tolerance, into commercially important fish species. Researchers mentioned that the successful application of CRISPR-Cas9 in zebrafish to produce offspring with increased disease resistance.

The technology also offers the potential for controlling invasive species and preventing their negative impacts on native fish populations and ecosystems. Invasive species can disrupt natural habitats and compete with native species for resources. CRISPR-Cas9 can be employed to modify the genomes of invasive fish species, aiming to reduce their reproductive success or increase their susceptibility to environmental factors. Although still in its early stages, CRISPRbased approaches hold promise for managing and mitigating the impacts of invasive species.

Furthermore, CRISPR-Cas9 has the potential to contribute to fish health management. It can be used to develop disease-resistant strains of fish by targeting and modifying genes associated with susceptibility to viral, bacterial, or parasitic infections. By enhancing the innate immune responses of fish, CRISPR-Cas9 technology may offer new possibilities for controlling disease outbreaks in aquaculture settings. A study highlighted the potential of CRISPR-Cas9 for enhancing fish immunity against viral pathogens.

It is worth noting that the application of CRISPR-Cas9 technology in fisheries is still in the early stages, and further research and evaluations are needed to assess its safety, efficacy, and potential ecological impacts. Okoli et al. suggested the regulatory frameworks and guidelines must be established to ensure responsible and ethical use of CRISPR-Cas9 in fisheries. CRISPR-Cas9 technology offers exciting prospects for fisheries research and applications. Its potential includes genetic improvement of aquaculture species, management of invasive species, and enhancement of fish health and disease resistance. Continued research and responsible implementation of CRISPR-Cas9 in fisheries can contribute to sustainable aquaculture practices and the conservation of fish populations.

CRISPR-Cas9 technology in nanotechnology

CRISPR-Cas9 technology has the potential to synergize with nanotechnology in various ways. Nanotechnology offers precise control at the nanoscale, and when combined with the precision of CRISPR-Cas9, it opens up new possibilities for targeted delivery, imaging, and sensing applications. Here are a few potential areas where the integration of CRISPR-Cas9 and nanotechnology could be explored:

- Targeted delivery systems. Nanoparticles can be engineered to carry CRISPR-Cas9 components, such as guide RNA and Cas9 protein, to specific cells or tissues. This targeted delivery approach could enhance the efficiency and specificity of genome editing while minimizing off-target effects.
- Imaging and tracking. Nanoparticles with imaging agents can be used to monitor the delivery and activity of CRISPR-Cas9 components in real-time. This allows researchers to visualize and track the editing process, providing valuable insights into its efficacy and potential optimization.
- Gene regulation and epigenetic modifications.

Nanotechnology-based platforms can be employed to modulate gene expression and epigenetic modifications, either by directly targeting DNA or influencing regulatory factors. Integration of CRISPR-Cas9 with nanotechnology can enable precise control over gene expression and epigenetic regulation, offering new avenues for disease treatment and fundamental biological research.

While the specific references on CRISPR-Cas9 technology in nanotechnology are limited, some researchers published articles and reviews on CRISPR-based delivery systems, nanocarriers for gene editing, or nanotechnology-enabled gene regulation for further insights into the potential synergies between CRISPR-Cas9 technology and nanotechnology.

CRISPR-Cas9 technology in bioinformatics and machine learning

The integration of CRISPR-Cas9 technology with bioinformatics and machine learning has yielded significant advancements in genomic research and data analysis. CRISPR-Cas9, a powerful gene editing tool, combined with computational methods has enabled researchers to gain deeper insights into the structure, function, and regulation of genomes.

The application of machine learning algorithms to CRISPR-Cas9 data has further enhanced the analysis of largescale genomic datasets, enabling the discovery of novel patterns and relationships. One area where CRISPR-Cas9 technology has merged with bioinformatics and machine learning is in the prediction of off-target effects. By leveraging machine learning algorithms and genomic sequence data, researchers have developed computational models that can predict potential off-target sites of CRISPR-Cas9 gene editing. These models help guide experimental design and increase the specificity and accuracy of CRISPR-Cas9-mediated gene modifications. Furthermore, machine learning techniques have been employed to analyze CRISPR-Cas9 screening data. CRISPR screening involves perturbing genes and observing the resulting phenotypic changes, generating large-scale datasets. Machine learning algorithms, such as neural networks and support vector machines, have been used to extract meaningful information from these datasets, identify gene functions, and uncover genetic interactions. In addition, the integration of CRISPR-Cas9 technology with machine learning has facilitated the development of computational tools for predicting the functional consequences of genetic variants.

Machine learning algorithms can be trained on CRISPR-Cas9 experimental data to predict the impact of specific genetic variants on gene function or disease risk. These tools assist in the interpretation of genomic variations and aid in prioritizing potential disease-associated variants.

Moreover, machine learning has been utilized to improve the efficiency and accuracy of CRISPR-Cas9 experimental designs. By leveraging large genomic datasets and computational models, machine learning algorithms can guide the selection of optimal target sites for CRISPRCas9 gene editing, taking into account factors such as on-target activity and off-target effects. This approach streamlines the experimental process and increases the success rate of CRISPR-Cas9 experiments. The integration of CRISPR-Cas9 technology with bioinformatics and machine learning holds great promise for advancing our understanding of genomic biology and accelerating discoveries in personalized medicine. By combining experimental data, computational algorithms, and machine learning, researchers can unlock the potential of CRISPR-Cas9 for precision genomics and pave the way for targeted therapies.

INTERNATIONAL REGULATIONS AND GUIDELINES

In July 2018, the European Court of Justice ruled that plants derived by mutagenesis techniques are to be considered as GMO and that techniques developed and established later than 2001—like GEd—demand the application of all obligations for the approval of transgenic organisms. The increasing number of advanced research on Ged applications in many crops and deliberate regulations of Ged plant products worldwide, and with envisaged and first market releases, initiated a policy debate within the EU how to address law enforcement facing international trade as well as about the proportionality of regulatory demands.

The Council of the European Union asked the European Commission to conduct a study on the impact of the ruling. This study took into account the state-of-the-art knowledge and the views of the EU countries and stakeholders. It was published by the Commission end of April 2021. The findings of the study were the starting point for a further policy initiative by the Commission to explore the necessity and perspectives to update the GMO regulations with regards to GEd and cisgenesis in plants (only). A primary step was the so-called Inception Impact Assessment (IIA) outlining objectives and a roadmap toward the decision on a legislative action.

This IIA concluded that the given EU regulation on GMOs bears legal uncertainties, disproportionate or inadequate requirements, enforcement challenges and may hinder contribution of GEd to a more sustainable and resilient agricultural system (regarding Green Deal, Farm to Fork Strategy, Biodiversity strategy). The future legislation shall maintain a high level of protection, enable access to benefits, enhance the competitiveness of the EU and ensure the consistent functioning of the internal market. The key policy options for further reconsideration relate to a proportionate risk assessment and approval requirements of GEd and cisgenetic plants, sustainability analysis, appropriate traceability, and labeling provisions that are implementable and enforceable. In the course of the public consultation about the IIA in October and November 2021, the Commission clarified that it is not intending to deregulate GEd.

Until spring 2022, an in-depth impact assessment with a detailed suggestion for a legal initiative will follow with a public consultation prior to further legislative steps. Such impact assessment will explore economic, social, and environmental impacts of a revised or amendment regulation. Issues on administrative burdens and with regard to fundamental EU rights are to be considered.

There are two potentially interrelating policy actions that are linked with the legislation of GEd plants and worth following and tracking the further discussions. There is the "Sustainable food system framework initiative" that aims to make the EU food system sustainable and to integrate sustainability into all food-related policies based on a respective regulation. The topic relates to the discussion about the criteria for sustainability of GEd plants. The initiative for the "Revision of the plant and forest reproductive material legislation", is less observed in the broader public, but intensely discussed in the breeding sector. It addresses issues about the market introduction of plant reproductive material, IPR, and accessibility that are also key topics in the European debate about GEd plants.

In Hungary, which is one of the strongest opponents of genetically engineering in the EU, it came somehow as a surprise that the Ministry of Agriculture as well as financial, scientific, and agricultural organizations support "non-transgenic genome editing." Since the ECJ ruling, the ministry is somewhat in a passive mode and nothing has been released officially but statements, e.g. from the Hungarian minister of agriculture that the EU GMO legislation should be revised. If and when an official position will be released is to date still uncertain.

Other European and North American countries

In Switzerland, the association "Varieties for Tomorrow" which is composed of various actors from the Swiss agriculture and food industry called for an open and differentiated approach to NGT in plant breeding. Furthermore, the Science Commission of the States voted for the first time in favor of GEd. The Commission demands an exemption of GEd if no foreign DNA has been introduced into the plant. The current moratorium for commercial GM plant cultivation that is in place in Switzerland shall be prolonged for another 4 years but plants without foreign DNA shall be exempted.

The United Kingdom released new plans "to unlock the power of gene editing to help farmers to grow more resistant, more nutritious and more productive crops" on September 29, 2021. The plan is divided into several steps. In a first step, the government wants to use the existing Environmental Protection Act 1990 to lay a Statutory Instrument by the end of 2021. The aim is to make research and development easier for plants that have been produced by genetic technologies where the resulting genetic changes could have been developed using traditional breeding methods. This enables research performing field trials with GEd plants without obligation of an extended risk assessment. However, scientists will continue to be required to notify Department for Environment, Food and Rural Affairs (DEFRA) of any research trials. As a second step, DEFRA aims to review the regulatory definitions of genetically modified organisms, to exclude such ones produced by GEd and other techniques if the product could have been developed using traditional breeding methods. GMO regulations would continue to apply if gene editing introduces DNA from other species into an organism. Furthermore, appropriate measures enabling GEd products to be brought to market, including consumer choice and traceability, will be considered in the near future in England.

In January 2022, the parliament released a Post Note on Genome Editing clarifying that the government is proposing that genome-edited crop plants are exempted from GMOs regulations, provided the genetic changes could occur naturally or via existing conventional breeding techniques. This will be achieved in two steps: the first step, which enables field trials, is already being laid by a statutory instrument; the second step will change the definition of a GMO in England and exclude GEd from it. For the registration of plant varieties, a new category will be proposed.

In Canada, there is no separate GMO regulation, but there is a regulatory regime dealing with novel foods and with respect to risk assessment obligations with new substances notification. Following from a consultation from May 19 until September 16, 2021, Canada released a complementary draft guidance to Part V of the Seeds Regulations (Regulations Respecting the Quality of Seeds Including Seed Potatoes, and the Testing, Inspection, and Sale Thereof) for clarifying the regulatory status of seeds from plants derived with modern breeding techniques. Seed that is not substantially equivalent to the seed of that species that is already present in Canada, in terms of its specific use and safety for the environment and human health, is subject to Part V, and must be authorized before releasing into the environment. The guidance applies to all plants intended for release into the Canadian environment, including agricultural crops independent of the various technologies that may be used in the development of a plant. The draft guidance states:

Virtually all plants developed by conventional breeding techniques qualify for an exemption from Part V, on the basis of being substantially equivalent to the lines they are derived from. Similarly, genetic changes that do not include foreign DNA will, for the most part, resemble conventional breeding outcomes, and will also qualify for an exemption. The CFIA recognizes that gene editing techniques can introduce genetic changes that are comparable to conventional breeding outcomes, and will also qualify for an exemption. Plants derived from populations that have been previously grown in Canada qualify for an exemption, provided that they do not present new risks to the environment. Plants previously grown in Canada include those that were present prior to 1996 when Part V came into force, as well as those that were authorized after 1996.

The presence of foreign DNA in the final product triggers the novelty aspect of the Canadian regulatory regime and so for the Seed Regulation. The DNA that encodes the gene editing components (e.g. CRISPR Cas protein(s) and associated guide RNAs) are considered to be foreign DNA. If these sequences are removed from the final products by using plant breeding and selection, it is exempted, if no other point of the regulation is met. If the sequences stay within the plant material or if the plant species is new to Canada, it is still subject to Part V. If a plant is exempt from Part V, there is no requirement to submit information to the CFIA. The plant can be released in Canada, subject to any other applicable requirements; if needed the CFIA will issue an exemption opinion letter.

Policy activities outside the EU

Africa

Nigeria has authorized guidelines on gene editing through the National Biosafety Management Agency (NBMA) in December 2020. The Government of Nigeria sees science and technology (including "molecular breeding techniques") as a major driver of agricultural productivity and as part of the solution to provide food security and fight malnutrition in Nigeria and Africa as a whole. NBMA aimed to introduce adequate regulation that will ensure that products of GEd do not cause harm or prevent any adverse effect to the environment, humans, and plants. The released guidelines are directed to all "person(s), institutions or bodies wishing to carry out gene editing as it relates to plants, animals, and microorganisms ranging from containment, confined field trial, multi-locational trial, commercial/general release and imports intended for direct use as food or feed, or for processing".

Nigeria has adopted a case-by-case approach to regulate GEd and products thereof. Products of GEd that result in a "novel" combination of genetic material or contain recombinant DNA such as nuclease gene, present in the final product (InDels are not seen as novel combination) will be classified as GMO and regulated as such. However, products, which do not contain "recombinant DNA," or the recombinant DNA that has been removed in the final product is not seen as GMO and a non-GMO regulatory classification is applied. Nonetheless, every applicant wishing to carry out gene editing in Nigeria has to approach NBMA, which will come back to the applicant with feedback within 21 d.

South Africa which was long time undecided on how to deal with GEd has announced its regulatory approaches for "New breeding technologies" through a public notice after a discussion on the GMO status of "NBT products" on October 27, 2021. The executive council of South Africa has concluded that for "NBTs" the same risk assessment framework should apply as for GMOs based on the definition of a GMO in the South African GMO Act. Based on this decision, the application forms were updated and products of "NBTs" are seen as a GMO regardless of the type.

In contrast to this, *Kenya* has recently released its guidelines as well. The Kenyan guidelines rely on a case-by-case evaluation based on the presence of transgenes and are comparable to the ones which are in place in Nigeria including deregulation of cisgenesis and in cases where foreign DNA is absent ("all deletions/knock outs"). Furthermore, processed products whose inserted foreign genetic material cannot be detected (e.g. processed Oils) will also be excluded, the same holds true for "conventional" breeding methods, e.g. mutagenesis, polyploidy, and haploidy. The applicant in Kenya will receive feedback within 30 d. Kenya has already approved five GEd events including three which confer resistance toward plant–pests.

Some other countries in Africa are still in the debate on GEd and haven't come up with regulations so far. Most advanced drafts were prepared by Eswatini and Burkina Faso. Burkina Faso is running experiments with genome-edited rice (*Oryza sativa*) which is resistant to bacterial blight. Other countries in Africa like Mali, Senegal, Gambia, Niger, Mauritania, and Guinea-Bissau do have a regulatory system for GMOs in place or are in the process of releasing, when and whether this will specifically be adapted to GEd is uncertain, but discussions in Africa are being held intensively. However, the trend for current and future regulations in Africa tends toward open regulations just as in South American countries.

Asia

South Korea has informed the WTO in a short statement on the introduction of a preliminary review system for low-risk LMOs, covering "LMOs created through modern biotechnology" under Article 2 (Definition) of the LMO Act. In the short statement, South Korea notifies that "low-risk LMOs in which foreign gene has not been introduced to the final product shall be exempted from risk review and approval on the import, production and contained use." Further details will follow adoption of the LMO policy. A formal statement is expected in late 2022, clarifying which kinds of changes will be excluded from the regulation.

In *Japan*, amendments have been made to the current guidance on GEd in December 2020. The Ministry of Health, Labour and Welfare (MHLW) amended the handling procedures for food and food additives that are derived from cross-breeding varieties which have already been notified to MHLW, and these products are no longer subject to the MHLW consultation process. In April 2021, the Ministry of Agriculture, Forestry and Fisheries (MAFF) approved a proposal to amend the Feed Safety guidelines for handling of "genome edited feed and feed additives developed by crossbreeding already notified genome edited organisms with other allowed varieties." In the past, products derived from crossbreeding of genome-edited varieties had to undergo a consultation with MAFF. These guidelines have been adapted on April 20, 2021 notifying that no prior consultation is required anymore of crosses of "conventional varieties etc.*... with varieties that have been notified as genome edited feed." The etc.* category includes, but is not limited to products notified to MAFF as genome-edited, genetically engineered, etc. The same holds true for MHLW.

Indonesia drafted a regulation on CRISPR-based GEd and other GEd techniques in early 2021. According to the draft, GEd products will be regulated as GMOs when they contain a novel combination of DNA ("from outside the taxon") or if foreign DNA is present in the final product. If this is not the case the products will be deregulated, and further details should be published later in 2022.

The *Philippines* released a resolution in late 2020 declaring that products of "innovative breeding techniques" such as GEd are going to be regulated under GMO law in the case of novel combination of genetic material, which is not possible to achieve through conventional breeding. In the other cases, products should be regulated as non-GMO or conventional. The resolution is going to be set in a new regulation that will clarify further details. The regulation is expected in the mid-2022.

India has recently defined regulations for products of innovative technologies such as GEd. A draft document on a proposed regulation has been published and commented by the public by July 2020. The draft stated that products of SDN1 and SDN2 GEd approaches that do not carry or involve exogenous DNA and are comparable to naturally occurring events should be exempted from the Indian GMO legislation. However, an appropriate tiered-based risk assessment is foreseen categorizing GEd into three categories. The first category in which single or few base pair edits/deletions should be addressed, an assessment is made to confirm targeted edits as well as the absence of any biological relevant off-target genomic changes. Phenotypic equivalence will be checked if necessary on a case-by-case basis. The second category that covers targeted few/several base pair edits an assessment will be supplemented with phenotypic equivalence and trait efficacy through appropriate contained and/or confined field trials. The last category addresses products harboring targeted edit(s) synthetic/foreign DNA. The assessment is the same as for traditional GMOs. On March 30, the Indian Ministry of Environment, Forest and Climate Change accepted in many parts of the proposed regulation and exempted genome-edited plants of the category SDN1 and 2 from the Indian GMO regulation.

China has released guidelines for the safety evaluation of gene-edited plants for agricultural use. These guidelines cover all products of GEd that have not introduced exogeneous genes (all except SDN3 approaches). For products that have introduced exogenous genes the classical GMO guidelines are applicable. The guidelines for GEd provide four requirement categories, based on the risk profile of the target trait. The first category does not increase the risk of environmental and food safety, the second may increase environmental risks, the third may increase food safety risks and the fourth may increase both risks. For each category, different requirements need to be provided for production and or import. There are some general requirements that need to be provided in every category: (1) molecular characterization of the plant including targeted gene-related data, gene editing method applied, data on the targeted gene edit (on target data), potential vector sequence residue, and off-target analysis, as well as (2) genetic stability of the edit and the trait in at least three generations. Additionally, the need for environmental and/or food safety assessment needs to be provided if necessary. These requirements are in line with the ones requested in the guidelines for the safety evaluation of GMOs. However, the guidelines do not indicate how to classify a product into one of the four categories nor to which extent data have to be provided. This will presumably be claimed on a case-by-case basis, but no official statement on this has been released. Clarification will be granted after the first decision has been made in China.

Other countries in South-East Asia are still in debate including *Bangladesh, Vietnam, and Taiwan*. Taiwan's food and drug administration is working with research institutes in Taiwan to draft a regulatory guidance for innovative biotechnologies including gene editing. However, a public draft is still forthcoming, but considerations concerning risk assessment and management are being prepared.

Vietnam is amongst other countries (e.g. United States, Canada, Uruguay, etc.) a supporter of the International Statement on Agricultural Application of Precision Biotechnology submitted to the WTO, but when and how a draft will be published is still open.

Perception of GE in society

Aside from a shift and the clarification of the legal status of GEd products in many jurisdictions, as described above many countries have clarified their legal status of GEd products in the last 2 years. Besides this, some other countries, especially in jurisdictions that haven't decided yet, tried to image the public perception of NGTs in their respective countries through official surveys. The Norwegian collaboration GENEinnovate performed a sophisticated consumer survey to elaborate a more precise understanding of how consumers are informed and how is their opinion, with respect to GE. GENEinnovate is a collaboration of breeders, companies the Norwegian University of Life Sciences, and the Norwegian Biotechnology Advisory Board and therefore represents a broad collection of stakeholders. The results of the survey showed that there is more than just like or dislike of GE in the Norwegian population. Regarding the knowledge of consumers, the vast majority (96%) have heard about genetically modified foods but less than 50% have ever heard about GE. Regardless of this, the general attitude of more than 50% of the consumers (up to 70%) was positive with respect to using GEd for breeding purpose that result in traits with a clear sustainability or societal benefit (i.e. reducing pesticide use or climate adaption of plants). In contrast to this, if GEd is used for traits that are of no societal benefit (e.g. changing the appearance of a product), the consumers are more negative. In summary, the survey showed that the consumer attitude is nuanced and that the important fact is the new trait and not the technique used to create it. Furthermore, there is a need for knowledge building about new genetic techniques in the public.

A very similar survey was conducted in 2020 by The Swedish Technology advisory board. This report used the questions from the Norwegian survey and adapted it to the Swedish population. The report is not published so far but it showed that, as in Norway, almost half of the Swedish consumers never heard about gene scissors or GE. Again as in Norway, the main result of this survey was that there is a positive attitude of more than 60% toward GE plants if the trait would contribute to environmental or societal benefits. If this is not the case the attitude is negative.

Another survey was conducted in 2021 by The Food Standards Agency as a consumer poll. The key findings of this poll were that consumers in the United Kingdom find GEd more acceptable than GM and in general GM or GE applied to plants more acceptable than either of them applied to animals. As in the Norwegian survey, more informed consumers were more accepting of GE food than less informed ones. Interestingly, most consumers felt it appropriate to regulate GE and GM differently. Another claim of the poll was that United Kingdom consumers demand a transparent labeling if GE food reaches the market.

In parallel to this consumer poll, the DEFRA held a public consultation (January 7 to March 17, 2021) to collect views regarding the regulation of genetic technologies in England. As outcome of this consultation, DEFRA as first step wants to alleviate the burdens on developers doing research and development using genetic technologies. Field trials will be possible without risk assessment but still under notification duty. As a second step, DEFRA wants to elaborate an amendment regarding the regulatory definitions of a GMO in a way that organism "that could have been achieved through traditional breeding or occur naturally" but have been produced by genetic technologies are exempted from GMO regulation.

Eventually, The Australian National University performed a systematic literature review. This review highlighted that the existing literature is limited, riven by gaps, and often methodologically framed or biased in other aspects. A reasonable number of the reviewed literature focused solely on the consumer's "Willingness to pay" and neglected other aspects. Nevertheless, the key findings in the systematic literature review showed that there is a slightly more positive consumer attitude regarding food produced from GE techniques than toward older forms of genetic modification.

CHAPTER-9

FUTURE TRENDS IN BIOTECHNOLOGY AND CRISPR

Biotechnology holds significant promises for the future, but it must be carefully regulated to ensure that the benefits are maximized while risks are minimized. This article will provide an explorative journey through biotechnology by shedding light to leading biotech trends focusing on various biotechnological techniques and applications which are potentially beneficial to the future of research and humanity.

Fig. 1. Top 10 trends in biotechnology

Given that the benefits of this technology usually become apparent in the future, it is crucial to understand which trends have produced the most favorable results in order to emphasize the development and advancement of the technology. Below are some of the top 5 core trends that will propel breakthroughs and innovation in biotechnology:

The Top 5 Biotech Trends

Gene Editing

This technique in biotechnology inserts, removes, changes, or replaces specific pieces of a person's existing DNA. To treat diseases, scientists are exploring ways to edit pieces of DNA at precise spots along the gene. The goal of gene editing is to change the existing gene and correct mutations where they occur. CRISPR is the most common type of editing.

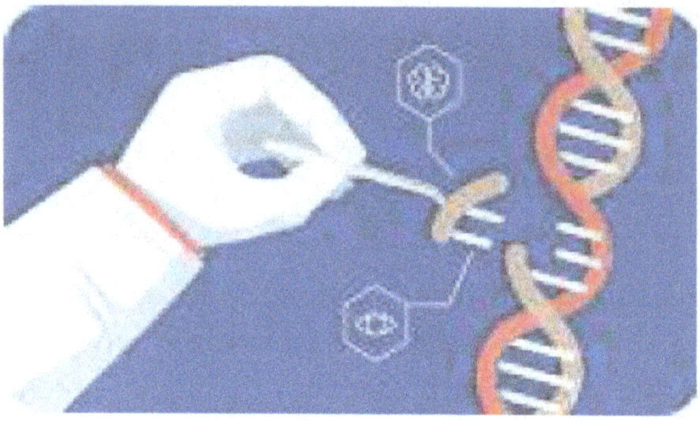

Fig. 2. Gene Editing

With this innovation, scientists can now modify genes or replace faulty genes with healthy ones to potentially treat, cure, or prevent a disease or medical condition.

Precision Medicine

The approach to precision medicine allows researchers and doctors alike to predict more accurate treatment and preventive strategies for a particular disease, usually in groups of people.

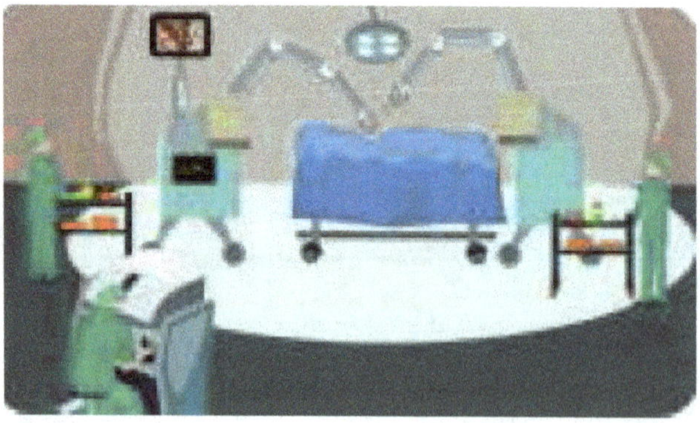

Fig. 3. Precision Medicine

The goal of this trend is to reduce the risk of human complications in the healthcare industry. Professionals in the field of science are holding on to the promise of precision medicine as it expands in the future.

Biomanufacturing

This trend relies on naturally occurring processes and reactions to produce outputs such as chemicals, material, etc. that are normally produced through a synthetic process. The most common biological process of biomanufacturing is fermentation.

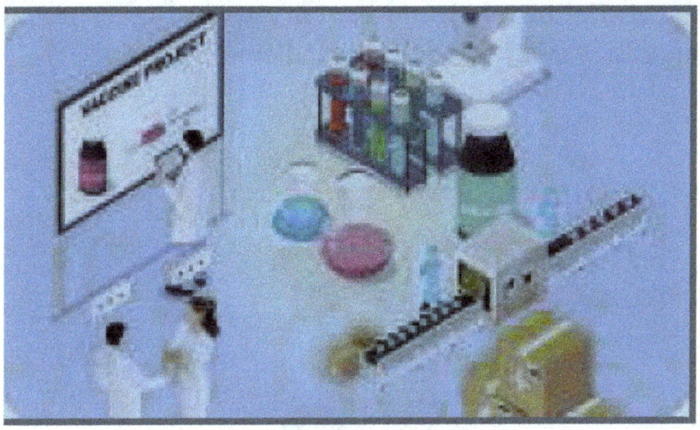

Fig. 4 Biomanufacturing

The goal of this application in biomanufacturing is to search for ways to boost sustainability, reduce energy consumption, and increase innovation and productivity. This trend is truly poised to continue advancing.

Synthetic Biology

The principle of this trend is to engineer new biological systems or re-design existing ones for useful purposes. It involves manipulation of biological compounds which are then integrated into cells which are chosen to provide suitable experimental strategy.

Fig. 5 Synthetic Biology

Although there are hurdles to overcome in this trend, its potential still outweighs the challenges. It is capable of delivering new solutions to agriculture, global healthcare, manufacturing, and many more.

Bioprinting

The relatively new technology is an additive manufacturing process like 3D printing. The only difference is that bioprinters print with cells or biomaterials to create organ-like structures that is beneficial to the healthcare industry. The grandiose potential of this trend benefits several industries including drug discovery, regenerative and personalized medicine.

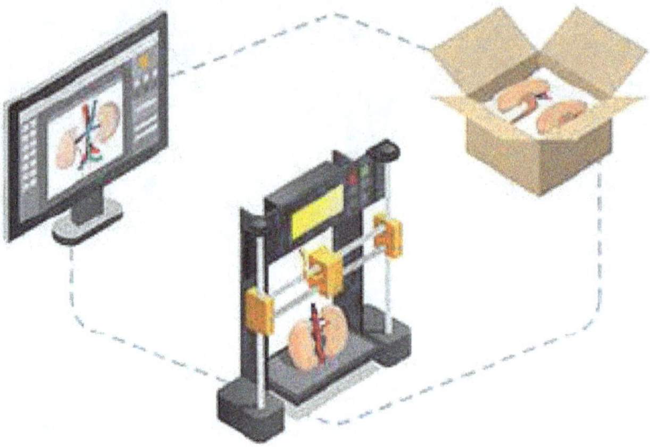

Fig. 6. Bioprinting

As more researchers gauge the latest bioprinting technology, the developments will surely break-through in the coming years.

Throughout the years following the discovery of the structure of DNA, scientists have endeavored to genetically manipulate organisms. Until recently, most of the genetic engineering tools developed were based on DNA:protein recognition principles, such as restriction enzymes, site-directed zinc finger nucleases (ZFs), and TAL effector nucleases (TALENs).

However, these tools are commonly experienced with difficulties in design, synthesis, and efficiency which altogether prevented a global widespread adoption, e.g. TALENs require 30–35 amino acids repeats, each only recognizing a single nucleotide (nt). On the other hand, the RNA-programmable CRISPR/Cas9 technology has led to a scientific revolution by solving all of the above-mentioned issues. The technology relies on two elements: a protein, the CRISPR associated protein (Cas9), and a RNA molecule, the guide RNA (gRNA). Cas9, the first Cas protein used in genome editing, is a large multi-domain enzyme interacting with the gRNA, the target DNA, and the Protospacer Adjacent Motif sequence (PAM) (Fig. 7a). The gRNA element is composed of two distinct elements: the spacer, a 20 nt domain that binds to the DNA; and the scaffold, a ~79 nt domain that interacts with Cas9 (Fig. 7a). Once guided to the target, Cas9 catalytically cleaves the DNA sequence 3 nt upstream the 52 -NGG PAM, resulting in the activation of endogenous repair mechanisms, such as homologous recombination (HR) or non-homologous end joining (NHEJ) (Fig. 7b).

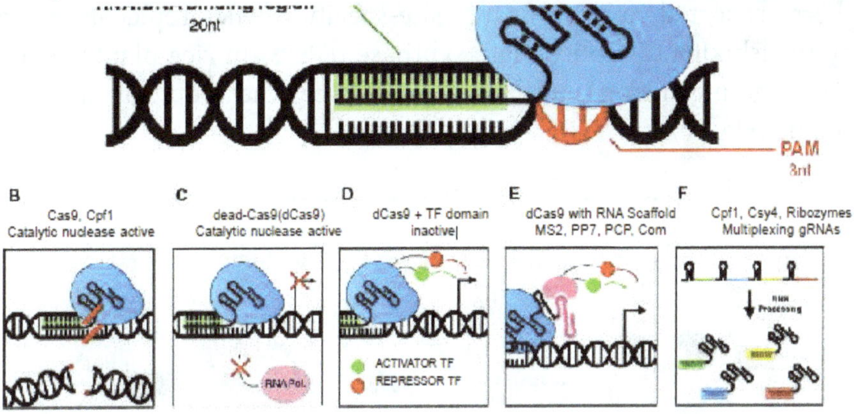

Fig. 7. CRISPR/Cas9 for genome editing and gene regulation. a The gRNA:Cas9 complex binding to the DNA target. In green, the spacer region which interacts with the DNA target. b DNA cut generated from the Cas9 nuclease activity. c Gene regulation with dCas9 physically blocking the RNA polymerase from binding to the promoter region. d CRISPR interference further enhanced with dCas9 fused with transcriptional regulators. e gRNA scaffold extended with stemloops recruiting regulator elements. f. Multiplexing gRNAs from a single transcript through endoribonuclease or self-processing elements

Industrial applications through metabolic engineering

Genome engineering

There has been an increasing interest in improving microbial cell factories through metabolic engineering approaches using CRISPR/Cas9 technology. The efficiency and versatility offered by CRISPR tools have shown great potential in rewiring the metabolic network of host cells to enhance their production of metabolites used in various areas of industrial biotech ranging from applications as biofuels to chemical building blocks and pharmaceuticals.

Metabolic pathway optimization towards the product of interest commonly requires deletions of multiple genes, e.g. competitive metabolic pathways, which is traditionally performed through iterative cycles of genetic marker integration and removal.

Conversely, the CRISPR technology does not necessitate integrative markers, and several efficient marker-free approaches were developed to perform multiplexed genome editing, e.g. knockouts, point-mutations and gene integration, which extensively reduced the time and effort required to perform targeted strain engineering.

The CRISPR technology has also improved genetic engineering in difficult-to-engineer industrial organisms, such as food crops. Among several examples, researchers reported a significant site-specific gene replacement of the 5-enolpyruvylshikimate-3-phosphate synthase (*EPSPS*) in rice plants using a pair of gRNAs targeting introns, ultimately converting the crop into a glyphosate-resistant one (Fig. 8c).

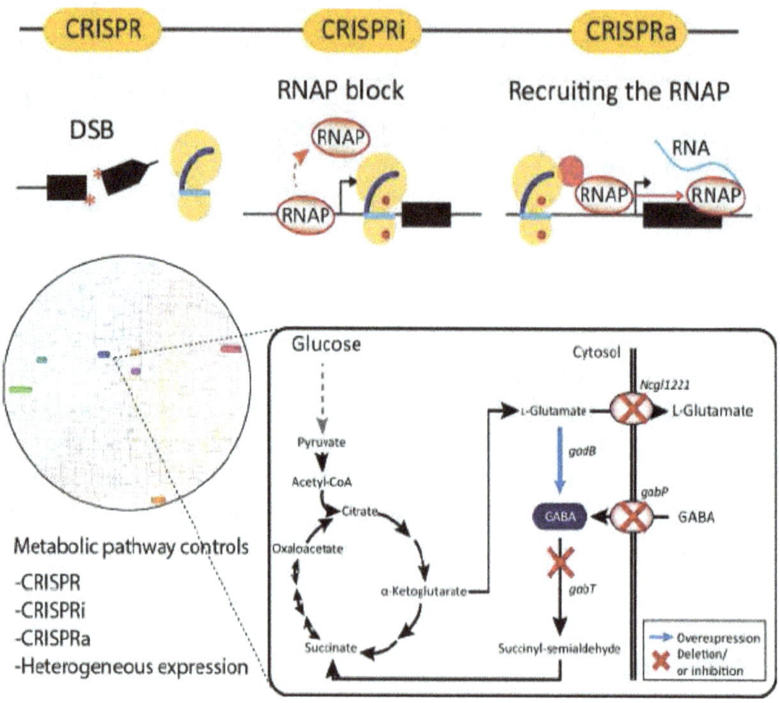

Fig. 8 CRISPR applications in metabolic engineering. a CRISPR enabled trackable genome engineering (CREATE) strategy for optimal expression of 4 genes involved in isopropanol biosynthetic pathway. Integration of DNA libraries composed of variant RBS sequences. b Multiple integration of the xylose and BDO pathway into retroposons sites in S. cerevisiae. Retrotransposons are composed of similar DNA sequences, which ultimately allows to generate a promiscuous gRNA able to target several of these targets at once. c Genetic engineering in rice using dual gRNAs targeting ESPS introns for double amino acid substitution [T102I + P106S (TIPS)]. d Systematic testing of enzyme perturbation sensitivities (STEPS) approach to iteratively find bottleneck

More recently, several studies have highlighted significant improvements in genome editing in plants using DNA-free CRISPR/Cas9 ribonucleoproteins.

Genome editing technology

"Genome editing" is a recent technology that can be considered as an ultramodern microsurgery on genes. It is an innovative technology that is booming in molecular biology laboratories; experts considered it the "breakthrough of the year" and, given its precision, low costs and ease of use, it has become the model of the "democratisation of science".

Genome editing is a technology that generates DNA sequence variants at defined positions in a genome. This can happen either in the coding region for a protein, thus affecting its function, or in the promoter region to impact cell type specificity or timing of promoter activity.

The best-known system in the genome editing toolbox is CRISPR/Cas9, for which Emmanuelle Charpentier and Jennifer Doudna, the inventors of the genetic scissors, received the Nobel Prize in 2020. Alternative systems are transcription activator-like effector nucleases (TALENs) or Zinc-finger nucleases (ZFNs).

All these editing tools target a specific sequence in the genome and induce a DNA double strand break at the target site. Once the DNA is cut, the cell uses its own DNA repair machinery, consisting in two major mechanisms that occur in almost all cell types and organisms: homology-directed repair (HDR) and nonhomologous end-joining (NHEJ), resulting in targeted integration or gene disruption, respectively.

The ease of use, accuracy and efficiency of genome-editing tools has led to their broad adoption in research, as well as to proof of concept applications in gene therapies involving non-reproductive (somatic) cells. It is also possible to deploy genome editing in human germline cells (sperm and eggs) as well as in early embryos. Many stakeholder groups are debating this issue, which, for some, remains a line not to be crossed whereas for others it provides possibilities for improving the human condition through the repair of deleterious genetic mutations.

However, the promise that genome-editing offers in diverse areas of human health is immense – and will make SDG3, Health and Well Being, attainable for many diseases which hitherto could not have been cured or prevented. The use of this frontier technology in ICGEB and the capacity building towards its constituency, particularly developing countries is also perfectly aligned with SDG10 by reducing inequalities across the globe.

In plants, the situation with regard to genome editing appears similar to the issue of genetically modified organisms where certain techniques applied to living, replicating organisms may result in genetically modified products that are risk assessed on a case-bycase basis before they are released into the environment or placed on the market.

However, there has been no universal consensus yet as to whether products resulting from genome editing should be treated in the same manner as GMOs or whether another regulatory framework is to be applied.

Over the past years, different countries have taken different paths along these two possible regulations. Therefore, before the edited lines can be introduced into breeding programs and used as a product, the country has to develop a legal regulatory framework. However, there is huge potential for the technology to make massive impacts across the world, both in addressing SDG2- Zero Hunger where such technology will rapidly overtake traditional means of crop improvement. In addition, many countries face incredible challenges as a result of climate change. This technology offers unique opportunities for mitigating the effects of climate change in various aspects of the agricultural sector, thereby addressing SDGs13 and 15.

Uses of genome edition in human health

Potential benefits of genome editing for human health include faster and more accurate diagnosis, more targeted treatments and prevention of genetic disorders. So far, ex vivo genome editing has been the most widely used approach. This entails genetic engineering of cells in vitro and their re-engraftment back to patients. In recent years, teams in China and in the United States have performed a series of clinical trials of gene editing, aimed at producing more effective CAR T cells for cancer treatment and the knockout of the erythroid specific enhancer of BCL11A to upregulate gamma globulin in autologous hematopoietic erythroid progenitors as a therapy for sickle cell disease and â-thalassemia. In addition to the many benefits of genome editing, there arc some technical challenges in translating these treatments to clinical disease therapy, primarily in terms of accuracy, efficacy and delivery hurdles.

Being a novel and sophisticated technology, not ready to be deployed in all countries, human genome editing has the intrinsic risk of fueling more health inequity between countries. It will therefore be essential to promote systems-level improvements to build capacity in all countries and ensure that human genome editing is used safely, effectively, and ethically. Indeed, this can only be achieved through appropriate enhancement of capacity in such technologies, meeting SDG4 for quality scientific education and SDG9 to build industrial capability to make use of these technologies and foster appropriate innovation in the countries most in need.

Uses of genome editing in plants

A new agricultural transformation is underway which aims at improving crop yields while using reduced levels of agrochemicals; developing new abiotic and biotic resistant varieties and finding alternatives for plant disease control and plant growth promotion are very important targets 8. With the global population projected to arrive at 10 billion by 2050, todays' agriculture will face enormous challenges, requiring crops with higher yields and of improved nutritional value, and needing fewer inputs being more sustainable. In addition, climate change is affecting crop yield due to the increase in temperatures as well as creating water shortages. At present, conventional breeding is the most widely used approach in crop improvement, however it is labor intensive and it takes a long time to go from the screens of the first crosses to arriving to commercial varieties.

Genetically modified (GMO) crops have been and are being created with beneficial traits via the transfer of genes (transgenes) or gene elements of known function into elite crop varieties. Despite the success in certain countries and usefulness that GMO crops have for global food security, their use is affected by largely unproven health and environmental safety concerns. Consequently, the tangible advantages of GMO traits have been few and these are restricted to a small number of crops. Genome editing involves advanced molecular biology techniques that allows for precise, efficient, inexpensive and targeted modifications at the specific position in the plant genome. This technique can be used for gene knockout mutants as well as gene replacement (knockin) and insertion mutants which can be generated in a wide variety of plants, and many of these genome modifications can be useful for crop improvement. The potential and negligent risks involved in modifying genomes through genome editing technology are significantly lower than those of GMO crops because most changes consist of only a few nucleotides, producing changes similar to ones that routinely happen in naturally occurring populations. Therefore, the use of genome editing into modern breeding programs can revolutionize plant breeding via rapid, cheap and precise crop improvement.

CRISPR-BASED THERAPEUTICS ON THE HORIZON

Traditional gene therapies have been in development for decades, and typically involve the delivery of a therapeutic gene to a patient via a viral vector. In contrast, gene editing therapies are designed to tweak the DNA that is already in a patient. This might involve deleting and adding bits of DNA, or a flexible mixture of both.

Earlier generations of CRISPR-based therapies are limited to tweaking cells removed from the patient and reinfused, referred to as an "ex vivo" model. Newer so-called "*in vivo*" candidates are in development to be delivered directly into the patient like a gene therapy, and could permanently correct mutations in a patient's DNA.

One limitation of traditional gene therapy is that it is most effective in tissue where cells aren't dividing and reproducing themselves, like nerve cells in the brain. Meanwhile, CRISPR-based edits to the genome can be transmitted to new generations of cells, so gene editing therapies can be used in tissues with a higher turnover rate, such as the liver in young children.

"You are almost dependent on being able to go in and correct the cell at the DNA level in order to get that long term durability," said Albert Seymour, president and CEO of Homology Medicines.

Safety and regulations for gene editing

As the gene editing space is still very new, there are many unknowns regarding the safety of the technology.

In the last year, LogicBio faced its own safety issues in a phase 1/2 trial in children with the life-threatening disorder methylmalonic acidemia. The firm recorded two cases of serious side effects known as thrombotic microangiopathy, leading the U.S. Food and Drug Administration (FDA) to put a clinical hold on the trial.

Unfortunately, we faced some safety issues in one of our clinical trials. He added that it is reassuring when companies communicate with each other at events and share how they can overcome these issues.

Another key point was that the field is so early-stage that gene editing companies must stay in close contact with the FDA to navigate the shifting regulatory scene and learn from each other.

The onus will also be on the regulatory agencies to determine what the safety threshold will be for gene editing therapies. For example, demanding data on very rare adverse events like off-target edits might sometimes be counter-productive.

Investor and payer attitudes to gene-edited therapies

Gene editing technology is seen as a new, shiny object in the greater landscape of gene and cell therapy investments. This has led to some impressive funding rounds backing gene editing players, with one example being Arbor Biotechnologies' $215 million Series B round in November 2021.

Additionally, in spite of the gloomy outlook in public biotech markets, gene editing players have been spared the worst of the uncertainty.

The valuations have held up remarkably well considering what's happened in the market. Investors get it, but I think it's incumbent on all of us in this space to continue to execute and hopefully generate that positive clinical data so that momentum continues.

However, there are headwinds for investors in the gene editing space to come. The emerging field needs to answer many questions regarding safety, monitoring the technology and manufacturing.

In addition, the way that gene editing therapies will be paid for is currently unclear. Companies developing traditional gene therapies such as Orchard

Therapeutics and bluebird bio are currently blazing a trail. Their ongoing negotiations with payers in multiple nations demonstrate how difficult it is to set a fair price for their potentially life-saving products.

From rare diseases to bigger indications

One reason that gene editing companies treat rare diseases is that their target indications in this camp are often caused by a single mutation that can be corrected easily. However, once the technology advances, the companies can begin to expand to treat diseases with more patients, but that have more complicated genetic underpinnings.

There are companies within the gene editing space that are swinging for the fences and they're going after common cardiovascular disorders, for example.

However, more common diseases such as cancer will take a lot more study, as the genetic factors driving the disease are much less clear.

Additionally, the question of reimbursement will need to be addressed when gene editing is applied to common indications affecting a million patients or more. As the first antibody treatments get approved for slowing down Alzheimer's disease, this will be watched closely by the gene editing community as a case study on how to decide the pricing strategy in this very common indication.

In vivo: how to deliver the therapy into the body

As more and more gene editing companies develop *in vivo* approaches, the question of how to deliver the gene editing machinery to where it needs to go in the patient's body opens up. Additionally, designing the gene editing machinery to work in human cells makes it tough to test them in animal models first.

Many companies are using viral vectors to get the gene editing tool into the patient's body. However, great control is required to make sure that the editing only happens in the tissue of interest. For example, Editas is using cell type-specific "promoter" regions in its therapy to make sure the genetic material is only expressed in the correct cells.

Other challenges for delivering *in vivo* gene editing therapies will be making sure that the immune system doesn't attack the therapy and blunt its effects. Another will be crossing the blood-brain barrier to treat conditions in the central nervous system.

According to researcher, there has been massive progress in the development of gene therapy vectors, such as viral vectors and the emergence of RNA therapies that can be delivered in lipid nanoparticles. However, it will still take time to find ways for these therapies to cross from the blood into the brain.

Gene editing will find its niche

There is a tradeoff developing for gene editing players as traditional gene therapies gain momentum. If the early generations of gene therapy could theoretically cure rare diseases, would that leave less room for the gene editing players to shine?

The trick will be for gene editing companies to carefully select indications that are less covered by the competition, or to find their edge over traditional gene therapy. For example, traditional gene therapy might suit adult patients with hemophilia, whereas gene editing might be better for younger patients with the same disease.

Development of Genome Editing Tools from the Prokaryotic Immune System

Experimental results evidenced that a chimeric single-guide RNA (sgRNA) (containing the features of crRNA and tracrRNA) can program the Cas9 to a specific locus. A small (~20 nucleotide) portion of the sgRNA directs or guides the binding of sgRNA-Cas9 complex to a complementary DNA sequence and subsequent DNA break by Cas9. So, customizing the 20 nucleotides from the 50 end of the sgRNA would enable the sgRNA to bind with the sequence of interest. It allows researchers to retarget Cas9 to virtually any genomic sequence that precedes a PAM sequence (50 -NGG-30). Cas9 makes a double-strand break (DSB) in the target DNA mostly 3 bp upstream of the PAM sequence.

It is clear that we can make a targeted and precise DSB in the DNA using the CRISPR-Cas9 system. But, how does it help in genome editing? The answer lies in the cellular repair mechanism. Since DSBs are the most toxic class of DNA damage, cells immediately deploy its various DSB repair pathways. In higher eukaryotes, nonhomologous end-joining (NHEJ) is a predominant repair pathway, which is error-prone. Due to the error during NHEJmediated repair, random small insertions or deletions (indels) or nucleotide substitutions are incorporated at the cut point while rejoining the two ends. Those incorporated errors frequently result in the disruption of the function of the target gene. By simply designing a 20-nucleotide guide RNA, researcher can develop a mutant organism with a targeted gene knocked out with high efficiency.

Orthologous and Engineered Cas Proteins for Diverse Genome Editing Platform

Presence of "NGG" protospacer adjacent motif (PAM) in the target site is a prerequisite for SpCas9 binding and cleavage. Therefore, a genomic site lacking an NGG sequence cannot be targeted by SpCas9. For those sites to be targeted, orthologues of SpCas9 which has a distinct PAM requirement could be utilized. Many Cas9 orthologues have been discovered and successfully demonstrated as genome editing tools. Another advantage of using SpCas9 orthologues is to avoid the large size of the protein which is a constraint in successful cellular delivery. SpCas9 is 1368 amino acid (AA) long (4104 bp), while the orthologue from Streptococcus aureus (SaCas9) is only 1053 AA in length. SaCas9 can efficiently cut the target sites having "NNGRRT" (R ¼ A/G) PAM sequence. Similarly, Nme2Cas9 has a compact size (1082 AA) and alternative PAM (NNNNCC) requirement. Cas12a (previously termed as cpf1) recognizes a T-rich PAM, 50 -TTTV-30. The tool is expanded further to RNA targeting by the discovery of Cas13 (previously termed as C2c2) that binds and cut RNA. For a further detail on different orthologues of Cas9 proteins discovered.

SpCas9 and other Cas proteins have been engineered to alter PAM compatibility, impair the cleavage activity, and increase specificity. SpCas9 has two nuclease domains, HNH and RuvC. HNH domain cuts the complementary or target strand, while RuvC cleaves the noncomplementary or nontarget strand (Fig. 7A). Fully active Cas9 generates double-strand break in the target DNA. However, engineered Cas9 containing a mutation in the catalytic residues of either RuvC or HNH domains generates single-strand nick. This type of engineered Cas9 is known as Cas9 nickase (nCas9). D10A mutation in the RuvC domain converts Cas9 into nCas9 that nicks in the target strand. Similarly, H840A mutation in the HNH domain of Cas9 generates a nCas9 that nicks in the nontarget strand. Incorporation of both the mutations turns an active Cas9 to a dead Cas9 (dCas9), which can bind to the target locus but cannot cleave.

ADVANCES IN SYNTHETIC BIOLOGY

Introduction

Synthetic biology is seen as an emerging field which builds on the work in genetics and associated fields over the last few decades. Like most new fields at this stage of development, it is subject to a number of possible definitions. For the purposes of this report, the following definition has been adopted: "Synthetic biology aims to design and engineer biologically based parts, novel devices and systems as well as redesigning existing, natural biological systems." This is the simplest and perhaps the most widely accepted definition and, in addition, is consistent with the Academy's view of the importance of synthetic biology to the future of engineering, industry and the general economy. Synthetic biology strives to make the engineering of biology easier and more predictable.

Biological systems

In the context of synthetic biology, it is perhaps important to make some points about biological systems. Living systems are composed of a number of key components (cells, genes, proteins) that allow biological organisms to grow and replicate. Understanding how the orchestration of these components leads to the complex nature of cellular and physiological behaviour exhibited by living systems has formed much of the basis of biological and biomedical research over the last 100 years

The research has resulted in the realisation that living systems operate at different scale levels: for example, from populations to individual organisms; and from the physiological level to the individual cell.

Over the last 60 years our understanding has extended to the molecular scale. Today the functions of individual cells are known in detail - from individual biochemical reactions to metabolic pathways, gene regulation to the control of cell division and cell-cell signalling.

This level of detail also extends to visualisation, where the macromolecular components such as proteins, sugars, lipids and nucleic acids that perform all of the necessary biochemical functions for cellular life can be imaged at atomic resolution.

The central tenet that underpins much of current molecular biology lies in the relationship between DNA, RNA and proteins. The genetic code (the DNA) instructs cells to produce proteins by translating the DNA sequence into an intermediary messenger RNA (mRNA) – this process is known as transcription. The mRNA is then translated into a polypeptide chain comprising a defined sequence of the 20 naturally occurring amino acids (a process known as translation) to produce proteins that carry out most of the cellular functions and activities within organisms.

The revolution in molecular biology which has occurred over the last sixty years has been underpinned by parallel technology developments. Examples of such developments are: more and more powerful computers (including the widespread use of personal computers); the rise of the internet; powerful search engines (e.g. Google); and broadband networks. This has resulted in a massive explosion in data - including the genome sequences of hundreds of different organisms including the human genome - where the full complement of protein components can now be identified and studied.

Systems approach

The traditional approach to biological research has been to isolate a small number of biological components in order to understand their structure and function. The realisation that biological systems are multi-scale and multi-level has led to a deepening realisation that biological systems can no longer be studied using a reductionist approach. This approach assumed that single biochemical events resulted in single effects, in a simple cause and effect relationship.

In reality, most genes, proteins and other components carry out their functions within a complex network of interactions, with positive and negative feedback loops that regulate their operation. Consequently, a single component (such as a gene) rarely specifically controls any particular biological function or disease, and conversely any given component may influence many different functions.

Such a simple cause and effect framework posses severe restrictions on the ability to understand, manipulate and design increasingly complex biological systems. This realisation has led to the emergence of systems biology (which was the subject of a previous report by The Royal Academy of Engineering and The Academy of Medical Sciences which tightly couples 'state-of-the-art' biological measurements with sophisticated mathematical and computational modelling.

This facilitates the understanding of how networks of interactions between the components of a biological system give rise to its observed properties. The systems biology approach allows the study of multi-scale, multi-level (including multi-cellular) organisms. One example of this approach has been the creation of accurate models for the human heart, by combining physiology with computer modelling.

Over the last decade, advances in molecular biology technologies have allowed researchers to manipulate, or re-engineer, DNA coding regions in many different organisms from bacteria, to plants, to animals. This has led to a revolution in molecular and cell biology – driven, primarily, by the genome projects. Animal models of human disease can be created, e.g. mice cancer models; cell-based systems can be manipulated to address fundamental biological problems, e.g. mutant yeast and bacterial strains; and organisms can be altered in beneficial ways, e.g. transgenic mosquitoes for the control of malaria. However, all of these approaches are primarily based on hypothesis driven questions - where researchers aim to address specific questions with often single and specific experimental solutions. By contrast, synthetic biology aims to establish a rational framework for DNA manipulation, based on the application of engineering design principles.

The emergence of synthetic biology

Having reviewed some of the key concepts of living systems, the question of how and why synthetic biology has emerged as a discipline and what differentiates it from other related subjects will now be addressed.

Developments in Information and Communication Technology (ICT)

Shannon's work established the basis of the information and communication revolution which has taken place over the last sixty years. The reason for this is that digital computers can only work with what is known as sampled data. In general, sampled data is uncommon in the natural world. Data tends to be continuous, that is it exists at all points in time over the period for which it is being measured.

Shannon developed his sampling theory which allows data to be converted from its continuous form to its sampled form without loss of information and vice versa. Shannon's sampling theory is the basis of modern information and communication systems.

This sampling theory naturally leads biological research to produce large amounts of data requiring ever more powerful computers to analyse. As such, the rapid increase in the power and availability of computers is another important building block in the development of the quantitative techniques which underlie synthetic biology. Indeed, as data mining becomes increasingly intensive, the computational power that will be required for more sophisticated applications is likely to exceed the capabilities of local computers.

Furthermore, the demand for computational operations involving data sources that may be distributed across many sites, where they are maintained and updated on a regular basis, is increasing - along with interactions between researchers working in different and often distant locations. Hence, high speed telecommunications networks and access to high performance computing are very important in modern biological research - and are an essential element in the development of synthetic biology.

The relationship between systems biology and synthetic biology

Systems biology aims to study natural biological systems as a whole, often with a biomedical focus, and uses simulation and modelling tools in comparisons with experimental information. Synthetic biology aims to build novel and artificial biological parts, devices and systems. Many of the same methods are used and as such there is a close relationship between synthetic biology and systems biology.

But in synthetic biology, the methods are used as the basis for engineering applications. The basis of quantitative systems biology lies in the application of engineering systems and signal theory to the analysis of biological systems. This allows the definition of systems in terms of mathematical equations. Once a system, or part of a system, has been described in this way, then synthetic biology allows the reduction of the system to biological parts (bioparts) whose function is expressed in terms of input/output characteristics. These characteristics are then presented on a standard specification sheet, so that a system designer can understand the functional characteristics of the part. The parts are then entered into an inventory.

The parts defined in an inventory (or registry) can then be combined into devices and, finally, into systems. Tolerances are built into the design of any engineering part, device or system to compensate for imperfections in the manufacturing. Bioparts tend to have wider tolerances than standard engineering parts, so biologically-based devices are designed to accommodate such features.

Hence, synthetic biology incorporates the classic reductionist method whereby complex systems or processes are built from defined parts and devices. This means that if a biologically based system is being developed, it is developed on the basis of standard parts and devices which form part of a reference repository or registry.

Gene Editing Techniques for the *C. glutamicum*

Traditional Gene Editing Techniques

With the advancements in genomics and molecular biology, there are now many tools available that can be used to engineer microbial strains. These tools can be broadly categorized into two groups: traditional gene editing technologies and CRISPR technologies. Traditional gene editing technologies refer to the introduction of target genes into the host cell through homologous recombination, reverse screening with screening markers, and then editing target genes through an allelic exchange.

Currently, traditional gene editing techniques applied to *C. glutamicum* can be categorized into non-replicating plasmids (suicide plasmids) and replicating plasmids-mediated homologous recombination, Cre/loxP-mediated site-specific recombination, and RecT-mediated single-stranded recombination based on the method of recombination.

Homologous Recombination Mediated by Non-Replicating Plasmids

Non-replicating plasmids, also known as suicide plasmids, can be categorized based on their mediated homologous recombination into single-exchange and double-exchange recombination. Single exchange is a method of inactivating the target gene by introducing resistance markers into the host strain. Since the introduction of the resistance gene may alter some physiological properties of the host strain or influence the gene expression or inactivation of other genes, another allelic exchange is performed on top of this to discard other parts of the gene editing process, and this method is called double exchange.

In the 1990s, researchers developed a gene knockout method for *C. glutamicum* by utilizing the *sacB* gene. Since then, the pk18*mobsacB* system, a secondary exchange homologous recombination mechanism mediated by a suicide plasmid, has become the most widely used gene editing system for *C. glutamicum* to date. The *sacB* gene in this system functions as a negative screening marker gene by encoding a secreted sucrose fructanase.

This enzyme catalyzes the hydrolysis of sucrose into glucose and fructose while also facilitating the synthesis of fructose into a polymeric fructan that is lethal to the strain. By applying a reverse screening strategy, the appropriate colonies can be obtained when cultivating the suicide plasmid in a medium containing sucrose. Researchers used the pk18*mobsacB* system to adjust the metabolic pathway associated with 4-hydroxyisoleucine biosynthesis in *C. glutamicum* by manipulating gene expression through knockout or overexpression strategies. Currently, the classical pk18*mobsacB* system is a commonly used tool for researchers to engineer strains.

However, the practical application still presents some challenges, such as prolonged recombination time, complex operation, and susceptibility to false positives. In response to issues with the pk18*mobsacB* system, certain scholars have proposed alternative anti-selection markers and nuclease systems. Researchers used the *rpsL* gene (streptomycin resistance) as a counter-screening marker for *C. glutamicum*, which not only improved the editing efficiency but also improved the growth of the cells. Researchers used the *upp* gene as a reverse screening marker in combination with the '!-Sce'!-mediated recombination system to reduce the likelihood of false positive strains and achieve efficient mutant screening. Researchers attempted to mutate the *rpsL* gene encoding small ribosomal protein S12P in *C. glutamicum*, resulting in streptomycin resistance and significantly enhancing counter-selection efficiency during suicide plasmid-mediated genome editing. This approach simplifies strain screening and improves overall efficiency.

Site-Specific and Single-Stranded Recombination

Cre Recombinase (Cyclization recombinase) was first identified in 1981 from P1 phage and is a site-specific recombinase that exhibits catalytic activity. Cre recombinase mediates specific recombination between two LoxP sites (a palindromic sequence of 34 base pairs), which enables the deletion or recombination of gene sequences located between the LoxP sites. Compared with homologous recombination methods, this technique has been applied to gene knockout, exogenous gene integration, and screening of efficiently expressed motifs due to its high efficiency, specificity, and accuracy. As such, it is a powerful tool for gene editing technology.

Researchers developed a Cre-loxP-mediated deletion system to achieve the deletion of large fragments without affecting the normal growth of *C. glutamicum* and improved the efficiency of gene knockout. However, the knockout efficiency of this method is unsatisfactory due to prolonged secondary recombination time, reversible LoxP site mediated by Cre, and gene residues present in the Cre/loxP system.

Therefore, the RecT recombination system was subsequently introduced in a later study by the researchers. RecT is a single-stranded DNA (ssDNA) annealing protein (SSAP) that utilizes the RecT recombination system derived from the phage Rac's *recT* gene to facilitate single-stranded recombination, which is both straightforward and independent of DNA sequence and length.

In 2013, the first RecT-based recombinant system was developed and applied in *C. glutamicum*, which was combined with nanosensing technology to achieve rapid and efficient strain isolation. Later on in the research, RecT-mediated single-stranded recombination was gradually applied to engineer strain. However, it was found that this system was inadequate for editing multiple genes simultaneously in *C. glutamicum*. Therefore, researchers have developed gene editing systems such as the RecET-Cre/loxP system and the CRISPR/Cpf1-RecT system, which have been applied in *C. glutamicum*, further enriching the arsenal of microbial gene editing tools.

CRISPR/Cas Gene Editing Techniques

CRISPR/Cas9 System

CRISPR technology was first proposed in the 1990s and has since been widely used for gene editing across various fields, including life sciences, medicine, and agriculture. Among these applications, the CRISPR/Cas9 system stands out as a simpler yet highly effective tool that is currently favored by researchers worldwide. CRISPR/Cas9 technology includes two fundamental processes: gRNA-guided Cas9-targeted DNA cleavage and subsequent DNA repair. Additionally, DNA repair mechanisms include non-homologous end joining (NHEJ) and homology-directed repair (HDR).

Since non-homologous end repair is absent in *C. glutamicum* and homologous end repair is inefficient, the researchers introduced a CRISPR-based base editing system into the organism. Base editing technology is constructed by fusing the expression of Cas9 nuclease and deaminase protein, which enables targeted mutations without inducing DNA double-strand breaks, exogenous DNA templates, or homologous recombination repair mechanisms. There are two primary categories of base editors, namely the CG-to-TA cytosine base editor (CBE) and the AT-to-GC adenine base editor (ABE), both of which have been applied in *C. glutamicum*.

Among them, researchers fused CRISPR/Cas-based Cas9 and cytidine deaminase (AID)-based cytosine base editor (CBE) tools to achieve multi-locus gene editing in *C. glutamicum*. The BETTER technology was subsequently developed by the researchers based on the base editing system, which allows for simultaneous editing of up to 10 genes with varying expression levels, thereby facilitating gene editing diversification. A study revealed that the leakage of the Cas9 toxin posed challenges in transferring plasmids containing the *Cas9* gene into *C. glutamicum*.

Therefore, to mitigate the toxicity of Cas9, the researchers integrated the *Cas9* gene and rec-ET-encoded recombinase into the genome. They also constructed plasmids carrying donor DNA and sgRNA expression cassettes to mitigate Cas9-induced toxicity. Ultimately, they successfully produced a mutant strain of *C. glutamicum* that was highly efficient in producing L-homoserine. The development and application of these gene editing technologies offer novel approaches for multi-locus gene editing in *C. glutamicum*, expedite the genetic engineering of *C. glutamicum*, and provide additional tools for fundamental gene editing.

CRISPR/Cpf1 System

Compared to CRISPR/Cas9, the CRISPR-Cpf1 system has attracted widespread attention as a novel gene editing tool with minimal off-target effects and no cellular toxicity. Cpf1 (Cas12a) is a type V, class II CRISPR effector protein with specific double-stranded ribonucleic acid endonuclease activity. It was initially characterized in *Francisella novicida* U112. Compared to Cas9 protein, Cpf1 protein exhibits several advantages: it can recognize and cleave the AT-rich proto-spacer adjacent motif (PAM) without requiring tracrRNA; its own RNase activity enables processing of CRISPR sequences into mature crRNA; its smaller molecular weight facilitates easier entry into tissues and cells.

In addition, the Cpf1 protein generates sticky ends upon cleavage, while the Cas9 protein produces blunt ends, as illustrated in Fig. 9a,b. The presence of sticky ends can facilitate the insertion of target genes into specific loci through non-homologous recombination. Given the advantages mentioned above, the CRISPR/Cpf1 system indicates excellent potential for application in the field of synthetic microbial biology.

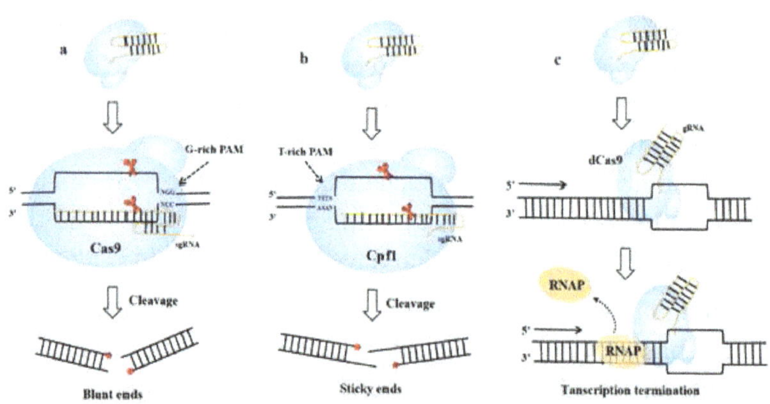

Fig. 9. CRISPR- Cas9/Cpf1, CRISPRi gene editing characteristics. (a) Cas9 protein shears to form blunt ends; (b) Cpf1 protein shears to form sticky ends; (c) the dCas9/gRNA complex binds to target genes to suppress gene expression. RNAP: RNA polymerase.

Elements of Gene Expression and Regulatory Strategies

Promoters

A promoter is a DNA sequence located upstream of the 52 end of a structural gene that specifically activates RNAP to bind precisely to template DNA for transcription initiation. As the promoter serves as the initiation point for controlling gene transcriptional expression, it stands as a crucial element in enhancing the yield of the target product. To date, compared to *E. coli*, *B. subtilis*, and *Saccharomyces cerevisiae*, there are relatively few mature promoters available for *C. glutamicum*, which limits both the use of strategies to precisely regulate the expression levels of *C. glutamicum*-associated genes via promoters and the development of various synthetic biology tools that rely on promoters. More studies in *C. glutamicum* focus on inducible promoters, such as IPTG-inducible (isopropyl-â-D-thiogalactopyranoside) promoter tac, arabinose-inducible promoter araBAD, etc.

Most of the inducible promoters currently characterized are stronger than endogenous constitutive promoters. However, their use requires expensive inducers, which increases costs, and some of these inducers may be toxic or unsuitable for industrial applications. As a result, the use of constitutive promoters has been increasingly favored in recent studies. In addition to screening for more robust and more efficient promoters, replacing or modifying existing promoters with efficient alternatives represents a direct and effective approach to enhancing target product expression. In subsequent research, a series of synthetic promoters were gradually developed to achieve precise regulation of microbial metabolism. In summary, there are three main sources of promoters currently used in *C. glutamicum*: (1) endogenous promoters derived from itself, (2) promoters obtained from other hosts, and (3) promoter libraries constructed through random mutations.

Non-Coding Regions in the 52 Untranslated Region (UTR)

Promoter optimization is a widely used approach for optimizing transcriptional strategies. However, screening for strong promoter-52 UTR complexes to achieve precise regulation of metabolic pathways in microorganisms has been applied to numerous model strains, including *E. coli* and *B. subtilis*. The 52 untranslated region (UTR) is a non-coding segment located at both ends of the messenger RNA (mRNA) molecule, and its secondary structure plays a crucial role in regulating the efficiency of internal ribosome binding site (RBS) recognition and start codon identification. The initiation of translation in the 52 UTR region depends on recognition by 16S rRNA of the RBS ribosome binding site, and the efficiency of this interaction affects gene expression levels. Therefore, optimizing the non-coding 52 UTR region, developing robust complexes in the 52 UTR, and improving RBS binding efficiency are crucial strategies for improving cellular efficacy.

Researcgers chose 90 potentially promoter-52 UTR-containing sequences of *C. glutamicum* species and obtained 17 strong PUTRs by screening with fluorescent proteins as reporter genes. Researcgers found a 3.2-fold increase in ScFv secretion from *C. glutamicum* upon substitution of the original Shine–Dalgarno (SD) sequence with tpi-SD.Researcgers optimized the RBS sequence and the spacing between the start codon and RBS core sequence based on a series of natural promoters, which ultimately resulted in an 82% increase in ã-aminobutyric acid (GABA) production.

Similar to expression elements such as promoters, in most cases, RBSs are often incompatible across different hosts. Therefore, it is important and necessary to develop RBS elements that can be universally applied across species. After constructing a library of RBS mutants and characterizing them individually in *C. glutamicum* and *E. coli*, the researchers found that 18 RBSs exhibited similar intensities between the two species. After analyzing the correlation of RBS intensities between the two hosts, it was finally determined that these 18 RBSs could be used across species. In addition, some sequences, including GAAAGGCGA, GAAAGGA, AGAGGGG, GTTATTAG, and AAAGGGA, were also identified as strong RBS sequences for the expression of target genes in *C. glutamicum*.

Signal Peptides (SPs)

SPs, typically composed of 20–40 amino acids, facilitate the translocation of secreted proteins to the extracellular space and are subsequently cleaved by signal peptidase (SPase) during or after transmembrane transport. In *C. glutamicum*, secreted proteins are transported through two distinct pathways: the Sec pathway for unfolded proteins and the Tat pathway for folded proteins, as shown in Fig. 10b,c. According to different transport pathways, SPs can be classified into Sec pathway and Tat pathway SPs. The SPs of proteins in the Sec or Tat pathway exhibit similar characteristics but also have their specificities, as illustrated in Fig. 11.

A range of SPs, including cg1514, CspB2, PorB, CspA, cgR1176, and cgR2070, have been successfully isolated and identified in *C. glutamicum*. To enhance the production of extracellular secreted proteins, the screening of an optimal signal peptide (SP) is a primary and equally crucial consideration. Different production strains may exhibit varying levels of secretory expression for target proteins with different SPs, and the best choice for one strain may not necessarily be optimal for another. Of course, previous studies have demonstrated that the optimal SP varies for different secreted proteins.

Currently, the limited knowledge of SP properties required for target secreted proteins and the unpredictability of optimal SPs for specific secreted proteins necessitate the construction and screening of high-volume SP libraries combined with high-throughput screening. This strategy is by far the most popular and successfully applied method for identifying optimal SPs [90,92]. Jin et al. [91] constructed an SP library based on previously validated SPs to enhance enzyme secretion efficiency. Subsequently, they identified the optimal Cg2196 SP for secretion in *C. glutamicum* through SDS-PAGE and Western blot analysis. Other researchers have effectively utilized the Sec SP library of *B. subtilis* to enhance protein secretion in *C. glutamicum*. Using keratinase as a model enzyme, researchers studied the effect of *B. subtilis* Sec SPs and process conditions on the performance of heterologous protein secretion in *C. glutamicum* and demonstrated that optimizing biological process conditions could improve the efficiency of heterologous protein secretion.

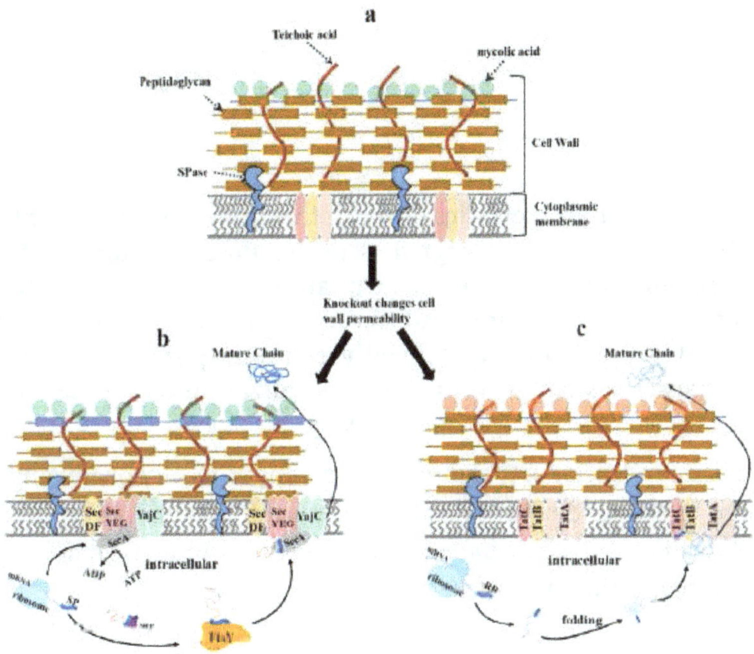

Fig. 10. The cell wall structure of *C. glutamicum* and its protein transmembrane secretion pathway. (a) Cell wall structure of C. glutamicum. The cell wall is composed of an outer layer primarily consisting of mycolic acid and an inner layer mainly comprising peptidoglycan. (b) Sec secretion pathway. The SecA protein facilitates ATP hydrolysis to generate energy and interacts with the Sec-YEG transporter complex, and Sec-YEG is composed of three subunits: SecY, SecE, and SecG. SecDF (consisting of SecD and SecF) forms a transporter protein-associated complex, while YajC participates in forming this complex. Additionally, YidC is involved in the integration and folding of membrane proteins. The signal recognition particle (SRP) binds to the SP at the initiation of translation, followed by the translocation of SRP and nascent protein to membrane protein FtsY. The unfolded protein then associates with SecA and is secreted extracellularly via SecYEG. (c) Tat secretion pathway. The protein export channel is composed of TatA, TatB, and TatC. The folded protein binds to the complex formed by TatB and TatC, translocates with the assistance of a TatA-targeted chaperone that is cleaved by SPase after translocation, and ultimately secretes the mature protein into the extracellular space [95]. (a)'!(b)\(c) indicates the disruption of either the mycolic acid or peptidoglycan layer in the cell wall, which leads to the change in cell permeability and facilitate the secretion of protein into the extracellular.

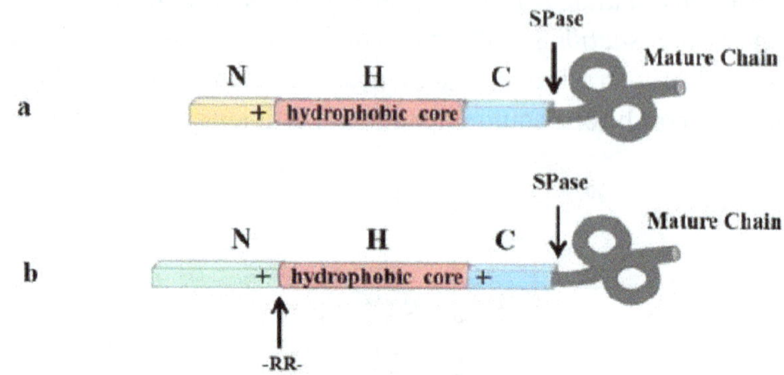

Fig. 11. (a) Structural characteristics of Sec pathway signal peptide. (b) Structural characteristics of Tat pathway signal peptide. Both contain an N-region, a hydrophobic region, and a C-region. Compared to Sec-type SPs, Tat-type SPs possess a longer N-region, a highly conserved arginine residue "-RR-" near the N-terminus, and a C-region that typically contains positively charged amino acids. These characteristics prevent the mislocalization of Tat substrates to the Sec pathway

In industrial applications, *C. glutamicum* presents a highly promising option for the expression of exogenous proteins due to its safety profile. However, the tough and thick cell wall of *C. glutamicum* poses challenges in industrial applications due to difficult fragmentation and increased cost, as shown in Fig. 10a. Although the addition of SPs for extracellular secretory expression vector construction has shown some potential in *C. glutamicum*, further research is needed before industrialization can be achieved. Therefore, altering the structure of the cell wall and enhancing its permeability is crucial in addressing this issue.

When the synthesis of mycolic acid or peptidoglycan layer on the cell wall is impaired or absent, it leads to an increase in cell wall permeability and extracellular secretion of target products. In a previous study, researchers treated *C. glutamicum* with Triton X-100. Although it increased cell wall permeability and improved cell activity, the non-food grade nature of Triton X-100 limited its industrial application for producing food-grade tagatose. The researchers later disrupted the mycolic acid layer to inhibit cell wall synthesis and increase cell wall permeability, but this also significantly impeded strain growth and division. Without affecting cell growth, researchers knocked down the *ponA* gene to induce structural changes in the peptidoglycan layer and partially inhibit cell wall synthesis. This led to a nearly five-fold increase in the efficiency of electro-conversion of exogenous DNA due to the resulting change in cell wall permeability. Because the host's proteases degrade or misfold proteins (including heterologous protein), researchers delete the endogenous protease *CLPS* gene from the genome of *C. glutamicum*, resulting in an increase in heterologous protein secretion. Although enhancing cell wall permeability and suppressing endogenous protein-degrading enzymes can boost extracellular secretion of the protein, further research is still required to apply this approach to *C. glutamicu*'s extracellular secretion production. In future industrial application research, the strategy of modifying the cell wall structure of the chassis can be combined with SP engineering to construct microbial cells with enhanced permeability and extracellular secretion capacity.

Terminators

The terminator, situated downstream of the 32 regulatory sequences of the target gene, plays a crucial role in regulating gene expression along with the promoter. The primary function of this is to effectively terminate transcription, which prevents interference with downstream genes and enhances the metabolic competition pathway of microorganisms. In addition, it is crucial in maintaining the stability of gene transcription products. Currently, the study of terminators has not been as extensive as that of promoters. However, as expression systems become increasingly complex and effectively terminating transcription becomes more important to avoid interference with downstream regions, research on terminators is gradually advancing.

The metabolic pathway from glutamic acid to proline competes with that from ornithine for precursors. To address this issue, researchers constructed two mutant strains of *C. glutamicum* S9114: orn9, in which a terminator was inserted upstream of gene *ncgl2228*, and orn10, in which a terminator was inserted upstream of gene *proB*. The findings indicate that both mutant strains upregulated ornithine expression and improved the status of the two competing metabolic pathways. The gene *argF* encodes ornithine carbamoyltransferase, which is a crucial enzyme in the degradation pathway of L-ornithine. However, the disruption of this gene to increase L-ornithine production results in growth defects within the strain.

Thus, researchers introduced a terminator (rrnB) between the *arg D* and *arg F* genes, resulting in an 11.2-fold increase in L-ornithine production (6.1 g/L) compared to the original strain and a 42.8% increase compared to the *argF* knockout strain (4.27 g/L). Additionally, growth levels were increased by 10.6% compared to the *argF* knockout strain. The strategy of incorporating terminators to attenuate gene expression and enhance metabolic pathways has also been applied in the production of other compounds, such as amino acids. This approach obviates the need for gene knockouts that impact target gene expression, simplifying experimental procedures and avoiding physiological and metabolic repercussions on host cells. It represents a straightforward and convenient strategy for engineering chassis cells.

Expression Vectors

The transcriptional expression levels of genes are not solely determined by the expression elements but are also influenced by their respective expression vectors. Currently, the primary expression vectors utilized in *C. glutamicum* include its endogenous vector, as well as *E. coli-C. glutamicum* and *B. subtilis-C. glutamicum* shuttle vectors. Further research is necessary to optimize expression vectors of *C. glutamicum* for industrial-scale production.

Currently, the majority of expression vectors used for efficient overexpression of target genes in *C. glutamicum* require the supplementation of IPTG and antibiotics, which are unsuitable for large-scale industrial production. Optimizing a specific element or combination of elements based on the aforementioned factors can enhance the yield of the target product. However, this process is time-consuming, laborious, and has limited applicability. Therefore, researchers developed a bicistronic protein expression vector that enhances the performance of inducible promoters and enables efficient production of various recombinant and exogenous proteins. Moreover, the research team has previously demonstrated that a bicistronic expression system in *C. glutamicum* provides distinct advantages compared to its monocistronic system.

In a subsequent investigation, researchers devised an antibiotic-free *E. coli-C. glutamicum* shuttle expression vector pLY-4, which exhibits the advantage of stable replication in nutrient-deficient host strains without resorting to antibiotics. This led to a 33.96% increase in L-methionine production via heterologous expression of the target gene. In addition, increasing the plasmid copy number represents a feasible strategy for constructing efficient expression vectors. Researchers conducted laboratory adaptive evolution experiments to screen a high-copy-number strain and successfully enhanced the endo-xylanase secretion in *C. glutamicum*.

However, it has been demonstrated that high copy number plasmids are unstable, which increases the likelihood of coding sequence mutations. Additionally, the maintenance of high copy number plasmids in bacteria imposes a significant metabolic burden on the host organism, leading to poor cellular growth and reduced production of recombinant proteins. Subsequently, researchers developed an industrial strain carrying the genome without plasmids, which has been successfully utilized for producing 5-keto-fructose by *C. glutamicum*. This breakthrough provides a powerful tool for constructing expression vectors in industrial strains.

CRISPR AND PERSONALIZED MEDICINE

Personalized medicine

Personalized or precision medicine aims in developing the treatment procedure tailored to the individual and eliminate the constraints of "one-size-fits-all" therapy. In addition to broadening our understanding of cancer, NGS aided the development of personalized medicine, providing oncologists with a powerful tool to comprehend each patient's disease and its unique genetic traits and whole-genome mutational status. NGS can detect tumor-specific mutations with the single-nucleotide resolution, allowing us to take advantage of this characteristic for target analysis.

The functional characterization of all annotated genetic elements in normal biological processes and disease has been a primary priority since the conclusion of the Human Genome Project. The individual's genetic profile can be analyzed using NGS and the mutations identified help us in identifying the therapeutic targets which can be edited with the help of CRISPR gene editing. The biomarkers from liquid biopsy make it easy to study the progression of the tumor.

The main objective of personalized molecular medicine is to target very specific disease-causing genes while minimizing the danger of off-target consequences. CRISPR as a genome-editing technique could aid in the development of more efficient gene-targeted modification technology. Figure 12 represents the application of Next Generation Sequencing (NGS) and CRISPR/Cas9 in personalized medicine.

To measure efficacy and determine treatment decisions, precision medicine approaches analyze patients' circulating DNA (liquid biopsy), immunological markers, and other biologic aspects. Tumor and cell-free DNA profiling, immune markers, proteomic and RNA analyses can be used as diagnosis methods to determine the personalized treatment options. Sequencing with NGS helps in genetic characterization, which can aid in target identification.

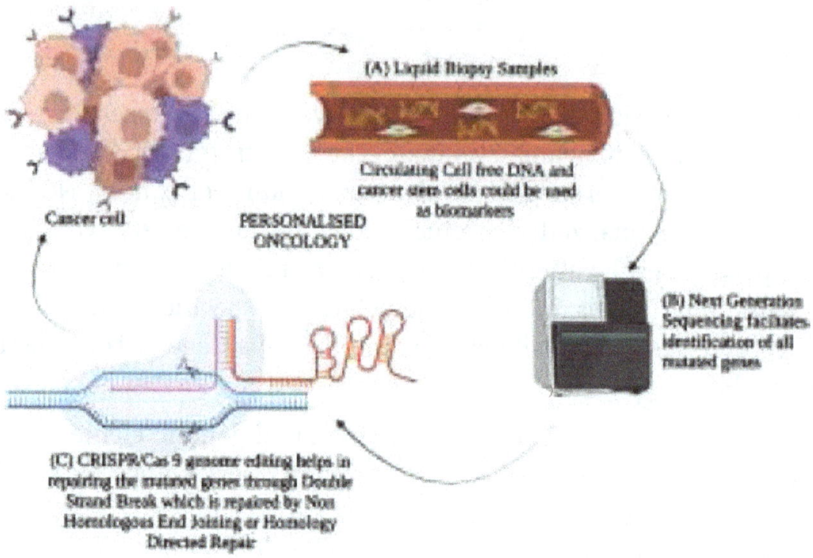

Fig. 12. Represents the application of Next Generation Sequencing (NGS) and CRISPR/Cas9 in personalised medicine: A Cancer can be diagnosed without non-invasive biopsy samples with the help of circulation biomarkers (liquid biopsy) like cell free DNA and cancer stem cells. B The advancement of NGS has helped in identification of various mutation in the cancer cells that cannot be identified by other methods like PCR. This helps in personalising the treatment for cancer. C CRISPR/Cas9 genome editing tool locates the mutated gene and modifies it by creating double strand breaks which is corrected by non-homologous end joining or homology directed repair which is now being studied for personalised oncology.

One of the important reasons for opting alternative therapeutic approaches and personalized medicine is drug resistance in cancer. Many tumors are initially responsive to chemotherapy, but they can develop resistance over time due to DNA mutations and metabolic changes that enhance drug inhibition and degradation. Reduced drug activation can potentially lead to cancer cells developing resistance to such treatments [18, 19]. In these cases, the mutation could be analyzed and treatment could be planned according to the genetic analysis in personalized medicine.

The efficacy of a pharmacotherapeutic drug is influenced by its molecular target and changes to that target, such as mutations or changes in expression levels. Signaling kinases like members of the epidermal growth factor receptor (EGFR) family are targeted by several anticancer drugs. In some malignancies, several of these kinases are constitutively active, which encourages uncontrolled cell proliferation. Over-activation of these kinases is usually caused by mutations; however, over-expression of these genes can sometimes have the same effect.

HER2, a receptor tyrosine kinase in the EGFR family, is overexpressed in 30% of breast cancer patients, and treatment resistance can develop following long-term usage of inhibitors targeting this kinase. Factors like this make the need for an alternative therapeutic approach significant. Drug resistance can also be achieved via altering the signal transduction pathway that mediates drug activation, in addition to alterations in specific drug targets. Drug resistance is a serious impediment to advancement in the field of targeted therapy and it is worth noting that gene mutations causing medication resistance are now being identified by NGS analysis, which could lead to ways to restore sensitivity such as editing of drug-resistant genes using gene-editing tools.

Liquid biopsy in diagnosis and monitoring of tumor

Liquid biopsies have ushered in a new era of precision medicine in the treatment of human cancer. Liquid biopsies may better reflect the genetic characteristics of all tumor subclones in a patient than tissue samples collected from only one tumor location. Liquid biopsies have the potential to guide cancer treatment and provide the optimal strategy for personalizing treatment in precision medicine.

Furthermore, liquid biopsies can be taken at predetermined intervals to track therapy responses, medication resistance, cancer recurrence, and metastasis. The most successful biomarkers from such biopsies so far have been genomic biomarkers, although other biomarkers, such as protein assays and transcriptomics, are being developed and tested.

Circulating cell-free tumor DNA (ctDNA) is one such promising biomarker from liquid biopsies that can support personalized treatment. Conventionally ctDNA can be assessed from various easily available physiological fluids, including blood, urine, and CSF. The two main methodologies for analyzing ctDNA are the detection of tumor-associated mutations or DNA methylation, and chromosome instability evaluation. NGS of gene panels, whole-genome sequencing (WGS), whole-exome sequencing (WES), RT-PCR, and other diagnostic platforms for ctDNA analysis are among the alternatives. Treatment-selected mutations, such as *EGFR T790M* in non–small-cell lung cancer, *KRAS G12V* in colorectal cancer, and *BRAF V600E/ V600K* in melanoma have been detected using ctDNA, thus paving way for personalized treatment options. Tumor NGS analyses genomes in the tissue biopsy, whereas ctDNA assesses DNA shed from numerous places, and ctDNA is associated with tumor load and can be detected at low levels.

ctDNA could be released into body fluids through passive mechanisms like tumor cell death and necrosis and active mechanisms like lymphocytes or complete tumors spontaneously releasing DNA, like the tumor DNA in exosomes. Many factors influence the amount of ctDNA in a patient's bloodstream, including clinical stage, tumor size, cancer type, and cell turnover rate. Tumors without solid shape, necrosis, or enhanced mitotic activity may not be shedding ctDNA into the bloodstream or maybe shedding ctDNA at levels below current detection thresholds which is the main drawback in liquid biopsy samples.

The first blood-based colorectal cancer screening test authorized by the US Food and Drug Administration (FDA) is SEPT9 gene methylation detection. The European Medicines Agency and FDA have approved ctDNA-based epidermal growth factor receptor (EGFR) mutation testing for therapeutic recommendations in patients with non-small cell lung cancer (NSCLC). Thus, biomarkers from liquid biopsy aid in more frequent analysis, monitoring, early cancer screening, diagnosis, and prognosis because of their minimally invasive or non-invasive properties and high public acceptance.

Identification of new therapeutic targets

Oncogene addiction makes cancer cells vulnerable to therapies targeting specific driver genes. Due to the variety of driver genes, CRISPR-cas9 provides a powerful tool to identify these genes, which are potential drug targets. CRISPR-cas9 screening has been utilized to identify new targets in a variety of tumors in various cellular contexts. Edwin et al. conducted a genome-wide LOF genetic screen in colorectal cancer HCT116 cells harboring mutant or wild-type KRAS.

They found that several genes involved in metabolism, such as NAD kinase or ketohexokinase, were potential therapeutic targets, and pharmacological inhibition of these enzymes attenuated the growth of colorectal cancer cells. Konstantinos el al. utilized the CRISPR-cas9 screening platform to identify genetic vulnerabilities in five types of acute myeloid leukemia (AML) cells (MOLM-13, MV4-11, HL-60, OCI-AML-2, and OCI-AML-3). As a result, they identified 492 AML-specific essential genes, including several established therapeutic targets, such as DOT1L, BCL2, and MEN1. KAT2A was proposed as a new therapeutic target for AML and validated in vitro and *in vivo*.

Recently, genome-scale CRISPR-CAS9 screening in 324 cancer cell lines from 30 cancer types was performed to develop a comprehensive database to prioritize candidate cancer therapeutic targets. For example, Werner syndrome ATP-dependent helicase (WRN) was identified as a promising new synthetic lethal target in microsatellite instability tumors.

Because CRISPR libraries are delivered by lentiviral vectors, it is difficult to operate *in vivo*. However, large-scale screening based on CRISPR-cas9 is not limited to the cellular system, and researchers managed to conduct screening *in vivo*. Researchers first introduced piggyBac (PB) transposons for sgRNA delivery for *in vivo* screening. PHD finger protein 5A was identified as a critical apoptotic suppressor involved in cancer progression and as a promising epigenetic target in breast cancer by using an *in vivo* CRISPR screen targeting RNA-binding proteins. Although CRISPR screening has contributed significantly to the discovery of new targets, further improvements to reduce the false positive rate caused by the off-target effects of cas9 is warranted.

Identification of biomarkers in cancer therapy

Biomarkers are widely used for disease diagnosis, staging, and prediction or for monitoring the efficacy of therapeutics. Thus, the discovery of biomarkers is an important factor in precision medicine. Screening based on CRISPR-cas9 is very helpful in searching for biomarkers to identify populations sensitive to a given treatment. Researchers found that BRD4 promotes the proliferation of colon cancer RKO cells through LOF screening.

Combined transcriptomic and genomic analysis further revealed that the long noncoding RNA (LncRNA) colon cancer-associated transcript 1 (CCAT1) was a biomarker that could indicate the sensitivity of colon cancer patients to BET inhibitors. Small cell lung cancer (SCLC) is generally considered to be associated with RB1 inactivation and TP53 mutations. CRISPR-cas9 screening revealed that RB1-negative SCLC NCI-H82 cells are hyperdependent on chromatin separation-related proteins, such as Aurora B kinase. Therefore, the loss of RB1 may be a potential biomarker for sensitivity to Aurora B kinase inhibitors. Responses to the first-line chemotherapeutic drug cisplatin have not been improved in the treatment of bladder cancer. Researchers performed whole-genome CRISPR screening to confirm that MSH2, as the most significantly enriched gene, promoted resistance to cisplatin. Therefore, MSH2 is a potential biomarker that may indicate the response of bladder cancer patients to cisplatin. A number of genes have been identified as potential biomarkers based on CRISPR-cas9 screening, which deserves further validation in clinical studies.

Identification of genes involved in drug resistance

Drug resistance is prone to occur during drug treatment and may ultimately lead to tumor recurrence. Therefore, elucidation of the resistance mechanism and discovery of combinatorial regimens are important strategies to overcome drug resistance. With the emergence of CRISPR-cas9 technology, researchers have been devoted to searching for essential genes causing drug resistance or synergistic lethality through large-scale screening. The Zhang lab constructed a Genome-Scale CRISPR Knock-Out library targeting 18 080 genes with 64,751 unique guide sequences to identify vemurafenib resistant genes in A375 melanoma cells.

Since then, unbiased LOF screening based on CRISPR-cas9 has been widely used in the search for synergistic lethal or resistant genes. Researchers utilized a genome-wide CRISPR-cas9 screen to determine that the knockout of glutaminase, which acts in the first step in glutamine metabolism, has a synergistic lethal effect when combined with FLT3-TKI inhibitors in FLT3ITD AML patients. Thus, by inhibiting the feedback-activated survival signaling pathway, the recurrence of tumors caused by single drug therapy can be effectively prevented.

Genome-wide phenotypic changes can be achieved by transcriptional activation, and high-throughput screening based on GOF has also been introduced to identify genes inducing drug resistance. Konermann et al. synthesized a transcriptionally activated CRISPR library consisting of 70290 sgRNAs. They used a synergistic activation mediator system to screen genes potentially mediating resistance to BRAF inhibitors in melanoma A375 cells and patient-derived melanoma samples.

Unlike pooled libraries used for knockout screening, sgRNAs in CRISPR activation (CRISPRa) libraries are designed to target promoter sites in genes of interest. Moreover, investigators have developed comprehensive genome-wide CRISPRa screens targeting 14 701 lncRNA genes. By screening these libraries, transcriptional activation of the GAS6-AS2 lncRNA was found to lead to Ara-C resistance in multiple cancers, including AML. Although transcriptional activation CRISPR libraries avoid the deficiencies of cDNA libraries by capturing the complexity of transcript isoform variance and by utilizing cloning into size-limited viral expression vectors, function-acquired screening is still less widely used than knockout-based screening.

In recent years, cancer immunotherapy has achieved great success. However, the response rate is relatively low. In vivo CRISPR-cas9-based screening provides a powerful tool to identify genes mediating resistance to immunotherapy. Such an approach has been conducted in mice transplanted with melanoma, and defects in interferon-ã signaling have been identified to cause resistance to checkpoint blockade.

Recently, a genome-scale CRISPR-Cas9 screen was used to identify the mechanisms inducing resistance to the killing of tumor cells by cytotoxic T cells. The loss of PBAF function caused by inactivating genes, including Pbrm1, Arid2, and Brd7, sensitized mouse melanoma B16F10 cells to killing by T cells, and the inactivation of Pbrm1 rendered resistant tumors sensitive to immuno-therapy. In another study, the authors invented the CHIME (CHimeric IMmune Editing) method, which uses a CRISPR-CAS9 bone marrow delivery system that allows the rapid assessment of the genetic functions of innate and adaptive immune cells *in vivo*.

This approach was employed to perform an *in vivo* pooled genetic screen, and Ptpn2 was identified as a negative regulator of CD8+ T cell-mediated responses to LCMV clone 13 viral infection. Large-scale screening *in vivo* will be used primarily for the study of cancer immunotherapy, and the delivery system will be a critical factor for the efficient introduction of genetic alterations.

Challenges of CRISPR-cas9 screening in precision medicine

Large-scale screening based on CRISPR-cas9-mediated gene editing provides a powerful tool in the era of precision medicine. Although this approach has shown tremendous advantages, there are a few issues that need to be improved urgently. Minimization of the off-target effects of gene editing has been a significant challenge for the development of CRISPR technology. Several websites are available to help researchers design sgRNAs with improved specificity.

Scientists have also been working to optimize the design principles of sgRNAs to improve their specificity. Strategies for designing sgRNAs targeting noncoding RNA are also being improved. A lentiviral paired-guide RNA library for the large-scale knockout of noncoding RNA has been developed. Moreover, programmable single-base RNA editing has been realized by a RNA editing for specific C-to-U exchange system. Genetic perturbations can be introduced by either LOF or GOF. Therefore, both strategies are utilized in large-scale screening. However, the GOF method requires the participation of coactivator proteins together with sgRNAs, which makes the method less accessible. To allow GOF screening to be more widely used, the multiple-plasmid system should be streamlined. Another concern is that the knockout or overexpression effects caused by CRISPR-cas9 may be compensated for by paracrine pathways.

Neighboring cells can compensate for functional knockout or gain by secreting growth-promoting cytokines, which would interfere with the functional readout. Furthermore, the phenotype and microenvironment may fail to adhere to the desired pathological conditions when cells are cultured in 2D. Large-scale screening based on CRISPR experienced a process from in vitro to *in vivo*. The improvement of the delivery system and CRISPR-cas9 efficiency will be critical for improved screening *in vivo*. Alternatively, screening in organoids could better simulate the tumor microenvironment.

Finally, the analytical methods need to be optimized to reduce false positive or negative candidates during large-scale screening. A model-based analysis of genome-wide CRISPR/Cas9 knockout (MAGeCK) method was created to control the quality of original sequencing results and to establish model analysis so that 'seemingly relevant' genes can be removed. Researchers recently performed computational analysis of CRISPR screens using the MAGeCKFlute pipeline, which combines the MAGeCK and MAGeCK-VISPR algorithms and incorporates additional downstream analysis functionalities.

CHAPTER-10

CHALLENGES AND RISKS

Genome editing allows the specific modification of a genome; genes are modified within their respective location in the genome, making the changes often indistinguishable from natural mutations. Developments of this technology such as the use of gene drives, where specific genes are spread within populations, or the use of viral vector systems, are enabling additional applications in environmental engineering and disease treatment. There are substantial individual and societal benefits from applying genome editing; nonetheless the technology also poses significant risks to individuals, society as a whole and the environment.

The central focus of this case study is on the unresolved ethical issues related to safety and security that pose both short-term and long-term challenges to international research partnerships. As such, the case study focuses not on a single incident but on the risks in the proliferation of a new and very powerful technology at a time when accepted and tailored ethical and legal frameworks at the international, national and local level are missing.

In the case study two areas of *safety* risks are mapped and existing governance approaches described: first, risks to humans, for example in relation to therapeutic applications of genome editing; second, risks to the environment in relation to the use of genome editing on animals, plants and microbes. In addition, two aspects of *security* risks are also assessed: first, the creation of harmful agents relevant in the bioweapons context; second, human enhancement in a military context and its medium- and long-term implications for international security.

Risks of gene editing technology

In today's world, the increasing importance of the knowledge of biology as a basic science is undeniable. As a result of deep studies and many investigations, the boundaries of biology and the findings related to the knowledge of nature have been expanded tremendously. The volume of resulting information and its increasing growth cannot be compared to any era. Today, biotechnology, as a branch of biological applications, has progressed more than at any other time.

Due to its applications in health, hygiene and the economy, its importance and value have increased daily. These important advances in biotechnology are mainly due to advances in instrumentation and their application in developing the frontiers of biology. The most remarkable developments of this knowledge and technology have been achieved in ecology, genetics, microbiology, molecular biology, biochemistry, cell culture technologies and process engineering.

The emergence of the new sciences of genomics, proteomics, bioinformatics, systems biology, synthetic biology and gene editing has also resulted from these developments. On the other hand, today, biology is exposed to hostile abuse as much as chemistry in World War I and physics in World War II. The enormous force of international trade that underpins this basic science has driven it to innovations that, along with its marketable medical value, may also be used for destructive purposes. If a country exploits science and technology in biological fields, it can reveal one of the most serious problems of humanity that it has not faced so far. If the production of a new generation of bioweapons is pursued with force, especially if used to control and conquer humans, it can cause dangerous technological competition.

If the force of biotechnology is not politically contained, it will be able to invent scientific methods that will change the way war is conducted and increase the means of civilian sacrifice. It should be noted that the discussions discussed below do not in any way attempt to magnify the possible risks of biotechnology; because such a possibility exists with any other technology, both those that are widely used in societies today (such as IT and telecommunications) and those that will emerge in the future in the field of health (such as gene editing, artificial synthesis, and human creation).

Such studies can also show the potential capabilities of biotechnology in promoting new defense capabilities; so that the guardians of the country's biological defense pay special attention to the potential of this technology for defense purposes (such as providing new treatment trends and new diagnostic methods in the direction of biological defense).

Among the various technologies proposed in the field of life and health today, gene editing technology can cause irreparable risks to the biological security of societies. Now genome editing is much easier, faster, cheaper and more efficient than ever before (just like editing an article on a computer) and helps researchers in various scientific fields. The fourth generation of gene editing, CRISPR-Cas9, can handle new editing programs, from viruses and bacteria to animals, plants and humans.

However, as this technology develops, how should it be controlled? While expressing concern about the negative consequences of the development of gene editing technology, *Professor Jennifer Doudna*, the discoverer of Cas9 nuclease and 2020 Nobel Prize winner, has received $3.3 million in funding from the US Defense Advanced Research Projects Agency (DARPA) to investigate anti-CRISPR solutions. The red line of this technology is the manipulation of the human embryo.

It should be remembered that CRISPR-based genome editing technology has caused a tremendous revolution in medical sciences and other scientific fields. This technology, developed for about a decade, has fascinated scientists. In a way that it allows a person with less than a high school education to edit the genome of any animal or plant. High school students are now using this technology to perform experiments that were previously only dreams for most scientists. Much of the scientists' research has focused on the tremendous potential of using genome editing to treat cancer. This work is based on the familiarity and understanding of the new model of the cancer network, which examines how cancer cells are controlled. New gene-based approaches for cancer treatment have suggested using genome editing as an effective tool. Unfortunately, while we can cure cancer using CRISPR, we can also create cancer using the same technology.

Based on this, researchers will have been able to design cancer models on the computer. The possibility of artificially creating cancerous tumors will also be provided when they can treat cancer patients with CRISPR. Therefore, the cancer bomb can be implemented using CRISPR. As a result, today, CRISPR technology is proposed as a new biological weapon, and scientists warn about its biosecurity consequences.

The very simple application of CRISPR makes this technology potentially very dangerous. This technology has features that make it an ideal military and bioterrorist weapon and has been favored by biohackers. Mr. Josiah Zayner, one of the pioneers of biohacking and the founder of Odin Company since 2006, offers online training and simple gene therapy kits based on CRISPR technology to his customers around the world, who, in addition to the simple training of gene editing, can design and produce genetically engineered products (such as the green tree frog with the high growth rate).

Based on this, it seems that in the not-too-distant future, we will see the design and production of biological weapons (mass destruction) based on CRISPR, which will be even more dangerous than a nuclear bomb. According to some scientists, nuclear weapons are obsolete because they are complicated to maintain. From a military point of view, CRISPR weapons are considered far superior to nuclear weapons and will probably replace them. These CRISPR-based editing features make it very easy to weaponize.

It is possible to create an engineered virus that edits CRISPR in a completely controlled manner so that only humans whose genomes have specific characteristics are killed or disabled by the virus. The importance of CRISPR biosecurity is so high that scientists have warned about the design and production of programmed humans and editing on human embryos and are drafting national and international laws in this regard. It is necessary to explain that the Chinese scientist, He Jiankui, officially announced the birth of Chinese girl twins with Crisper technology. In these twins, who were born from a father with AIDS and a healthy mother, the CCR5 gene, which is related to the entry of HIV into cells, has been edited and deleted.

This scientific activity, carried out illegally and confidentially, shocked the scientists in this field and led to a fine and three years of imprisonment for this Chinese scientist. Jiankui had to present the result of his research at the Second International Conference on Human Genome Editing held in Hong Kong in 2018, which was widely criticized by the scientific community, even Chinese scientists.

The question raised today is why the Crisper weapon can be so dangerous. The answer to this question can be found in the following:

- CRISPR-Cas based editing can be precisely designed to edit a specific part of the target genome.
- Viruses can deliver CRISPR-Cas-based editing to a given host.
- Mathematical rules can fully control edits. In other words, target genome editing is applied only if certain precise conditions are present in the target person's genome.

For example, two people can be infected by a CRISPR-based editing virus, but only the person with the prerequisites will have their genome edited. In addition to these, unfortunately, many more characteristics of a potential CRISPR-based weapon make them ideal weapons for future precision and targeted mass destruction. For example, its effects can remain hidden for months, and CRISPR bombs do not have the long-term toxic effects of nuclear weapons. Also, it can be implemented for most organisms and programmed as an effective weapon for the nervous system. As a result, CRISPR as a weapon of mass destruction for genocide is significant and considered.

Also, from a biosecurity perspective, CRISPR technology could potentially create and induce precise cancers that would kill people within months. Unfortunately, it is much easier to cause cancer than to cure it with genome editing. The main appeal of a deadly CRISPR-engineered virus as a bioterrorism weapon is the precision in mass incapacitation and destruction of people. Unfortunately, these topics are unrelated to science fiction and fantasy horror movies, but this is a real danger in humanity's present and future. Notably, there are probably labs worldwide developing CRISPR technology to develop the next generation of bioweapons. As mentioned, Crisper also has genocidal capabilities. Given that a particular generation has unique genetic traits that distinguish it from others, all members are potential targets for a CRISPR-engineered killer virus.

For example, if the prerequisite is that one must have brown eyes, then anyone with brown eyes is a potential target for a deadly CRISPR-engineered virus. Using the Crisper weapon, you can create diseases that cause the target person to die slowly or quickly. Some scientists who have provided significant analysis about the possible possibilities have raised various dangerous issues. Some of these reports indicate major future concerns that could result from secret research programs.

Especially these reports have not ruled out the possibility of hidden viruses that can secretly enter the genome of a population and later be activated by a signal. Another example is "programmed cell death". This ability to insert a genome into the genetic reserve of a specific population and attack it at will, or to create a completely new pathogenic agent, indicates a change in capabilities.

Challenges and disadvantages of CRISPR in agriculture

Despite the significant advantages of using CRISPR to genetically modify crops, there are also considerable challenges. These include:

- Precise delivery. Delivering CRISPR/Cas9 components to the correct cellular location remains a significant challenge, particularly *in vivo* in rigid plant cells. This may require novel delivery strategies or adaptation of existing ones
- Off-target effects. There is significant concern about gene editing techniques, including CRISPR, having off-target effects on non-target genes. Although the fidelity of CRISPR is relatively high, more work is required to improve this.
- Efficiency. CRISPR is not efficient in all species of plants or at targeting some traits that are controlled by several different genes. Some crops or gene targets show lower transformation and editing efficiencies and do not obtain the desired genetic changes.
- Loss of genetic diversity. Many people are concerned that genome editing using CRISPR and other techniques will lead to the narrowing of the genetic pool, making crops more susceptible to pests and disease in the long term and harming biodiversity
- Legal and regulatory challenges. Being a relatively new and rapidly advancing area of research, genetic modification of crops using CRISPR is

subject to legal and regulatory challenges stemming from the lag in the time taken for laws and regulations to be updated in line with technological advances. Moreover, to gain approval, robust evidence must demonstrate safety and efficacy

OFF-TARGET EFFECTS AND PRECISION CHALLENGES

Challenges with ZFNs include design and engineering of the ZFP for high-affinity binding of the desired sequence, which can prove non-trivial. Also, not all sequences are available for ZFP binding, so site selection is limited. Using open-source ZFP domains, sites could be targeted only every 200-bps in a random DNA sequence. This may not be a concern if the objective is gene knockout or deletion; however, this may be an obstacle if the objective is a gene correction or addition product.

Another significant challenge is off-target cutting. ZFN design improvements addressing off-target concerns have included ZFNs that work in pairs, with each pair recognizing two sequences that flank the target cleavage site. One ZFN binds the forward strand, and the second ZFN binds the reverse strand, increasing the total number of recognized bps to between 18 and 36. Further, *Fok*I domains that are obligate heterodimers with opposite charge have been fused to ZFPs such that only properly paired ZFNs will result in *Fok*I dimerization/activity and the formation of a DSB.

Transcription activator-like effector nucleases (TALENS)

Four years following the development of ZFNs, a new class of natural DNA-binding proteins was discovered in the plant pathogenic bacteria *Xanthomonas* sp. The proteins, termed transcription activator-like effectors (TALEs), contain 33–35 amino acid repeats that flank a central DNA binding region (amino acids 12 and 13). This DNA binding region, known as the repeat variable di-residues (RVDs). Shortly after the discovery of TALEs, TALE nucleases (TALENs) were developed that, like ZFNs, are a fusion protein comprised of a TALE and a *Fok*I nuclease. For an in-depth TALEN function review.

Unlike ZFNs, design and engineering of TALENs is much simpler and can be done in a shorter time. TALENs are also not as limited as ZFNs in target site selection due to the 1:1 TALE–DNA binding. While off-target cutting remains a concern, TALENs have been shown in one side-by-side comparison study to be more specific and less cytotoxic than ZFNs. TALENs, however, are substantially larger than ZFNs, requiring 3 kb of cDNA encoding for one TALEN versus just 1 kb for a single ZFN. This makes delivery of a pair of TALENs more challenging than a pair of ZFNs due to delivery vehicle cargo size limitations. Further, packaging and delivery of TALENs in some viral vectors may be problematic due to the high level of repetition in the TALENs sequence.

CRISPR/Cas9

The most recently developed site-specific gene editing tool, CRISPR/Cas9, is a naturally occurring RNA-guided endonuclease. A methodical investigation by the scientific community has deciphered the natural function of the CRISPR/Cas9 gene editing system. Based on this work, several laboratories developed CRISPR/Cas9 as a tool that has now been applied in much of modern molecular biology. A key difference of this system from the protein-based binding to DNA of ZFNs and TALENs is the use of a short RNA sequence as the specificity-determining element to drive the formation of a DSB at the targeted site. The use of CRISPR/Cas9 avoids the need for protein engineering to develop a site-specific nuclease against a specific DNA target sequence, requiring only the synthesis of a new piece of RNA. This dramatically simplifies and greatly reduces the time needed for gene editing design and implementation.

Off-target cutting

In addition to rational design of sgRNA, efforts to improve specificity and reduce off-target cutting have resulted in the design of mutant Cas9 systems. While Cas9 itself does not cause off-target effects – these exist solely due to the sgRNA – improvements to the Cas9 protein can limit these effects nonetheless. One mutant system disrupts the Cas9 protein so that it introduces only single-stranded DNA nicks. The nickase CRISPR/Cas9 is then used as a pair with one Cas9 binding to the forward DNA sequence and another Cas9 binding to the reverse DNA sequence flanking the target site. Only when binding in this configuration is a DSB formed through cooperative nicks. Off-target cutting results in only a single-stranded nick that is repaired with simple DNA ligases. The use of this system in mammalian cells reduced off-target cutting by three orders of magnitude with little to no reduction in on-target cutting efficacy.

ZFN and TALEN systems served as inspiration for another mutant Cas9 system, a fusion protein of inactive dCas9 and a *Fok*I nuclease dimer. Again, sgRNAs are designed to bind both the forward and reverse sequences flanking the target, and only when bound in this configuration will the *Fok*I nuclease dimers reconstitute into a functional *Fok*I and form a DSB. While these approaches increase specificity and reduce off-target cutting, the number of potential target sites is lower due to PAM and other sgRNA design constraints.

This system also significantly increased the size of the gene editing tool, providing greater constraints on *in vivo* delivery approaches. Another reported mutant Cas9 systems designed to reduce off-target effects include fusions of Cas9 with ZFPs or TALEs that can target nearly any genomic locus with improved precision. Cas9 mutants have also been designed to reduce non-specific DNA contacts by weakening binding of the target DNA strand or the non-target DNA strand to Cas9 while maintaining robust on-target cleavage.

Cas9 activity

Several Cas9 proteins from differing species have been identified and used for gene editing, including *Staphylococcus aureus* (SaCas9), *Neisseria meningitidis* (NmCas9), and *S. thermophiles* (St1Cas9). Each has differing PAM sequences and variable activity. Thus, selection of a specific Cas9 ortholog may provide improved gene editing efficiency for a given target sequence and should be considered as part of gene editing system design.

In addition to the inherent activity of a given Cas9 protein, other factors have been shown to influence activity. For gene editing in eukaryotic cells, Cas9 must translocate into the nucleus. In these systems, the nuclear location signal (NLS) is connected to the Cas9 protein. Increasing access to the NLS by adding a 32 amino acid spacer between the NLS and Cas9 was shown to increase DNA cleavage activity. Increasing the relative concentration of sgRNA to Cas9 protein was also shown in increase on-target cutting activity, presumably by ensuring all Cas9 proteins formed the active ribonucleoprotein complex. However, excessive sgRNA was also shown to increase off-target effects.

Finally, in comparison with other enzymes, the activity of Cas9 is quite low, with a single turnover rate of <-0.3–1.0 min. And, once bound to the target DNA sequence, displacement of Cas9 from the DNA strand, even after DSB formation, is challenging –1 nM Cas9 cleaved <-2.5 nM plasmid DNA after 120 min. Thus, Cas9 is less like a catalytic enzyme and more like a single-shot actuator. While this characteristic may be useful in some instances, such as gene activation/inhibition or short-lived activity for gene editing with lower off-target effects, it may be undesirable for other applications where catalytic activity is useful.

Physical delivery methods

Microinjection

Microinjection is considered the 'gold standard' for introducing CRISPR components into cells, with efficiencies approaching 100%. In this method, either plasmid DNA encoding both the Cas9 protein and the sgRNA, mRNA encoding Cas9 and sgRNA, or Cas9 protein with sgRNA, can be directly injected into individual cells. Using a microscope and a 0.5–5.0 ìm diameter needle, a cell membrane is pierced and cargoes are delivered directly to a target site within the cell. This process circumnavigates barriers associated with delivery through extracellular matrices, cell membranes, and cytoplasmic components.

Further, microinjection is not limited by the molecular weight of the cargo, which is a significant limiting factor with viral vector delivery systems. This method also allows for the controlled delivery of known quantities of the cargo, improving control over off-target effects. Naturally, microinjection is best suited for *in vitro* and *ex vivo* work only, as the use of a microscope to target individual cells (and precisely inject cargoes to specific locations within them) precludes the use of microinjection in a true *in vivo* setting.

Nucleic acids are by far the most common cargo for microinjection delivery. There are three primary methods for injection of these components: (1) as DNA directly delivered to the cell nucleus, (2) as *in vitro*-transcribed mRNA molecules delivered to the nucleus, or (3) as *in vitro*-transcribed mRNA molecules delivered to the cytoplasm. These different methods have benefits and drawbacks.

By placing the DNA encoding both Cas9 and the sgRNA into the nucleus, the cell is free to transcribe and translate the components. This method is preferred by some groups, due to the ability to omit lengthy *in vitro* transcription reactions from the overall process. However, single-stranded DNA is prone to random integration into the host genome, which may disrupt genes, result in constitutive expression of Cas9, and lead to greater off-target effects. Even circularized plasmid DNA can undergo this phenomenon.

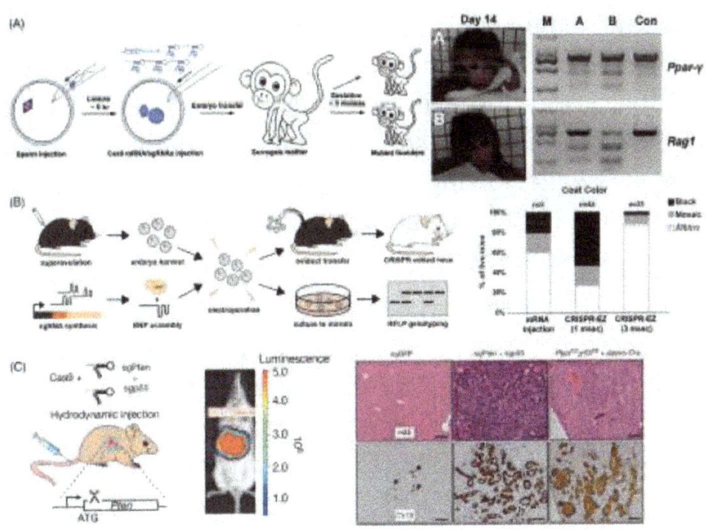

Fig. 1. Microinjection is also the most commonly used method for generating animal models. Injection of the gene editing cargo into zygotes allows for efficient germline modification. In addition, there is evidence that injection of Cas9 mRNA and sgRNA into the zygote cytoplasm is the most efficient method for yielding normal embryos and fullterm mouse pups harboring the desired modification Microinjection can also be useful for CRISPRa and CRISPRi to provide transient up- or downregulation of a specific gene within the genome of a mature cell.

When delivering mRNA, the ideal case is to deliver the sgRNA directly to the nucleus and the Cas9-encoding mRNA to the cytoplasm, facilitating translation and shuttling of Cas9 to the nucleus. Unfortunately, microinjection is a technically challenging and laborious process, making two different micro-injections into a single cell impractical. Further, two microinjections, even when separated by several hours, typically results in non-viable cells. Therefore, microinjections of CRISPR mRNA components often occurs directly into the cytoplasm of the cell. This method has the advantage of putting the Cas9 mRNA directly into the cytoplasm, where it can be translated by the cell. sgRNA in the cytoplasm is then bound by Cas9 while being shuttled into the nucleus, allowing for modification of the host DNA. The vast majority of studies using microinjection to deliver CRISPR use this approach, including simultaneous knock-out of four genes from a single injection into rat zygotes, disruption of two genes in cynomolgus monkeys from a single injection into one-cell-stage embryos.

Electroporation

One of the long-standing physical methods for delivery of gene editing tools into a population of cells is electroporation. This technique utilizes pulsed high-voltage electrical currents to transiently open nanometer-sized pores within the cellular membrane of cells suspended in buffer, allowing for components with hydrodynamic diameters of tens of nanometers to flow into the cell. Electroporation is less dependent on cell type than other delivery techniques and can efficiently transfer cargo into cells that are traditionally difficult to manipulate. Electroporation is most commonly used in an *in vitro* setting, though as with microinjection, *ex vivo* applications are also valid. Owing to the oftentimes-large amounts of voltage needed to be applied across cell membranes, however, electroporation is typically not suitable for *in vivo* applications.

There are many published methods for electroporation of mammalian cells. While these protocols can provide a starting point, mammalian cells are often quite sensitive to precise voltages and current application times. This contrasts with bacterial cells, which are often more tolerant of electroporation. This problem becomes even more prominent when studying zygotes rather than immortalized cell lines.

Several groups have developed technological solutions to increase the prominence of electroporation within the CRISPR/Cas9 community. For example, Hashimoto and Takemoto built a custom electroporation chamber for 40–50 zygotes which allowed them to achieve both very high levels of CRISPR/Cas9 entry into cells and viable embryos. Other groups have used more standard electroporation cuvettes and methods to deliver CRISPR/Cas9 components with high efficiency to zygotes. In other examples, electroporation was used to deliver plasmid DNA encoding both Cas9 and mRNA to generate colorectal cancer models from Cas9-engineered human intestinal organoids, generate an early-onset Alzheimer's disease model in human cells, and correct mutations that cause DMD. Electroporation for delivery of Cas9:sgRNA ribonucleoprotein (RNP) to primary human cells has also been reported. The use of RNP was shown to reduce off-target effects versus plasmid transfection and be less stressful on the cells, producing two-fold more embryonic stem cell colonies than with plasmid transfection.

A specialized electroporation method designed to deliver cargoes directly into the nuclei of mammalian cells has also been used to deliver CRISPR/Cas9. Termed nucleofection, this technique does not require breaking down the nuclear envelope, or cells in a state of division, for cargo to enter the nucleus. Plasmid DNA encoding Cas9 and sgRNA has been delivered via nucleofection to correct a cataract-causing mutation in mouse spermatogonial stem cells, confer resistance to HIV infection by adding the natural CCR5Ä32 mutation to human cells, generate lung cancer cell models in human lung epithelial cells, and cure latent herpesvirus infection by Cas9-based cleavage and destruction of latent viral genomes.

The use of electroporation and CRISPR/Cas9 to edit genes *in vivo* was also recently reported. Researchers developed a model of a childhood malignant brain cancer, Sonic hedgehog medulloblastoma, via *in utero* electroporation of a developing mouse. Plasmids encoding both Cas9 and sgRNA were injected into the embryo cerebral ventricular zone, followed by electroporation using forceps-like electrodes.

Because of the attractiveness of high-throughput and high-efficiency CRISPR/Cas9 transformation utilizing commonly available laboratory resources, electroporation/nucleofection will likely continue to be used and refined as a major technique to efficiently deliver CRISPR/Cas9.

Hydrodynamic delivery

Hydrodynamic delivery is an *in vivo* delivery technique that involves rapidly pushing a large volume (8–10% body weight) solution containing the gene editing cargo into the bloodstream of an animal, typically using the tail vein in mice. As blood is incompressible, the large bolus of liquid results in an increase in hydrodynamic pressure that temporarily enhances permeability into endothelial and parenchymal cells, allowing for cargo not normally capable of crossing a cellular membrane to pass into cells. This includes naked DNA plasmids and proteins. Delivery of cargo using this method is significantly enriched in the liver, but also includes cells of the kidneys, lung, muscles, and heart.

Hydrodynamic delivery is attractive because it is technically simple and does not require any exogenous delivery components to successfully introduce gene editing components into cells. Hydrodynamic delivery is typically used for *in vivo* applications only, as the premise of delivery relies on transiently increasing the pressure in a closed system and forcing cargo through otherwise-impermeable barriers.

Researchers demonstrated successful delivery of DNA plasmid encoding Cas9 and sgRNA to liver cells using hydrodynamic delivery, resulting in *in vivo* correction of the *Fah* mutation in mouse hepatocytes modeling hereditary tyrosinemia. Although initial delivery efficiency was only one in 250 liver cells, the liver's regenerative capacity allowed for the expansion of the modified cells and phenotype rescue. Soon after, Guan et al. also used hydrodynamic delivery of naked plasmid DNA encoding CRISPR/Cas9 components to a mouse model of hemophilia B. Again, targeting the liver, they showed restored hemostasis in treated mice.

They also showed that an adenovirus (AdV) delivery system resulted in higher corrective efficiency, but no therapeutic effects due to severe hepatic toxicity, presumably a result of the high immunogenicity of the viral vector. Other examples of hydrodynamic injection of plasmid DNA encoding Cas9 and sgRNA include indel mutation of two tumor suppressor genes and oncogenes resulting in generation of liver tumors in mice, inhibiting hepatitis B virus (HBV) replication and gene expression in HBV-infected mice, and specific targeting of the HBV genome in the nucleus of HBV-infected mice, showing the potential of CRISPR/Cas9 as a therapeutic against chronic HBV infection.

Despite these successes, hydrodynamic delivery is not currently being considered for clinical applications. The process of hydrodynamic delivery can be quite traumatic, resulting in potential physiological complications, including cardiac dysfunction, elevated blood pressure, and liver expansion. It is relatively easy to cause accidental mortality with this method. Also, as discussed previously, transfection rates are very low, and only certain cell types are amenable to successful delivery.

Lentivirus (LV) and adenovirus (AdV)

While LVs and AdVs are clearly distinct, the way they are utilized for delivery of CRISPR/Cas9 components is quite similar. In the case of LV delivery, the backbone virus is a provirus of HIV (Naldini et al., Citation1996); for AdV delivery, the backbone virus is one of the many different serotypes of known AdVs. As in the case of AAV, these are plentiful, and finding a useful AdV to a desired target is relatively facile.

The serotype most commonly used is AdV type 5. LV is particularly useful because it can be pseudotyped with other viral proteins, such as the G-protein of vesicular stomatitis virus. In doing so, the cellular tropism of the LV can be altered to be as broad or narrow as desired. And, to improve safety, second- and third-generation LV systems split essential genes across three plasmids, significantly reducing the likelihood of accidental reconstitution of viable viral particles within cells.

Both LV and AdV can infect dividing and non-dividing cells; however, unlike LV, AdV does not integrate into the genome. This is advantageous in the case of CRISPR/Cas9-based editing for limiting off-target effects. As is the case with AAV particles, both LV and AdV can be used in *in vitro*, *ex vivo*, and *in vivo* applications, which eases both efficacy and safety testing.

In terms of mechanism, this class of CRISPR/Cas9 delivery is like AAV delivery described above. Full viral particles containing the desired Cas9 and sgRNA are created via transformation of HEK 293 T cells. These viral particles are then used to infect the target cell type. The biggest difference between LV/AdV delivery and AAV delivery is the size of the particle; both LVs and AdVs are roughly 80–100 nm in diameter. Compared with the <“20 nm diameter of AAV, larger insertions are better tolerated in these systems. When considering CRISPR/Cas9, additional packaging space for differently-sized Cas9 constructs or several sgRNAs for multiplex genome editing is a significant advantage over the AAV delivery system.

Many groups are currently using AdV or LV vectors for delivery of CRISPR/Cas9 components. Researchers recently used an AdV vector to inactivate genes in normal human lung fibroblasts and bronchial epithelial cells, wherein they reported efficient silencing of genes at MOIs of AdV as low as 20. Additionally, researchers created a unique lentiviral CRISPR/Cas9 system via Golden Gate synthesis. Their construct allowed for the expression of one Cas9 and four different sgRNAs, each under the control of a different promoter, to allow for the editing of several different types of human cells. The work done by these two groups, especially on primary human cells, allows for exciting possibilities for the use of these delivery systems *in vivo*.

Researchers reported the generation of a model of *Eml4-Alk* oncogene-driven lung cancer in adult mice by intratracheal instillation of AdV-delivered CRISPR/Cas9. Researchers used AdV delivery of Cas9 and sgRNA to target *Pten*, a gene involved in the liver disease non-alcoholic steatohepatitis (NASH). Four months post treatment, *Pten* gene-edited mice showed massive hepatomegaly and features of NASH. Importantly, in addition to displaying AdV vector-associated immunotoxicity in the liver, humoral immunity against SpCas9 was detected, as was a potential SpCas9-specific cellular immune response. This indicates the importance of also studying the immunogenicity of specific Cas9s for *in vivo* delivery of CRISPR.

The use of AdV for CRISPR/Cas9 delivery has been reported targeting loss-of-function *PCSK9* mutation in mouse liver, *in vitro* partial restoration of muscle function in a DMD model mice, resistance to HIV-1 infection of primary CD4 + T-cells by adding the cell membrane receptor *CCR5Ä32* variant, and tissue-specific gene knockout in mouse liver.

Examples of LV delivery of CRISPR/Cas9 include modification of up to five genes using a single LV to deliver plasmid DNA encoding Cas9, sgRNA and a fluorescent marker to develop a mouse model of acute myeloid leukemia, target of herpes simplex virus-1 genome regions essential to viral protein expression during early and late phases of viral infection/reinfection to suppress infection and prevent new infection, and the use of LV to mutate genes in mouse primary immune cells.

Leveraging the integration ability of LV, this system has also been used to create gene libraries for studying mechanisms of disease. Researchers used LV to deliver a pool of 73,000 sgRNAs to two human cells lines that had been previously engineered to express Cas9. Researchers delivered 64,751 unique sgRNA sequences by LV to human cancer cells, and researchers delivered 87,897 sgRNAs targeting 19,150 protein coding genes in mouse embryonic stem cells, both again constitutively expressing Cas9. These, and other similar studies, will enable identification of new therapeutic targets and the design of next-generation drugs. For example, genomics screening using CRISPR/Cas9 and LV was used to identify the genes essential for West-Nile-virus-induced cell death and define a signal peptide pathway required by flaviviruses.

However, there are drawbacks to LV or AdV delivery systems, as typical AdVs and LVs elicit strong immune responses. In addition, although care is taken to make the HIV provirus as integration-deficient as possible (Chen and Goncalves, Citation2016) and AdVs are naturally very low integrators into the cell genome, it is not currently possible to completely eliminate the chances of integration into the host. Additionally, although steps can be taken to make this integration targeted, one cannot guarantee that the viral payload goes to the same precise location every time. This can result in an increase of expression and off-target effects, or even potential damage to the cell if the insertion randomly occurs within an important cellular protein. Care must always be taken with LVs and AdVs when utilizing them for genome editing.

Methods of Off-Target Detection

Detecting off-target sites in a highly sensitive and comprehensive manner remains a key challenge in the field of gene editing.

The T7 endonuclease I assay was initially used to detect off-target mutations, but this assay suffers poor sensitivity (it cannot detect off-target mutations that occur at frequencies <1%), and it is neither practical nor cost-effective for large-scale screening.

Various advanced methods for off-targeting detecting including deep sequencing (measure off-target mutations at frequencies ranging from 0.01 to 0.1%, web-based prediction tools, and ChIP-seq have been recently developed and widely adapted.

The Web-based algorithms have an innate limitation as the tools assume that off-target sequences are closely related to the on-target site, which may miss detrimental off-target sites with less sequence similarity. ChIP-seq has also been used to identify off-target binding sites for sgRNAs complexed with catalytically inactive Cas9 (dCas9). Encouragingly, majority of published works suggest that very few, if any, off-target cleavage sites were caused by active Cas9 nuclease.

Genome-wide, unbiased identification of DSBs enabled by sequencing (GUIDE-seq) that has been reported recently, based on global capturing of DSBs introduced by RNA-guided endonucleases (RGEN), enables genome-wide profiling of off-target cleavage by CRISPR-Cas nucleases.

GUIDE-seq consists of a two-stage process. First, DSBs induced by RGEN in the genomes of living human cells are tagged by integration of a blunt, 34 bp double-stranded oligodeoxynucleotide at these breaks via an end-joining process consistent with NHEJ. Second, dsODN integration sites in genomic DNA are precisely mapped at the nucleotide level using unbiased amplification and next-generation sequencing.

This method extends the detecting range of off-target mutagenesis frequencies to as low as 0.12%, far beyond the existing computational methods or ChIP-seq. Of note, some sgRNAs may induce extremely low frequency of, or perhaps no, undesired mutations (perhaps beyond the detecting limitation of GUIDE-seq technology). Another newly developed, high-throughput, genome-wide, translocation sequencing (HTGTS) method, which identifies translocations of yeast I-SceImeganuclease-generated "bait" DSBs at target sites, introduced into the genome of mouse cells to other "prey" cellular DSBs genome-wide. This HTGTS method not only reveals 10-folds more off-target sites caused by certain previously characterized nucleases, but also provides nucleotide-level resolution of junctions.

This method further improves the detection of off-target sites through the identification of translocations between homologous chromosomes, an undesired collateral effect that has not been detected previously. The strategies outlined here can be used as part of a rigorous preclinical pathway for objectively assessing the potential off-target effects of any RGEN proposed for therapeutic use, thereby substantially improving the prospects for eventual translation of these reagents to the clinic.

Although *in silico* and *in vitro* methods are used to screen for potential off-target sites, they cannot precisely predict mutations that occur *in vivo*.

Thus, efficient methods to detect where the off-target mutations occur *in vivo* need to be explored. Linear double-stranded integrase-defective lentiviral vectors (IDLV) preferentially incorporate into DNA DSBs by NHEJ, making it possible to tag nuclease-generated DSBs *in vivo* and has been used routinely to detect the off-target cleavage sites of zinc finger nucleases.

A modified IDLV system was able to identify off-target cleavage sites with a frequency as low as 1% when CRISPR/Cas9 and transcription activator-like effector nucleases (TALEN) systems were used to target the Wiskott-Aldrich Syndrome (*WAS*) and tyrosine aminotransferase (*TAT*) genes. Therefore, this method can be an alternative tool to validate the target specificity of CRISPR/Cas9 and other gene-editing technologies in preclinical applications.

Although various approaches have been shown to reveal the frequency of off-target mutations by at least an order of magnitude at several sites, it still remains unknown whether these RGEN variants are free of off-target effects in the entire genome. Recently, Digenome-seq, *in vitro* Cas9-digested whole-genome sequencing, has emerged as a robust, sensitive, unbiased, and cost-effective method for profiling genome-wide off-target effects of programmable nucleases, including Cas9, in human cells.

Digenome-seq also consists of a two-stage process. First, intact genomic DNA was isolated from transfected cells without RGEN and transfected cells with RGEN. Then DNA isolated from RGEN-transfected was digested *in vitro* with RGEN or not and RGEN transfected cells were digested *in vitro* with RGEN, both followed by whole-genome sequencing. Second, these digests can produce many DNA fragments with identical 52 ends, which give rise to sequence reads that are vertically aligned at cleavage sites. In contrast, all other sequence reads would be aligned in a staggered manner. After mapping sequence reads to the reference genome, the authors used the Integrative Genomics Viewer (IGV) to observe patterns of sequence alignments at the on-target and the off-target sites.

This method has demonstrated that Cas9 off-target effects can be avoided by replacing "promiscuous" sgRNA gX19 or GX19 sgRNAs with ggX20 sgRNAs ("g" and "G" represent a mismatched guanine and matched guanine, respectively), improving target specificity for gene and cell therapy applications.

In addition, off-target effects of dCas9-FokI, paired Cas9 nickases, or other programmable nucleases could be monitored carefully via Digenome-seq to avoid unwanted mutations. Of course, the verification of animal experiments *in vivo* using Digenome-seq need to be further studied. Although different methods of off-target detection have their own characteristic advantages (Table 1), Digenome-seq is currently considered to be the gold standard for validating the specificity of next-generation genome-editing tools.

Table 1 Methods of off-target detection

Method	Advantage	Disadvantage
T7E1 assay	Simple	Poor sensitivity, not cost-effective
Deep sequencing	Precise	Biased, misses potential off-target sites elsewhere in the genome
In silico prediction	Predicts some off target mutation sites-	Fails to predict bona-fide off-target sites
ChIP-seq	Unbiased detection of Cas9 binding sites genome-wide	Most off-target DNA-binding sites recognized by dCas9 are not cleaved at all by Cas9 in cells
GUIDE-seq	Unbiased, sensitive (0.1%), qualitative translocations, identifies breakpoint hotspots	False negatives present, limited by chromatin accessibility.
HTGTS	Identifies trans-locations	False negatives present, limited by chromatin accessibility.
IDLV	Programmable, sensitive (1%)	Many bona-fide off-target sites cannot be captured
Digenome-seq	Sensitive (0.1% or lower), unbiased and cost-effective	Not widely used
FISH	Quick	Less precise

FISH, fluorescence in situ hybridization; IDLV, integrase-defective lentiviral vectors.

Off-target assessment can be time consuming and expensive, especially for simple *in vitro* projects. Yet some reports in the literature suggest that the frequency of off-target mutations is higher in *in vitro* cellular experiments compared to *in vivo* animal experiments. When performing *in vitro* cellular experiments, one can choose one of the unbiased methods for off-target detection as listed in Table 1, depending on the requirement of specific experiment, to reveal potential off-target sites, which may confound the interpretation of experimental results. The fluorescence *in situ* hybridization (FISH)-based methods for off-target identification, which is fast but less precise, can be also used as an alternative.

Interestingly, when it comes to animal experiments, for many model species modified by CRIPSR/Cas9 including mouse, pig, or monkey, no off-target mutations, and no detectable damage at known off-target sites have been reported so far. There are a few possibilities that might explain the no reported off-target cleavage in animals. Firstly, it is possible to control the expression level of Cas9 protein by injection of Cas9 mRNA or protein, which not only acts much faster than plasmid transfection, but also degrades rapidly. Secondly, researchers tend to keep the founder animals with right on-target mutations, and do not maintain the animals without on-target mutations, leading to off-target mutations being diluted over generations in small animals, even though the founder (F0) might harbor possible off-target mutations.

In addition, the off-target sites may be in a noncoding region,so, the likelihood of generating an off-target phenotype is small. Thirdly, the methods and experiments employed for detecting off-target cleavage in most of the published animal model studies were not performed in genome-wide, high resolution, and unbiased way, which may mask some low frequency off-target editing sites. Thus, much more comprehensive and stringent examination of off-target cleavage sites at the animal level should be performed for future studies.

Strategies for Minimizing Off-Target Effects

Various strategies have been reported to reduce RGEN off-target effects.

First, the sgRNA sequence can be altered. Truncation of the 32 end of sgRNA (derived from tracrRNA domain that interacts with Cas9), shortening the region complementary to the target site at the 52 end of the sgRNA by as many as 3 nt (tru-gRNA) or addition of two guanine nucleotides to the 52 end of the sgRNA (just before the 20-nt complementary region) improves target specificity, decreasing undesired mutagenesis at some off-target sites by 5,000-fold.

Meanwhile, RGENs using these altered sgRNAs also have decreased on-target activity. Efforts toward modifications of sgRNA sequences to enhance the specificity of sgRNAs without compromising on-target efficiency have not provided consistent results.

In addition, chromatin accessibility has been reported to be one major determinant of *in vivo* binding. As Zhang and Sharp reported recently, there are hundreds of thousands of "seed+NGG" sites in the genome, yet <1% are actually bound by dCas9, and most of the matches are in promoters, enhancers and genes.

Therefore, when we design sgRNA, we should choose sgRNAs in promoters, enhancers, and genes as far as possible to improve the target cleavage efficiency. However, those sites cleaved by CRISPR/Cas9 system can be both on-target and off-targets sites and we need to balance it according to different experimental purposes.

Second, one potential strategy for minimizing off-target effects is to control the concentration of the Cas9-sgRNA complex by titrating the amount of Cas9 and sgRNA delivered. However, increasing specificity by reducing the amount of transfected DNA also leads to a reduction in on-target cleavage. Therefore, a balance between on-target cleavage efficiency and off-target effects has to be considered. Nonetheless, future optimization of both Cas9 and sgRNA design may improve Cas9 specificity without sacrificing cleavage efficiency.

Third, the wild-type Cas9 nuclease can be replaced with D10 mutant nickase version of Cas9 and paired with two sgRNAs that each cleaves only one strand. Paired nicking strategy substantially reduces the off-target activity by 50–1,500-fold in cell lines and facilitates gene knockout in mouse zygotes without sacriûcing on-target cleavage efûciency.

This versatile strategy enables a wide variety of genome-editing applications that require high speciûcity. Additionally, the repair of two nicks generated by double nicking produces 5' overhangs, leading to the formation of indels much more frequently than 3' overhang.

Comicroinjection of mouse embryos with wild-type Cas9 and sgRNAs induces on-target and off-target mutations that are transmissible to the offspring. Conversely, double nicking can be used *in vivo* to efficiently mutate single or multiple genes with minimal off-target mutations while preserving on-target cleavage efûciency. There is also an established framework for the optimal design of paired sgRNAs and computational tools for predicting potential off-target sites for sgRNA pairs.

The optimal pair of Cas9 sites are in a tail-to-tail orientation, separated by “10 bp to +30 bp, with the sequence 5'-CCN(32–72)GG-3'. This method is applicable for editing the genome of any model organism and minimizes confounding problems of off-target mutations, thereby holds great application potential in clinical gene therapy.

Fourth, to further improve DNA cleavage specificity, fusions of catalytically inactive Cas9 with FokI nuclease domain (fCas9) have been generated, which edits target DNA sites with >140-fold higher specificity than wild-type Cas9 and at least fourfold than that of paired nickases at loci with highly similar off-target sites.

This work provides the foundation for the further characterization and improvement of Cas9 specificity and cleavage activity *in vitro* and *in vivo*. Recently, tru-RFN with 19 bp half-site complementarity lengths that combines truncated guide RNAs with dimerization-dependent RNA-guided FokI-dCas9 nucleases decreases off-target cleavage by 40% relative to standard RFNs, providing a useful and further improved tool for high-precision genome-editing applications in human cells.

We believe that the synergistic use of double nicking and fCas9 offers a promising route for mitigating the effects of off-target CRISPR/Cas9 activity, and may perhaps be a useful approach for exploring therapeutic applications.

Recently, a strategy that improves the efficiency of precise genome editing with CRISPR-Cas9 via inhibition of nonhomologous end joining using the inhibitor Scr7 to target DNA ligase IV has also been reported, providing another opportunity to improve the target specificity.

Moreover, a simple and effective way to codeliver chemically modified sgRNAs with Cas9 mRNA or protein has been recently described. It is an efficient RNA- or ribonucleoprotein-based delivery method for the CRISPR/Cas9 system, with lower cytotoxicity in primary cells than DNA plasmid-based systems, enabling the expansion and widespread application of this technology.

Programmable gene insertion

Programmable and multiplexed genome integration of large, diverse DNA cargo independent of DNA repair remains an unsolved challenge of genome editing. Current gene integration approaches require double-strand breaks that stimulate DNA damage responses. Furthermore, CRISPR-based methods that bypass double-strand breaks, such as prime editing, are limited to modification or insertion of short sequences.

In late 2021, Jonathan S. Gootenberg, PhD, and Omar O. Abudayyeh, PhD, both alumni of the MIT laboratory led by Feng Zhang, PhD, presented their latest CRISPR system, PASTE (Programmable Addition via Site-specific Targeting Elements). This technology achieves efficient and versatile gene integration at diverse loci by directing insertion with a CRISPR-Cas9 nickase fused to both a reverse transcriptase and a serine integrase. Without generating double-strand breaks, PASTE can integrate sequences as large as ~36 kb with rates of between 10 and 50% at multiple genomic loci in human cell lines, primary cells, and quiescent nondividing primary human cells.

One of the best applications of programmable gene insertion is in therapeutics. "There are companies that are using advanced CRISPR approaches that make single-nucleotide or small edits," Gootenberg remarked. These approaches are fascinating, he noted, but he added that they may not work with genes that harbor many mutations.

Cell-type-specific delivery

For diseases that require systemic administration of CRISPR-Cas9 systems, cell-type-specific delivery is ideal to avoid delivery-mediated off-target effects. Although viral and nanoparticle delivery systems have shown promise, especially for delivery to the liver, there's a lot to be done in other cells or tissues, especially to enhance selectivity. Efforts to achieve cell-type-specific gene delivery with nonviral vectors have spanned surface modification by various targeting moieties, such as small-molecule ligands, peptides, and antibodies. One route is to use aptamers, single-stranded oligonucleotides that fold into defined architectures and bind to targets such as proteins.

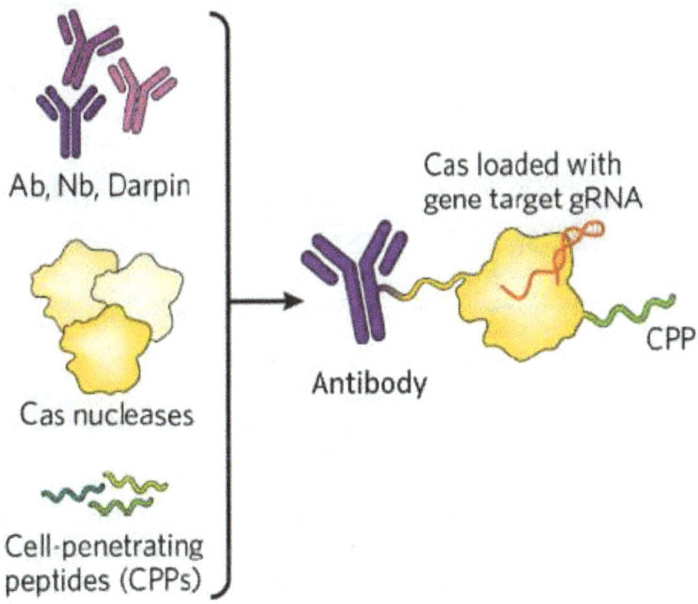

Fig. 2 .Spotlight Therapeutics develops cell-targeted in vivo CRISPR gene editing biologics by employing its Targeted Active Gene Editor (TAGE) platform. Each TAGE is a CRISPR effector that leverages a synergistic interaction between an antibody (Ab) and a cell-penetrating peptide (CPP) moiety. The Ab facilitates cell targeting, and the CPP moiety facilitates nuclear trafficking. The company's approach forgoes the use of complex viral, viral-like, and nanoparticle delivery systems.

Spotlight Therapeutics is addressing multiple limitations of the current in vivo delivery systems by developing engineered ribonucleoproteins with its Targeted Active Gene Editing (TAGE) platform. These ribonucleoproteins are created by combining transmembrane trafficking moieties (cell-penetrating peptides) and a cell-targeting moiety (ligand or antibody) with a CRISPR nuclease.

Gene silencing/activation

The epigenome, nature's gatekeeper to gene expression, consists of markers that are added or removed by very elegant and highly conserved mechanisms to determine the identity and function of each cell. For example, these markers do a great job of ensuring that our neurons stay neurons throughout our lifetimes. By targeted modulation of the epigenome, gene expression can be tweaked without ever cutting or breaking the DNA, which avoids some collateral damage that can come from using gene editing.

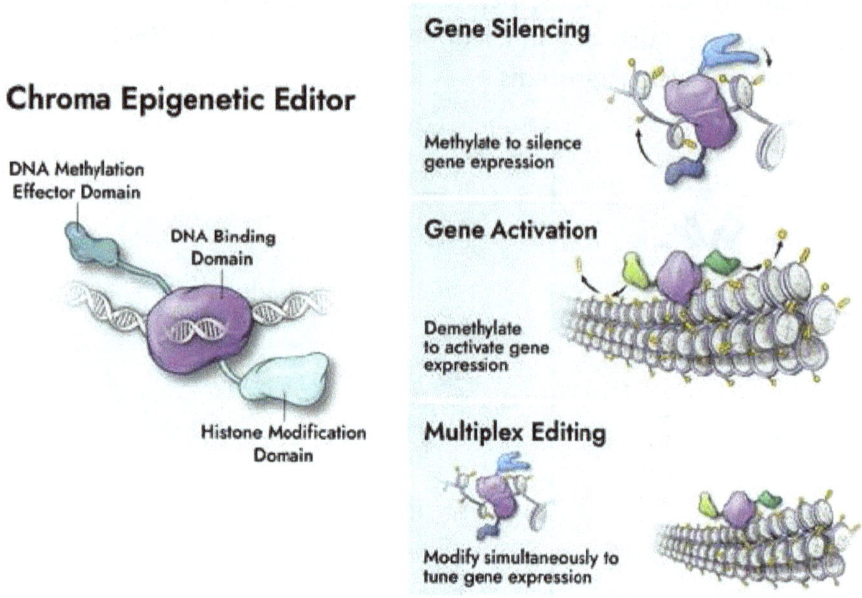

Fig. 3 Chroma Medicine develops programmable epigenetic editors that target genes and control chromatin conformation. These editors may be used to silence or activatc single genes. Also, they may be used to target multiple genes simultaneously. Each editor incorporates a DNA binding domain, a histone modification domain, and a DNA methylation effector domain.

Chroma Medicine is providing a modular, flexible, CRISPR-based platform for developing epigenetic medicines based on the silencing or activation of genes.

The platform couples a DNA binding domain with epigenetic effector domains. The DNA binding domain specifically targets the gene (or genes) to be silenced or activated. The effector domains create specific methylation patterns which control chromatin conformation and govern whether a gene is accessible or inaccessible for transcription. In simple applications, single genes may be targeted. In more complex applications, multiple genes may be targeted at the same time. It is even possible for some genes to be silenced while others are activated.

Microbiome modulation

A growing trend in the CRISPR world relates to the microbiome. CRISPR is being packaged into phages to affect bacteria in this setting, which is the opposite of how the CRISPR system exists in nature—bacteria use CRISPR to fight off phages. Irony aside, CRISPR technologies have the potential to enable unprecedented control over microbiome populations. For example, one may eliminate drug-resistant bacteria without targeting beneficial bacteria. This kind of control is critical to addressing antibiotic resistance, which is spreading rapidly around the world and seriously impeding efforts to control microbial infections.

Killing selected bacterial populations just one way to exert control over the microbiome. Another way is to genetically modify bacteria so that they express therapeutic proteins and peptides. The second form of control raises the prospect of hijacking bacteria to modulate critical processes such as metabolic and immunological processes.

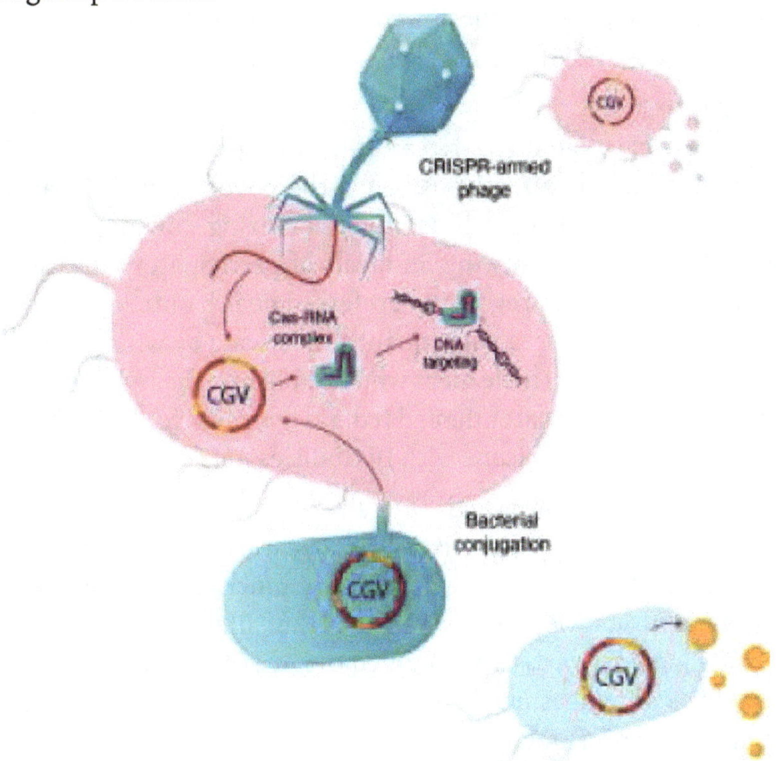

Fig. 4 SNIPR Biome, a developer of microbiome gene therapies, uses engineered phages or bacteria to deliver CRISPR Guided Vectors (CGVs) to bacterial hosts. The CGVs encode CRISPR RNAs that assemble into Cas-RNA complexes to drive therapeutic applications. These applications include highly selective and programable bacterial killing as well as modification of the human microbiome via gene modification and the expression of therapeutic proteins or peptides.

Safety is a big concern when making a chronic and live treatment. "If you introduce a gene," Grøndahl pointed out, "you also have to be able to control it and shut it off when you have obtained your treatment effect or when you want to lower the output." To do so, Snipr Biome is using its technology to install "safety switches" that allow inserted genes to be excised—a better alternative to harvesting the microbiome completely.

SOCIETAL AND ENVIRONMENTAL RISKS

Public participation is essential to policy making in democracies, as it captures dissenting views and helps to establish conduits of transparency and expanded understanding. Societal engagement in decision making promotes procedural justice—fairness and inclusion, as well as respect for autonomy—by incorporating those impacted into the process. Despite the inherent value and utility of public engagement, it is not without fundamental limitations. In aggregate, public opinion can reflect cultural biases and undervalue the perspectives of marginalized groups. Indeed, relying too heavily on public opinion can disregard essential human rights considerations. To illustrate this point, let us consider a notable health care policy case.

In 1989, the state of Oregon initiated an effort to expand eligibility for Medicaid health insurance to include all citizens below the poverty level. To achieve this, the state undertook to reallocate funds by prioritizing coverage of treatments based on their effectiveness. Drawing on ideals of participatory decision making, the state solicited public input regarding preferred health outcomes and quality of life for different conditions, creating a quantitative scale by which to evaluate the effectiveness of a treatment. The leaders who established the resulting policy were surprised when the state was later denied federal approval for the program, citing non-compliance with the Americans with Disabilities Act.

The Department of Health and Human Services charged that the state discriminated against people with disabilities. Research shows that able-bodied people rate the limitations of having a disability much more negatively than do people living with disabilities. By using quality of life ratings based on public opinion, the state had inadvertently devalued the lives of persons with disabilities. Using quality of life as a measure by which to prioritize the importance of health services and outcomes presumed an objective, shared understanding that does not exist.

Furthermore, analysts argued that under the Oregon plan, treatments for people with disabilities were more likely to fall to the bottom of the priority list. Thus, while public input is critical in decisions that affect the health of a population, it may not suffice to uphold the rights of vulnerable populations.

Given these challenges, simply querying the public suffers from key limitations. For decisions about GGE, public input on decision making about which conditions should be considered debilitating or undesirable enough to warrant possible future eradication may be primed for similar risks of inadvertent discrimination. Even the process by which public input is obtained may suffer from considerable limitations.

For example, while a powerful and beneficial tool, public polling can often lead to answers that are not the result of careful, informed deliberation but rather are influenced by readily available information in memory, automatically generated affect, and salient information in the environment. To prevent biasing respondents, efforts are often made to avoid framing and contextualization. Yet, contextual information about competing technical, ethical, or social considerations may actually lead to more useful answers. This tension underlines the limitations of polling in ethical policy making. Public forums, surveys, focus groups, and other methods for gathering public input may complement polling, but also suffer from insufficient reach, self-selection of participants, and inattention to differential power dynamics in presumed consensus building or democratic processes.

There are crucial ethical reasons to seek broad public participation in making such a large change in norms as permitting GGE: to respect individual and group autonomy and to be fair and inclusive. The call for public engagement issued after the 2015 International Summit on Human Genome Editing acknowledges the stake that everyone has in these decisions and the diversity of views and values that are often eclipsed by powerful scientific elites. In response, multiple efforts have been initiated to involve the public through varying approaches, including polling, town-hall style meetings, deliberative engagement, and online forums, and there is ongoing research into novel approaches that improve engagement.

Yet, while necessary, public participation alone is not sufficient to achieve policies that will genuinely protect individuals and society from changes that may not serve everyone's long-term well-being or rights. Clashing value systems and shifting public attitudes suggest additional approaches beyond public input must be incorporated when drafting robust policy. Establishing principles that protect vulnerable populations and value human rights may help address the limitations of public participation. Yet, despite the value of principles, below we also argue that these alone, or in combination with public opinion, are insufficient to establish ethical germline editing policy.

Building on Principles

Principles are crucial insofar as they involve reason-based arguments that underlie commitments to substantive moral claims. The Nuffield report, for example, asserts that germline editing cannot be ethically acceptable without weighing its use against two important principles:

- Protecting the welfare of future persons, and
- Upholding social justice and solidarity.

Germline editing will affect future generations, creating an impact on public health as well as individual welfare, thereby extending the moral duty of protection to people not yet born. Social justice demands that burdens and benefits be distributed fairly, so that socially disadvantaged groups do not bear an inequitable share of the burden of health interventions, and that benefits are not solely enjoyed by those already experiencing social and economic advantage. Solidarity recognizes that society as a whole is better off when all members do well.

But principles are usually decontextualized in that the reasoning behind the principle is usually based on philosophical commitments that are held to be universal in their scope. Delineating principles balances some limitations of public input. However, like public participation, principles are necessary but not sufficient for ethical decision making.

History shows that relying on principles does not always go far enough in actively determining ethical courses of action. During the 1980s, the U.S. President's Commission on Bioethics deliberated on seriously divisive bioethical issues, including developing criteria for brain death and allowing for the removal of life support. Experts could not agree on principles but did agree on specific cases (such as Karen Quinlan). Such case-based analysis is responsive to real—not abstract—human experience, and not bound by a single set of values or principles.

Further, a reliance on rules or principles to create a guide for ethical conduct has been called a "quest for certainty." This quest is futile when attempting to develop rules for technologies such as GGE where societal effects are so uncertain. Efforts to establish certainty about "permanent" solutions to human problems often reduce variegated human experiences to abstract standards that presume to apply in all situations. A central challenge is that a single agreed-upon set of principles presumes an ethical "universalism," minimizing social and cultural differences that result in unfair advantage for some groups over others. The Oregon case, where an assumption about common understanding of "quality of life" resulted in inequity, illuminates the danger when both public participation and principles omit critical societal contexts where power relations already create unfair advantages and disadvantages.

Where Do We Go From Here? A HRIA

Challenged by clashing value systems and in need of an inclusive shared basis for decision making that upholds crucial principles yet is responsive to dynamic societal contexts, we seek a way forward. We propose adding a third process for evaluating individual and societal harms: a HRIA.

While in medical ethics impacts of procedures are considered at the individual level, in public health ethics, impacts are considered at both individual and population levels. Because germline editing has the potential to impose burdens on future persons and societies, drawing on approaches from public health such as the HRIA widens considerations beyond individual impact.

The added value of a HRIA comes from providing an empirical assessment of power dynamics—in actual political and socioeconomic contexts—which may threaten human rights. Ethical principles lack context, and democratic processes may not protect the rights of vulnerable minorities. A HRIA provides the necessary third perspective and a way to apply principles to real situations. Empirical assessment of the likely influence of power dynamics on the implementation of new technologies is critical. As the former UN High Commissioner of Human Rights notes about new technologies such as artificial intelligence, there is no way for ethical guidelines to be meaningfully followed in societies where human rights conditions are not met. Public health assessments can evaluate the influence of racial, social, and economic inequities, and exclusion of people with disabilities, on health disparities. There are other contexts beyond public health where empirical assessments have been created to protect individual and societal rights. For example, for almost 50 years, the National Environmental Policy Act of 1969 (NEPA) has required the evaluation of environmental risks in impacted communities. More recently, a regulatory process involving race impact assessments has been proposed to examine risk of harm to racial minorities from new biotechnologies that may propagate the centrality of biological race as a primary health determinant, downplaying the role of discrimination. And the UN Guiding Principles on Business and Human Rights (2011) directs transnational corporations to use a HRIA to assess impacts on the rights of workers and communities.

We propose using a HRIA to guide ongoing decision making about GGE. This assessment can provide the socially shared transparent criteria necessary to structure inclusive productive democratic deliberation that aims to protect everyone's rights, irrespective of whether they participate in the deliberation. To achieve this, the HRIA uses a series of questions to help determine whether the potential benefit of a public health intervention or policy is outweighed by an undue burden on human rights.

A HRIA would consider the protection of human rights in the specific context of germline editing, operationalizing the Nuffield principles of "welfare of the future person" and "social justice and solidarity," and using a participatory process. The HRIA provides a way to examine how empirical conditions, including political and socioeconomic forces, combined with new technologies, may have the potential to undermine human rights.

Lawrence Gostin and Jonathan Mann introduced the HRIA in 1994 based on observations that public health interventions can ultimately invoke state power, and that protecting human rights promotes public health. They also built upon previously acknowledged rights, including the right to enjoy the benefits of scientific progress, which has been recently reaffirmed. While there are seven criteria in the HRIA, here we touch on a few of its priorities:

- Whether the proposed innovation or policy is well targeted;
- Whether it is the least restrictive/intrusive (in case of GGE) way to intervene; and
- What the human rights burden may be.

However, they are not sufficient for protecting the most vulnerable members of society who may not even be in a position to participate in democratic processes. We propose implementing a HRIA, which captures concerns of public health and offers an opportunity to evaluate and anticipate the societal impact of GGE iteratively as the technology advances, public sentiments evolve, and cultural contexts shift. We recognize that this will raise new challenges of how such assessments are shared and implemented and how they can be enforced. We urge regulatory bodies and policy makers to consider this assessment approach in helping to establish robust regulatory frameworks necessary for the global protection of human rights.

CRISPR can have environmental benefits and risks.

CRISPR can help:

- Limit agricultural emissions
- Make crops more resilient to climate change
- Use crops and soil microbes to store and stabilize carbon
- Reduce greenhouse gas emissions from agriculture and livestock
- Develop tomato and kiwi varieties that are more tolerant of saline lands while requiring less water

CRISPR can also have environmental risks, including:

- **Gene drives:** These can potentially decimate an entire species, eliminate a food source for other species, or promote the proliferation of invasive pests.
- **Loss of genetic diversity:** Genome editing using CRISPR and other techniques can lead to the narrowing of the genetic pool, making crops more susceptible to pests and disease in the long term and harming biodiversity.
- **Environmental risks:** These depend on the CRISPR/Cas9 delivery vector(s) and/or host cells(s) that will be used in conjunction with the CRISPR/Cas9 technology

BALANCING INNOVATION WITH SAFETY

CRISPR technology is still new, and any major safety issues could cause serious setbacks for the field as a whole. Ensuring the accuracy and safety of gene editing in different contexts remains crucial. Safety is essential for the widespread clinical application of the technology in the human therapeutics field. "Also, we believe that we must find more efficient ways to deliver gene editing tools to different cell types and tissues, and to identify solutions that lower off-target effects."

As the technology increasingly faces regulatory approval, safety and efficacy of CRISPR-based therapies will be validated, which should help generate public awareness and trust. But ethical considerations cannot remain unaddressed... One of the applications in which ethical questions emerge is evaluating the use of gene editing to resurrect extinct species or their traits, weighing the benefits against potential risks to ecosystems, and the moral implications of altering – even reversing – natural processes. But these ethical considerations can be addressed through the development of novel, more precise editing tools and by careful deliberation and open conversation between businesses, government, and regulatory institutions that would guide the ethical and responsible use of CRISPR technology.

Nevertheless, and as Paule confirmed, research will continue in Lithuania – and around the globe – with various stakeholders seeking to optimize current approaches to ensure the effective and precise delivery of CRISPR gene editing tools for a wide range of applications.

Some potential safety concerns include:

- Off-target effects
- Germline editing
- Unequal access
- Unintended consequences
- Mosaicism
- Long-term risks caused by genome editing

One strategy for increased safety is to select sgRNAs that yield high editing efficiency balanced with low off-target activity.

CRISPR technology has the potential to cure a range of genetic diseases, including neurodegenerative disease, blood disorders, cancer, and ocular disorders. However, one study found that the use of CRISPR-Cas9 in early human embryos carries significant risks.

CHAPTER-11

BIOTECHNOLOGY AND CRISPR IN EDUCATION

Biotechnology is a field of study that includes many areas that can be used in many occupations. Biotechnology education can help students understand the methods, achievements, and effects of biotechnology. It can also help students make their own decisions about the topic and act accordingly.

A decade after CRISPR started to become a major tool in genetic research, a new generation of scientists is growing up with the technology. Even high school students are able to run CRISPR experiments. Some specialized public high schools teach CRISPR as a hands-on lesson in biotechnology. These classes cover everything from molecular biology to pipetting to biomedical ethics and career options.

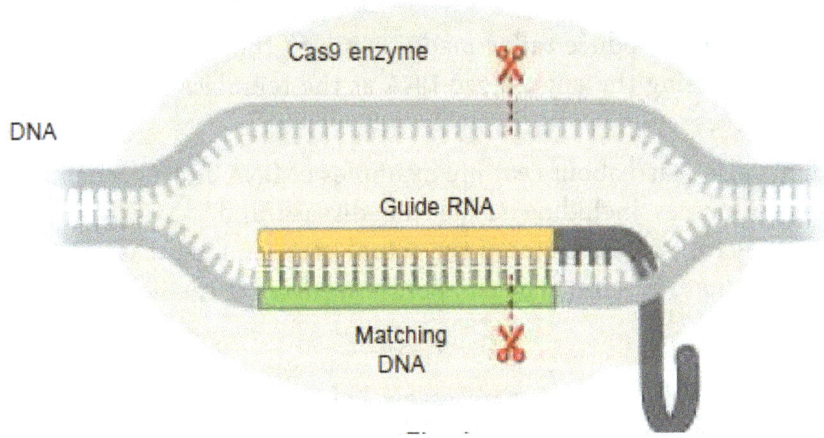

Fig. 1.

Visualizing Tiny Molecules

Visualizing and comprehending what's happening on the molecular level is usually always the challenge, a high school biotechnology teacher who taught CRISPR in her classroom last year. To help students better understand complex molecular movements, teachers use tools like paper models, 3-D printed models and online animations.

Target the right gene

Scientists engineer a piece of RNA that is a match for the DNA they want to edit. This is called the guide RNA. Students can practice designing their own guide RNA sequences using the same free bioinformatics search tool that scientists use for research.

Bind the target

An enzyme called Cas9 binds to a piece of DNA and temporarily unwinds a section of the DNA. Students can model the process with paper cutouts, pushing a paper DNA sequence along a printed guide RNA until they match.

Cut the DNA

If the guide RNA matches a section of the DNA, the Cas9 enzyme cuts both strands of the DNA double helix. An interactive animation from the Howard Hughes Medical Institute shows students how the Cas9 enzyme changes its shape and snips through the DNA (Fig. 2).

Repair and edit the DNA

Machinery inside the cell rushes to fix the broken DNA. One repair process uses a similar-looking, unbroken piece of DNA as a template to stitch the broken pieces back together.

Scientists can introduce tailor-made DNA into the cell — tricking the repair machinery into using the engineered DNA as the template for stitching together the broken pieces.

Students also learn about real-life examples of DNA editing, such as therapies for genetic illnesses including sickle cell disease and cystic fibrosis. In some exercises, they can see examples of the actual DNA sequence associated with each illness and evaluate specific gene edits proposed to cure the illness.

Hands-On Gene Editing With Bacteria

Ms. Gazda believes that hands-on lab lessons help students "open their minds to the idea that they can truly be a scientist." Several companies sell CRISPR curriculum kits to high schools and universities. One kit from Bio-Rad, a life science technology manufacturer, includes a prepackaged experiment using E. coli bacteria. (Fig. 3-4)

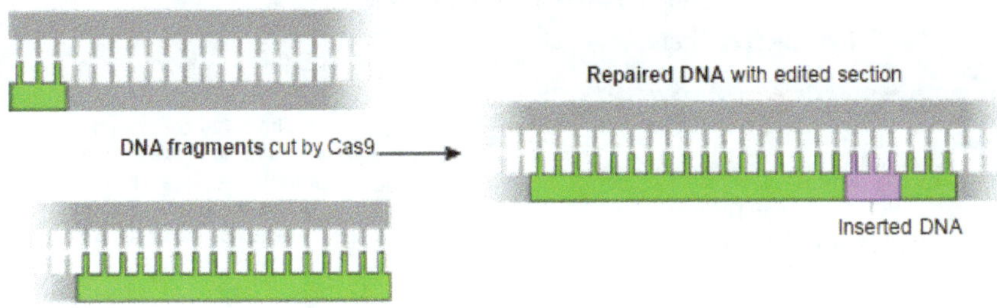

Fig. 2.

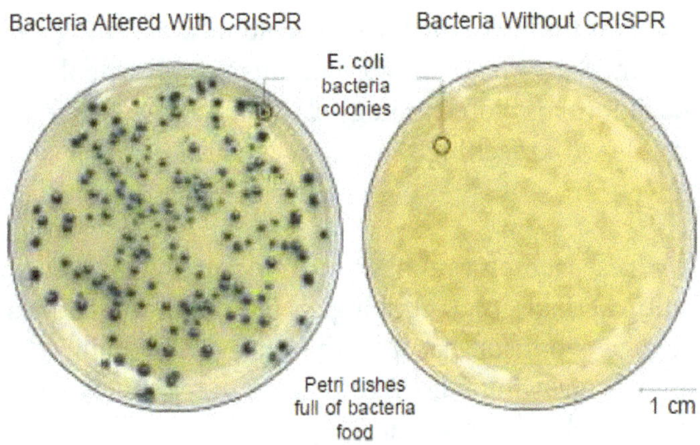

Fig. 3

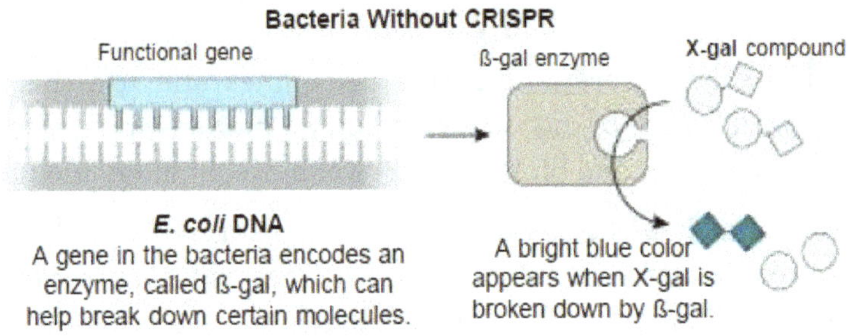

Fig. 4

The *E. coli* bacteria used in the Bio-Rad kit are grown on a food mixture that includes a compound called X-gal. Normally, these bacteria are able to use an enzyme to break down the compound into two parts: a sugar molecule and an indicator molecule.

The indicator molecule turns deep blue, showing students that the bacterial enzyme is working. This colorful display is an important part of the experiment. It shows students what to expect in an unaltered, or "control" group - a vital part of any scientific experiment (Fig. 5).

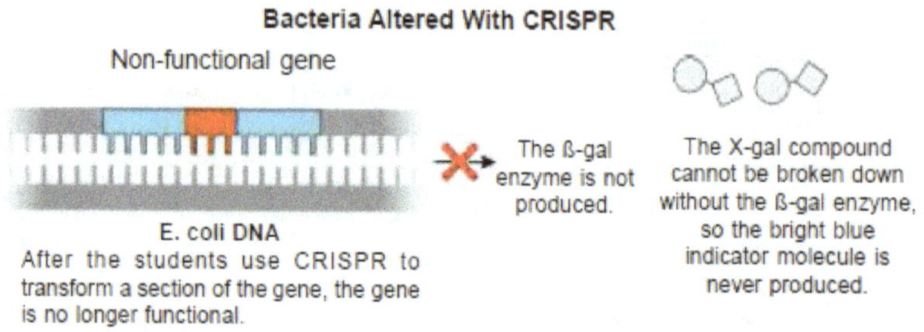

Fig. 5

The process of altering E. coli DNA with CRISPR involves lab techniques like pipetting liquids and carefully moving bacteria colonies. Teaching a new lab class like CRISPR can be intimidating, says Gregory Jubulis, a high school science teacher who uses the Bio-Rad kit in his biotechnology class. "It takes you a few years before you're real comfortable with teaching something," he said.

TRAINING THE NEXT GENERATION OF BIOTECHNOLOGISTS

Biotechnology is the research of a number of different fields, such as industrial, medical, and agricultural fields. Biotechnologists use techniques and knowledge to develop and create new products that help to improve the quality of human life. Biotechnology is an ever-growing field. As advances in technology and science are always happening, new subfields are emerging and opening new fields of growth and exploration.

Types of biotechnology and specialisations

Biotechnologists take different scientific disciplines and use them to improve processes in a wide range of industries, including biofuels, pharmaceuticals, food production, conservation and biofuels. Types of biotechnology include:

- **Medical and health:** The use of live organisms or biomolecular processes to improve or develop new drugs and treatments, cure certain disorders and illnesses and identify inherited diseases.
- **Marine and aquatic biotechnology:** Developing vaccines and medicines that help prevent certain diseases and viruses that may infect fish.
- **Environmental:** Pollution and contamination detection and control in the environment. Working towards the creation of renewable energy while also developing biodegradable materials that help reduce society's ecological footprint.
- **Biofuels:** Reducing the costs of bio-refining agents with the use of organic compounds. Creating chemicals from renewable biomass sources that reduce greenhouse gas emissions and put biofuels on the same level as fossil fuels.

- **Agricultural biotechnology:** Increasing productivity and pest resistance by genetically modifying crops and improving animal feeding.
- **Industrial:** The use of enzyme production and cloning to enhance the taste in food and drink and improve the preservation of food and drink. Using enzymes to improve the cleaning of clothes when washing at lower temperatures.

Biotechnologists usually specialise in areas like:

- Stem cell research for regenerative medicine.
- Cancer studies for the treatment and detection of cancer.
- Microbial sciences to study antibiotic-resistant bacteria.
- Virology for the study of viral diseases and viruses.
- Genetics to study DNA repair and detect heredity illness or genetic variation.
- Biochemistry to study plant science, microbiology, medicine and forensics.
- Pharmacology to study drug impact on biological systems.

Responsibilities of a biotechnologist

The amount of responsibility given to a biotechnologist depends on their level of experience and the type of company they are working with. Typically, biotechnologists:

- Create accurate reports and presentations using data and results gathered
- Maintain and operate laboratory equipment and computers for the company standard
- Ensure that health, safety, and environmental policies and regulations and met at a high standard
- Perform and monitor experiments involving live organisms to solve problems, develop innovative products and improve existing processes
- Follow new methodologies using scientific knowledge to produce quality results
- Continue with personal development by keeping up with new advances in developing products
- Work independently and as part of a team with other scientists

Skills of a biotechnologist

Biotechnology is a demanding field, so it's necessary to have skills such as:

- Problem-solving skills
- Attention to detail

- Excellent written and verbal communication skills
- Excellent organisation, planning and timekeeping skills to meet deadlines
- Good hand-eye coordination and the knowledge to use equipment accurately
- The ability to analyse and interpret data
- Excellent computer skills
- Analytical skills

Biotechnologists use scientific knowledge and techniques to develop new products that improve quality of life. They need a combination of technical and soft skills to excel in their roles.

Some skills and knowledge that biotechnologists need include:

- Knowledge of biology
- Knowledge of chemistry, including the safe use and disposal of chemicals
- Maths knowledge
- Analytical thinking skills
- Excellent verbal communication skills
- The ability to work well with others
- To be flexible and open to change
- Problem-solving skills

Biotechnologists use scientific knowledge and various methodologies to achieve results. They perform data analysis to interpret findings to support scientific investigations.

Biotechnologists face intricate challenges that demand sharp analytical and problem-solving abilities. They must be adept at critically analyzing data, identifying patterns, and devising creative solutions to complex biological problems.

Training programs

Training programs for biotechnologists can include:

- Technical competencies such as laboratory techniques, data analysis, regulatory compliance, and quality assurance
- New teaching methodologies to equip biotechnologists with data analysis abilities

In the fast-paced and innovative field of biotechnology, staying ahead of the curve is essential for success. As advancements continue to reshape the industry, organizations must invest in the continuous training and development of their biotech professionals to enhance their skills, foster innovation, and drive scientific breakthroughs. In this article, we will explore expert-level strategies to design and implement training and development programs specifically tailored to biotech roles.

- Assess Skills and Competencies: Before designing a training program, conduct a comprehensive assessment of the skills and competencies required for different biotech roles. Identify the specific technical expertise, scientific knowledge, and industry trends that are critical for success. This assessment will serve as a foundation for developing targeted training initiatives.
- Embrace a Blended Learning Approach: To ensure effective knowledge transfer and skill development, embrace a blended learning approach that combines various training methods. Offer a mix of classroom sessions, online courses, hands-on workshops, mentorship programs, and external conferences. This diverse approach caters to different learning styles and provides practical, real-world experiences.
- Collaborate with Industry Experts: Engage with industry experts, research institutions, and academia to bring cutting-edge knowledge and expertise into your training programs. Foster collaborations and partnerships to access specialized resources, guest speakers, and research insights. This collaboration will enrich the learning experience and expose biotech professionals to the latest advancements in their field.
- Provide Technical and Soft Skills Training: Biotech professionals require a combination of technical and soft skills to excel in their roles. Offer training programs that enhance technical competencies such as laboratory techniques, data analysis, regulatory compliance, and quality assurance. Additionally, prioritize the development of essential soft skills such as communication, teamwork, leadership, problem-solving, and adaptability.
- Foster Cross-Functional Experiences: Biotech professionals often work in multidisciplinary teams. Provide opportunities for cross-functional experiences, enabling them to gain exposure to different areas within the organization. Encourage rotations, job shadowing, and project collaborations to broaden their perspectives, foster collaboration, and develop a holistic understanding of the biotech landscape.
- Create a Culture of Continuous Learning: Instill a culture of continuous learning within your organization. Encourage biotech professionals to pursue ongoing education, attend conferences, and participate in industry-specific webinars. Support their enrollment in relevant certificate

programs, advanced degrees, or professional development courses. By valuing and investing in lifelong learning, you cultivate a workforce that is adaptable, innovative, and forward-thinking.

- Establish Mentorship Programs: Mentorship programs are invaluable for the growth and development of biotech professionals. Pair experienced mentors with early-career scientists to provide guidance, support, and industry insights. Mentors can share their knowledge, offer career advice, and help mentees navigate challenges and opportunities. Mentorship programs foster professional relationships and create a culture of knowledge sharing.
- Measure Training Effectiveness: To ensure that training programs are effective, establish mechanisms for measuring their impact. Use pre-and post-training assessments, surveys, and performance evaluations to gauge the knowledge gained, skills acquired, and behavioral changes observed. Gather feedback from participants to continuously improve the training initiatives and align them with organizational objectives.
- Encourage Continuous Professional Development: Training and development should extend beyond formal programs. Encourage biotech professionals to engage in self-directed learning and pursue their individual professional development goals. Provide access to learning resources, online platforms, and subscriptions to scientific journals. Support participation in industry conferences and encourage the sharing of knowledge within the organization.
- Foster a Learning Culture: Create an environment that values and promotes continuous learning. Recognize and celebrate employees' achievements in their learning journeys. Encourage knowledge sharing through internal seminars, brown bag sessions, and communities of practice. By fostering a learning culture, you create an atmosphere of collaboration, innovation, and growth.

How to become a biotechnologist

The complex nature of this field means that it's necessary for those who want to become a biotechnologist to complete many years of study. There are several steps to becoming a biotechnologist:

Earn a bachelor's degree

Holding a bachelor's degree in biotechnology allows you to become a biotechnologist. It's also possible to enter into the biotech sector by first completing a certification or diploma in general science before progressing on to a degree in biotechnology.

There are other degrees that act as doorways into the biotechnology industry, such as:

- Medicine
- Biology
- Microbiology
- Chemistry
- Process engineering
- Biochemistry
- Pharmaceutical science
- Environmental biology
- Chemical engineering

On some occasions, employers may look for a postgraduate qualification, such as a PhD or a master's. These postgraduate qualifications are important if you wish to pursue a career in academia or research.

Gain relevant work experience

Usually, while completing a bachelor's degree, most courses tend to offer work experience or internships within the biotech sector. Work experience allows you to get in-depth experience and learn more about the industry and also gives you the opportunity to interact with professions in the sector. The contacts you meet during work experience are invaluable, as they may lead to future career progression or may become mentors.If possible, try to secure your work experience in a laboratory setting. Gaining experience while working in a laboratory improves your chances of finding work after completing your degree. While lab experience is excellent, it's important to gain experience in your field of interest. There is also the possibility of gaining experience through work shadowing. Any experience that gives you commercial insight is a bonus, as some employers find it favourable to employ candidates with an understanding of business.

Continue professional development

When starting with a company, it's normal to complete a company induction. These inductions normally include regulation training, health, and safety training, and training on the control of hazardous substances. It's also likely that you may continue to get on-the-job training, which helps to keep up to date with new biotechnological techniques. Some companies may provide employees with permanent contracts with an annual personal development plan. Usually, a line manager creates this plan with each employee. This plan helps keep employees motivated while encouraging them to attend workshops and conferences to improve and learn new technical skills.

Career prospects and potential employers

As a biotechnologist, the opportunity for career progression may depend on several factors, such as area of speciality, company size, industry and qualifications. It's possible with enough experience to progress into more senior specialised positions. It may also be helpful to earn additional professional qualifications to help improve the chances of progression in a company. There is also the opportunity to move into a different area within the company or organisation, such as production, business development, information, and data technology or regulatory roles.Many larger and private-owned biotechnology companies may advertise vacancies with a focus on pharmaceutical, biochemical and medical disciplines. Smaller companies usually advertise roles under different job titles. There are a number of alternative job titles for the role of a biotechnologist, such as:

- biochemist
- genomic technologist
- bioengineer
- microbiologist
- bioprocessing engineer
- biophysicist

Biotechnology jobs are available in several areas:

- manufactures of food and drink
- agriculture organisations
- research institutes
- chemical and pharmaceutical companies

EDUCATIONAL INITIATIVES AND OUTREACH PROGRAMS

Biotechnology education is a broad field. Public education in modern biotechnology is a difficult and complicated endeavour. When biotechnology is discussed, usually two very different but interrelated areas come into context. On one hand, there is what can be termed as classical biotechnology, which covers activities such as fermentation and breeding by genetic selection, as two common examples. On the other hand is modern biotechnology, which takes into account the molecular systems that are involved in biology and these, will primarily cover recombinant DNA technology and its applications.

Classical biotechnology may have been around for thousands of years but is still relevant in this day and age of the post genomics era. Together, both technologies can work hand in hand, to provide a better quality of life for current and future generations. This is even truer in developing nations.

Many of these countries still suffer from undernourishment and a low quality of life in comparison to populations of developed nations. In many developing countries, biotechnology has become a source of economic development and social progress.

The policy makers are faced with the problem of responsibly educating the Malaysian public, while at the same time seeding the grassroots with the prospects of careers in biotechnology research and industry. It was deemed that one feasible and effective way of doing this was by embarking on a biotechnology awareness program targeted at students of Malaysian high schools.

This decision was made considering that the final years of high school is usually a time when students start thinking about true career directions as opposed to just childhood dreams. This age range was also selected when taking into account the maturity of the students, to begin comprehending modern biotechnology, and at the same time, this age range was also deemed suitable as students of this age group are generally vocal and critical enough to disperse the information to their peers as well as other family members.

APPLICATIONS OF CRISPER-CAS9 TECHNOLOGY

Human Science

Recently, CRISPR-Cas9 is being employed to study various genetic diseases, such as hemophilia, Duchenne muscular dystrophy, a1-antitrypsin deficiency, hearing loss, and hematopoietic diseases. Hematologic diseases have previously been difficult to treat via genome manipulation. Studies in recent years demonstrated that CRISPR-Cas9 can correct genetic errors in hematopoietic stem cells that give rise to hematologic diseases. These can then be applied to CRISPR-Cas9-based hematopoietic stem and progenitor cells (HSPCs) transplantation therapy.

One of the most promising methods to treat hematopoietic diseases is editing the HBB mutation with CRISPR-Cas9, and it has been successfully carried out in patient-derived induced pluripotent stem cells (iPSCs). In a survey, the CRISPR-Cas9 technology was used to rectify the *HBB* gene mutation in HSPCs. Final outcomes inferred a decrease in the amount of hemoglobin and sickle cells.

Since the CRISPR-Cas9 system originated from bacteria and acts as a bacterial immune system against invasive genetic agents, it has an inherent benefit in the treatment of bacterial and viral infection, and it is known as a new type of antiviral therapy against various incurable viral infections. Oncogenic viruses are highly related to carcinogenesis, including: Human papillomavirus (HPV), which could cause cervical cancer; Epstein Barr virus (EBV) that causes nasopharyngeal carcinoma; and hepatitis B virus that triggers liver cancer.

CRISPR-Cas9 was first used to cleave HBV genomes. It has previously reported that the production of the HBV core can decrease through the HBV expression vector. Among human genetic diseases, cancer is the most common cause of death universally. The tumor-suppressor genes have an important role in tumorigenesis. They can downregulate cancer progression by controlling cell proliferation and differentiation. CRISPR-Cas9 technology can target tumor-suppressor genes and restore them to interdict the tumorigenesis.

Clinical Trials Using CRISPR-Cas9 Technology

The United States clinical trial database contains useful information on experiments that have been used various genome editing tools such as ZFN, zinc finger, and CRISPR, Cas9, Cas12, and Cas13. Clinical trials using CRISPR-based therapies are still in their primary stages. It other words, if the treatments are safe and effective, they are likely to be several years away from legal authorities. New possibilities in precision medicine have arisen with the advent of CRISPR technology.

Recently, many experiments are underway for several diseases such as several types of cancer, blood disorders, chronic infections, eye diseases, and protein folding disorders. All CRISPR-related trials have been performed to edit specific cells or tissues without affecting eggs or sperm; hence no DNA changes can be transferred on to future generations during these experiments. The first application of CRISPR-based therapy refers to February 2019 in Germany. In this ex vivo trial, twelve patients were treated, and seven of them have been followed for at least three months. As a considerable result, they have observed that the patients had not needed blood transfusions even after treatment. Recently, researchers have provided a comprehensive information on application of CRISPR in the clinical trials.

Clinical Trials of the Eye Based on CRISPR-Cas9

Clinical trials of in vivo genome editors have also begun. In all trials, tissues such as the eyes, cervix, and liver have been subjected to experiment. Cas9 was delivered to the eye using Adeno-associated virus vectors (AAV) as a therapy for congenital amaurosis type 10 LBA (LCA10). Because the most frequent LCA10 produces a mutation in the intron of the Centrosomal Protein 290 gene (CEP290) and provides a new binding site that changes the mRNA to form an early stop codon, it is a good candidate for the therapy of Cas9. Leber's congenital amaurosis 10 therapy (LCA10) is reported as the first Cas9-related clinical trial. This therapy employs two sgRNAs that, when combined, cause partial deletion of the intron or in-version of the partial intron, resulting in normal CEP290 protein production in the patient's cells. Since the eye is an accessible tissue and subretinal injections in mice and mammals have provided stable gene editing, hence successful therapy may also be done in humans.

Limitations of CRISPR-Cas-Based Gene Therapy

There are many projects related to CRISPR-Cas9 for disease research, and recent results and reports summarize the benefits of CRISPR-Cas9. Before employing CRISPR/Cas9-based gene treatments in humans, their safety and efficacy should be assessed and adjusted. One of the main limitations of this technology is that not all mutation sites play a protospacer adjacent Motif (PAM) role, such that target detection depends entirely on this role.

In addition, DNA damage response, delivery vehicle, editing at off-target genomic sites, and immunogenicity are other challenges of using CRISPR-Cas9 for gene therapy. The limitations of CRISPR-Cas9 technology are mainly related to extra-target DNA fracture formation, extra-target mutations, and PAM sequence dependence. Off-target effects are a serious problem in genome editing trials. The CRISP-Cas9 technology has a great chance of causing off-target alternations in human cells compared to other editing approaches. Mutations outside the target location can induce gene dysfunction and, in rare cases, cell death. Choosing a suitable target locus possible during the bioinformatics analysis phase improves the effectiveness of the CRISPR-Cas9 system. Combining Cas9 D10A with sgRNA results in hydrolysis of one DNA strand at the target location.

Plant Science

Success in plant breeding depends on trait variability as well as overall genetic variation among plant populations. Recently, yield performance and several quality-related properties have been engineered by genome editing. CRISPR-Cas9 has proved useful in crop improvement to increase disease resistance, improve yield, improve nutrition, and aid domestication.

Plant Disease Resistance

Biotic stresses have usually been controlled by the spread of pathogen-resistant varieties and by applying agrochemicals. However, agrochemicals can cause contamination of the environment and negatively impact human health. Plant pathogens are also constantly evolving and can become unresponsive to these controls. Therefore, to create pathogen resistance in plants, resistance genes from wild species were introduced via breeding and established genetic transformation technologies involving large genomics regions.

Nonetheless, given the lack of specificity of the large genomic regions being introduced, other less-desirable traits may also be introduced using these approaches to create elite cultivars. Conversely, the CRISPR-Cas9 system provides a more accurate approach to genetic modification. Since first introduced to the field, it has proven to overcome several agricultural challenges including biotic stress resistance, fungal and bacterial disease resistance, and viral resistance. For instance, CRISPR-Cas9 was employed to knockout the mitogen-activated protein kinase-5 (*OsMPK5*) gene to enhance disease resistance in rice.

As the latest genome editing tool, CRISPR-Cas9 has a key potential in developing bacterial resistance in plants. In several studies, the considerable potential for CRISPR-Cas9 technology has demonstrated to counteract crop bacterial diseases. For example, mutagenesis of the ethylene-responsive factor (ERF) transcription factor gene *OsERF922* using this technology enhanced resistance in rice cope with the blast disease. In another case, researchers modified the effector-connection element in the promoter of the *Lateral Organ Boundaries-1* gene (LOB) to confer resistance to citrus bacterial canker (CBC) caused by *Xanthomanas citri* subsp. in Duncan grapefruit.

Examples for Biotechnology experiments and topics in the classroom

Already in elementary school traditional biotechnical methods can be dealt with, given that there is a relationship to the everyday life of the students, i.e. the application of biotechnology in food production (Table 1).

Table 1. Examples for biotechnological experiments in the classroom

School level	Example
Elementary school	Production of yoghurt and sauerkraut.
Middle school	Experiments with and microscopical investigations of yeast and yeast dough.
	Gradual formation of lactic acid in sour dough.
	From juice to wine, from wine to acid.
High school	Enzyme-Linked-Immunosorbent Assay - detection of the Pelargoium Flower Break Virus in Pelargonium leaves.
	Hydrolysis of l-DNA by restriction enzymes and separation of the DNA-fragments by gel electrophoresis.
	Production of citric acid.
	Comparative protein analysis.

By carrying out experiments with microorganisms students make the experience that these creatures play an important role in food production, and they learn to look onto microorganisms in a differentiated way. In addition the meaning of fungi, bacteria and of other small organisms for decomposition can be shown by investigations of compost.

On middle school level biotechnological processes can be integrated into the topics "ecology" and the education in environmental problems, i.e. in the meaning of microorganisms as reducers in the nitrogen cycle. In the context of health education bacteria, funghi and viruses are dealt with as pathogens that cause infections.

When talking about human physiology with about 15 year-old students, enzymes and enzyme technology can be picked out as a central theme. By easy experiments, i.e. the degradation of pectin in the context of juice production, a connection can be made to economically important fields of Biotechnology. On this school level also the topic breeding - especially modern ways of cultivation - can be a linked to Biotechnology.

On high school level the students ought to deal not only with applications of biotechnology but as well with the historical development of biotechnological knowledge. They are to know the main methods used in biotechnology and they must learn to judge in a reasoned way the (very often) ambivalent application possibilities of this technique. This means that ethical and social aspects of the complex topic biotechnology must be integrated in the science classes.

Research on preconceptions of bacteria and consequences for teaching

Microorganisms are indispensable for biotechnology. As we know pre-educational students conceptions of the topic to learn are very much influencing the learning process in the way that they hinder the acquisition of the scientific conceptions. Therefore it is necessary to find out students' preconceptions on microorganisms and to integrate them in teaching programmes about microorganisms and biotechnology.

Students' preconceptions of microorganisms in general and on bacteria in particular are hardly considered in literature. Researchers report that in general microorganisms are regarded as pathogens, and Researcher writes that in respect to bacteria the term mutation and mutant are being confused. Considering the important role bacteria have, regarding the education in the upper secondary level about biotechnology - including genetic engineering - we investigated preinstructional conceptions of 17-years-old-students about bacteria in detail.

The investigation was carried out by interviews following the method of White and Gunstone, 1992 for "interviews about concepts". This interview form was developed by the authors in order to sum up all the knowledge entities a person has concerning a certain term. For White and Gunstone it is decisive that the knowledge of a person concerning one term does not consist only of propositional parts but that it is determined by attitudes, opinions, own experiences and mental images.

For being able to take these aspects also into consideration in interviews the authors use a categorizing system of different types of knowledge in respect to the common cognitive-psychological presumptions about knowledge representation and thinking processes. This includes mainly propositions, images and episodes, and in addition intellectual and motor skills. Following researchers already the knowledge about a single term is a function of the multitude of the different types of knowledge that a person relates to it. In conclusion the sum of these elements constitutes the understanding of this term. Because of the subdivision into different categories of knowledge the 'interview about terms' suits also (and especially) for the analysis of preinstructional students' conceptions, i.e. for the pre-instructional knowledge the students have about a certain issue when they come to school; since students' conceptions include besides propositional knowledge about facts especially everyday experiences (like episodical knowledge) and mental images. Conclusively White and Gunstone use this technique for the investigation of the pre-instructional knowledge of students.

The interview guideline we used in the main study was developed on the basis of ten interviews with students of the 11th grade. In the main study interviews were carried out with fifteen students of the 11th grade about their knowledge on bacteria. The answers the students gave during the interviews can be categorized as follows: taxonomy of bacteria, comparison with other organisms, features of bacteria, occurrence of bacteria, fighting bacteria and the meaning of bacteria for biotechnology. In this context we will only outline the results concerning the occurrence of bacteria and the meaning of bacteria for biotechnology.

The students' ideas where bacteria occur in the human body are very diverse. The students let bacteria play a role in several organs of the human body. First of all there is their role in the intestine. 4 students make remarks about a role of bacteria in digestion. One student mentioned the idea that bacteria absorb noxious agents. A similar remark was made about the role of bacteria in the liver ('Bacteria in the liver process "unwanted" molecules'). Another student expresses the idea that bacteria in the liver play a role in the purification of the body.

The general as well as the detailed ideas about bacteria in the liver are incorrect. Bacteria only play a role during digestion by the production of vitamin K. Students not only made remarks about the role of bacteria as 'purification assistants' in the liver but about the role of bacteria to purify blood as well. These remarks show the lack of understanding about the human immune system.

According to the students' ideas bacteria also play a role in the production of bile. They think bacteria enhance the effectiveness of bile. A great number of remarks applies to the meaning of bacteria for the activity of the human stomach. 3 students give bacteria a role in the digestive process. In relation to gastric juice one student states that bacteria use it for nutrition. Other students think that bacteria can survive the gastric juice by the use of their own immune system.

Furthermore one student remarks that bacteria survive the acidity of the gastric juice by using the mucous membrane of the stomach. This is correct for Helicobacter pylori but it is not sure that Helicobacter pylori is meant. Altogether, according to the students interviewed, bacteria are present on the skin, in the eyes, blood, intestine, bile, liver, stomach, kidneys and 'organs in general' (Table 2).

Beside the human body students mention the soil as a place where bacteria live. This general idea is correct. Besides soil the following excorporal places are named: dirty places, swimming pools, glass, living creatures and air. The students name food that is prepared with the use of bacteria like yoghurt but as well they mention food that is not prepared with the use of bacteria like wine or beer. This finding suggests a partially correct general idea of the use of bacteria in the production of food.

Table 2. Overview of the students' remarks about the occurrence of bacteria in the human body and the role they play in different organs. (n=15).

Occurrence of bacteria in/on the human body	Total number of remarks	Number of students
Occurence of bacteria on the human body	49	10
Role in the human body	7	3
Role in the blood	6	5
Role in the intestine	34	10
Role in the liver	6	3
Role in the bile	2	1
Role in the stomach	13	8
Role in the spittle	3	3

Concerning bacteria and their relation to diseases, 2 students mentioned that bacteria secrete toxic substances. 2 students relate bacteria to diarrhoea. Another student makes bacteria responsible for the weakening of the immune system. The following diseases including corresponding toxic agents are mentioned: cholera – contaminated water, malaria fever – stablefly. Besides these illnesses the following are named: cold, flu, pneumonia, ulcer and tonsil inflammation. These remarks indicate detailed ideas about bacteria and their relation to illnesses.

As mentioned above students connect the production of particular food with bacteria. The first biotechnological process students connect to bacteria is the purification of water. They state, that bacteria purify water by decomposition of chemicals. This is a scientifically correct idea. Besides chemical purification one student mentions mechanical purification. This is correct as well since water is enriched with oxygen in a mechanical way. Two students mention chemical purification of water explaining that bacteria are killed by added chemicals. The remarks relating to molecular genetics are very diverse. One students names resistance, another one the restoration of races as a result of molecular genetics.

The statements are very general and only roughly correct. More detailed and correct is the statement that races are restored by the fusion of genetic factors. Four students name manipulation of genes as a description for molecular genetics. Two of these name the incorporation of resistance against a disease as characteristic. One student names the testing of medicines as characteristic for molecular genetics. One student makes detailed remarks about genetics talking about the extraction and modification of chromatine. Two remarks are made about methods for the transfer of DNA namely via bacteria and via viruses. These results indicate a partly deeper idea about molecular genetics.

Students' interest in Biotechnology topics

Interest in and attitudes towards the topics to be learned have a clear cut effect on the learning outcome of students. Researchers conducted an empirical study in which they captured students interest in and attitudes towards biotechnology and gene technology in particular. In this quantitative study they used questionnaires. The results showed that the interest of students in genetic engineering develops at an age of about 16. Girls of this age are more interested in social and ethical aspects of the topic, boys' interest, however, is directed towards economical and technical aspects. This interest is little knowledge-based. It is more a kind of curiosity and a general openness towards a new technology. The interest in genetic engineering develops parallel to the interest in other general social issues. Media seem to have more impact on the development of interest in this technique than science education at school. The students show a particular interest in advantages and disadvantages of gene technology, in its risks, in application possibilities in medical care, in the maintenance of world nutrition as well as in ethical questions. The youth shows a less interest in the application of gene technology in animal and plant production.

Between the interest and the negative attitudes towards gene technology there are more or less no meaningful (negative) correlations. In respect to the use of genetic engineering in environmental protection and for the diagnosis of hereditary diseases hopes prevail; in respect to the use of gene technology for plant and animal production as well as in connection with the release of genetically modified organisms students' fears are higher than their hopes. Students see risks in connection with the abuse of gene technological knowledge and in the possibility that gene technology could get out of control in research as well as in its different application fields.

The acceptance of genetic engineering is very differentiated. Most of the students interviewed ask for a sound public information about risks, about control mechanisms, and about ethical aspects before a decision on the genetically modification of an organism or species, respectively, is taken. Students show a serious interest in a sound informing on the topic gene technology. Todt and Götz, 1997 give a detailed overview on the hopes and fears of students concerning the various applications of this modern kind of biotechnology.

Material for teaching Biotechnology in schools – The European Initiative for Biotechnology Education (EIBE) as an example

Several investigations in Europe show that citizens are very critic whether their governments and other public committees can cope with the risks of biotechnological applications. As a reason for this findings a lack of sound information of the public is seen. As a reaction to this results in several countries of the European Union the topic "Biotechnology" has been integrated into the school curricula.

This fact led – among others – to a project called EIBE (European Initiative for Biotechnology Education) that was financed by the European Commission. The aim of EIBE is the information of the public – and of school students in particular – on Biotechnology.

About 25 teaching units on different biotechnology topics were developed by international groups of scientists and educators. They are published in the internet under the address http://www.rdg.ac.uk/EIBE/ENGLISH/INTRO.HTM.

In these materials information for teachers, student activities and materials for students as well as hints for literature etc. are combined. One characteristic of the units is their multidisciplinary approach. Not only the science aspects but as well ethical, social, legal and economic questions related to the topic in question are integrated in the materials.

Table 3. Examples for units developed by the European Initiative for Biotechnology Education (EIBE). Each unit is available in English and in several other languages as well.

Microbes and Molecules	Transgenic Plants I + II
DNA Fingerprinting	Transgenic Animals
Biscuits and Biotechnology	A Model European Council

Issues in Human Genetics	Novel Food
Fermentation Technology	Human Genome Project
DNA Model	Biotechnology and the 3rd World
Debate of a Personal Dilemma	Environmental Biotechnology
Practical Immunology	The EIBE-Family

Ethical aspects of biotechnology – how to teach them in biology classes?

As has been mentioned before when teaching biotechnology at school the topic can not be restricted to pure science aspects. Especially social and ethical questions related to the topic are the aspects students are mostly interested in and occupied with. When teaching a biotechnological issue in school we therefore must differentiate two dimensions: the descriptive-explicative dimension, that deals with scientific facts and methods on the one hand and that can be judged as right or wrong; and on the other hand the normative dimension that deals with the moral judgement of human actions that can be judged as morally good or bad. With this second dimension the teaching aim is connected, to improve students' ability of moral judgement.

This term means in this context the ability to decide in a dilemma situation between two values - i.e. the health of a human being versus the health of an animal - in a sound way. To give the students a structure for their argumentations we define to higher categories: the category of welfare of man and/or nature on the one hand and the category of human dignity on the other.

The question, however, is, how to deal with these two dimensions in the classroom with the students in a reasoned way. For this purpose we developed an ethical analysis to help teachers to structure their teaching on ethically ambivalent topics like biotechnology. By using the ethical analysis we try to help the students to develop a reasoned opinion towards a particular topic of biotechnology. In addition the students shall get a pattern at hand how he or she can develop a sound opinion in normative questions.

The ethical analysis can be structured in the following steps:

- Formulation of the dilemma;
- Finding out the possible options of action in this situation;
- Assignment of the optional actions to values that are touched by the action and reflection of the consequences;
- Taking a justified decision for one of the options of action in respect to the reflections in step 3;

- Assignment of the decision taken to one of the two higher categories of argumentation;
- Description and discussion of the consequences of the decision taken in step 4 for the individual and for the society.

A question that could be dealt with in the classroom – after teaching the students knowledge on DNA and basic molecular biological methods - by ethical analysis is for example the question whether human DNA should be made patentable or not. When using the ethical analysis it is indispensable that at first the students get information on the topic; this means the descriptive-explicative dimension must be considered as the basis for the ethical reflection. In this case the students must for example get information on patents.

The three options of action that are possible in this case are:

- All human DNA-sequences must be patentable, no matter whether their function is known already or not;
- Only DNA-sequences can be made patentable, which function is known;
- DNA-sequences may not be made patentable at all

In the next step the students assign values to these three options that are touched by the particular action respectively. In this example this could be the right of self-determination, economic profit or individual health. In the following the students take a decision for one of the options of action and assign it to one of the higher categories of argumentation.

The point of reference in this example for self- determination would be the dignity of man; the welfare of man would be the point of reference for the economical profit or for the right of individual health. Finally the students describe the consequences of their decision (see step 6). In conclusion the different opinions that have normally been developed by different student groups in the classroom and their way of argumentation are discussed in a comparative way.

ROLE OF BIOTECHNOLOGY IN STEM EDUCATION

Biotechnology is an academic subject and professional pursuit that is included in STEM. Biotechnology's interdisciplinary nature requires biotech professionals to be well-versed in multiple STEM areas, including: Biology, Chemistry, Environmental science, Physics, Engineering. STEM education is an interdisciplinary approach to learning, where students use science, technology, engineering, and mathematics in the real context that connects schools, community, work, and the global world to develop STEM literacy. Teachers can develop, apply, and support STEM literacy in the classroom learning by concerning three aspects. Firstly, Science, Technology, Engineering, Mathematics should not be seen as separated contents. Secondly, personal, social, and economic needs. Thirdly, domains of cognitive, affective/ attitude, and psychomotor. Eventually, this STEM literacy will make students being able to compete in a new economic era which is based on knowledge. Besides that, they are also trained to understand the complex issues so that they can find creative solutions. However, the separation of STEM literacy aspects particularly in computing the test results is needed to identify the weak points of students on each aspect of literacy.

Biotechnology constitutes a scientific study in science and technology which experiences rapid growth in recent years. Biotechnology can be applied in various fields and has social, ethical and economic implications in the form of risks and benefits for human life and the environment. Considering the impact of biotechnology on food, health, agriculture and the environment is very important; therefore, learning biotechnology in the school environment is needed for students as the sufficient provisions of scientific knowledge in developing a better understanding of the world around them. Biotechnology (especially food biotechnology) is one of science materials in secondary school, which contains STEM. Besides, the material is closed to students' daily life, especially the present of technology that makes students easier to access information related to biotechnology.

Experimental Method

The method that used in this research was the descriptive method. It was used for determining profiles of students' mastery in STEM literacy without any treatment previously. The design of the research can be seen in Fig. 6.

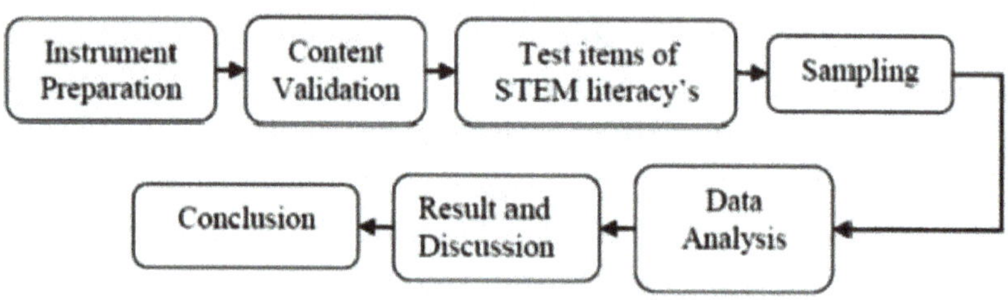

Fig. 6. Research design

The sample was 45 students of grade 9 in one of the junior high schools in Tasikmalaya district. The sampling technique used in this study was purposive sampling because cluster group has been formed in school and there was zero change by the researcher. The sample selection was based on particular purpose, which is because students have been trained in the biotechnology topic. The instruments used in this research were 24 multiple choice questions, consisting of 10 questions on science literacy, 8 questions on technology engineering literacy, and 6 questions on mathematic literacy problems. The scientific and mathematical literacy problems were developed using indicators which were developed by PISA

2015, while technological and engineering literacy was developed using indicators which were developed by NAEP 2014.

An instrument is said to be valid and reliable when it is able to measure what will be measured and can reveal data from the variables studied consistently. In this study, the validity testing was performed by the validator experts consisting of five people who are experts in the field of STEM, assessment, and biotechnology content. The validation process resulted in content validation index (CVI) of 0.907 using CVRcritical = 0.877 on the level of a = 0.05. Reliability analysis using Cronbach's alpha was processed using SPSS 20 resulted in the reliability value = 0.786. The results of test scores show appropriated validity and reliability.

Exploring Biotechnology Kits and Labs

Biotechnology education comes alive through hands-on experiences in well-equipped labs. Assuming adequate resources, a top-notch student biotech lab should feature instruments like thermal cyclers, gel electrophoresis apparatus, and microcentrifuges. These tools are used throughout the biotech field, and the process of using them will help students build better laboratory skills and gain greater confidence in undertaking advanced research.

Short of a fully-equipped, in-school lab, educators can turn to a wide range of prepared kits that introduce biotechnology concepts in a way that keeps students engaged.

For beginners, a DNA extraction kit provides an excellent starting point, helping students extract DNA from various sources to understand its importance in biotechnology. As they progress, intermediate kits can introduce techniques like PCR, which amplifies specific DNA sequences, and gel electrophoresis, which separates DNA fragments based on their size. Advanced kits might involve gene editing experiments using technologies such as CRISPR-Cas9, giving students a glimpse into the latest biotech advancements.

Equipping Educators to Teach Biotechnology

With biotechnology achievements moving forward so rapidly, professional development for biotech educators is critically important. By participating in specialized programs, teachers can enhance their knowledge, teaching methodologies, and interdisciplinary collaboration skills.

Here are some notable teacher training programs related to biotechnology:

- The Amgen Biotech Experience offers teachers hands-on molecular biology lab experiences. Participating educators gain practical skills in biotech techniques, enabling them to bring real-world applications to their students.

- BioBuilder focuses on synthetic biology and principles of genetic engineering. Educators are trained in a comprehensive curriculum aimed at introducing synthetic biology concepts in the classroom, such as by designing and building biological system.
- The Genome Project and similar programs provide comprehensive resources and curriculum guides. These online platforms help educators integrate genomics and personalized medicine into their teaching—with tips for delivering lessons remotely or in a blended learning environment.

Remember, professional development is not limited to specific programs. It extends to online courses, workshops, and other resources—wherever you passion for teaching biotechnology leads you.

Success Stories in Biotechnology Education

Many schools have implemented successful biotechnology education programs, igniting the curiosity and passion of their students. Here are some high school-based biotech programs in the United States that are said to be making a significant impact on biotechnology education:

- The Jefferson High School Biotechnology Academy (San Francisco, California) offers a comprehensive biotechnology curriculum that spans multiple years. Real-world applications of biotechnology are explored, including biofuel production and genetic engineering for medical purposes.
- The Biotechnology and Life Sciences Program at North Carolina School of Science and Mathematics (Durham, North Carolina) focuses on giving students a solid foundation in biotech and life sciences. Critical thinking, problem-solving, and the ethical considerations in biotechnology are covered in detail.
- The Biotechnology Signature Program at Thomas Jefferson High School for Science and Technology (Alexandria, Virginia)—recognized as one of the nation's leading STEM schools—is said to be especially rigorous. It emphasizes interdisciplinary collaboration and partnerships with universities and industry professionals.
- The Biotechnology Academy at Montgomery Blair High School (Silver Spring, Maryland) offers a comprehensive curriculum on key aspects of biotechnology, from basic lab techniques to much more. As with other successful programs, it stresses industry connections and real-world applications.

Navigating Ethical Considerations in Biotech

Biotechnology education goes beyond the laboratory, often stimulating deep discussions about the ethical considerations that can arise in the field. Many experts say topics such as biopiracy, genetic testing, informed consent, and social justice are essential in shaping responsible biotechnologists.

For example, educators can engage students in exploring the implications of biopiracy, whether it involves the unauthorized commercial use of living genetic resources or the traditional knowledge of Indigenous communities. By discussing the impact of so-called biopiracy on biodiversity and cultural heritage, students gain a deeper understanding of the ethical dimensions of biotechnology.

Genetic testing is another critical topic that prompts discussions on privacy, discrimination, and consent. Educators can guide students in examining the benefits and potential drawbacks of genetic testing, considering issues such as access to healthcare, genetic discrimination, and the importance of informed consent in research.

Some teachers believe biotechnology education also provides a platform for exploring social justice issues. By examining whether there is unequal distribution of biotechnology advancements and access to resources, students can develop a sense of social responsibility and advocate for equitable practices within the field.

Challenges and Solutions in Teaching Biotechnology

While biotechnology education offers tremendous opportunities, it also presents challenges.

Access to resources, including equipment, reagents, and laboratory facilities, can be a barrier for schools and educational institutions. However, innovative solutions such as virtual labs, open-source resources, and collaborations with research institutions can help overcome these issues, providing students with valuable experiences even in resource-constrained settings.

Another challenge lies in addressing safety concerns associated with working in a biotechnology lab. Educators must prioritize safety protocols and ensure that students understand the importance of following guidelines to prevent accidents and promote responsible laboratory practices.

Public perception of biotechnology can also pose challenges in the classroom. Addressing possible misconceptions and fostering a positive perception of the field requires effective communication and public engagement. Educators can encourage students to present research findings, participate in science fairs, and engage in community outreach to highlight the benefits and responsible use of biotechnology.

Benefits of STEM Courses

When questions like what is STEM and why to pursue these courses come up, there is not one definite answer. In the current scenario, 75% of the total jobs in such industries need STEM graduates. Since, this demand for qualifications will continue to rise in the coming future, being a STEM graduate would always have an edge over the thousands of job applicants overseas.

To cater to the above requirements, the STEM education system is also evolving at a high rate to produce more graduates every year. The programmes are more focused on enhancing the way of learning while making its global student body network better than before.

Its practical learning through real-world projects and workshops prepares the students with a strong theoretical and practical foundation. Additionally, the industrial exposure they are provided with as an essential part of this training equips them with the exact tools that are required to succeed in their challenging careers.

The best feature of the STEM education system is that it transforms students into professionals who can adapt to a changing workplace.

Importance of STEM Education

- Vocational Training
- Experimental Learning
- Real-World Exposure
- Better Networking Opportunities
- More Professional Demand

The more students exploit a STEM University's resources to conduct experimental learning, the more the chances of their academic success. Similarly, succeeding in their respective domains would increase their chances of being headhunted for industrial training by a global technological and scientific giant.

To be precise, the STEM Education System or a STEM Course prepares aspiring Scientists, Technologists, Engineers, or Mathematicians to live their dreams on better platforms with better recognition.

Benefits of STEM Education Abroad

With more overall exposure, STEM graduates are certain to enjoy more benefits than any other generic course. Hence, pursuing STEM, especially on the international front, has an endless list of benefits for its graduates at both levels, bachelor's and master's. International universities offer a multicultural environment wherein practising a discipline becomes more fun and fruitful.

Benefits of Pursuing STEM Education

More Length on Work Visa

- Especially in USA

Higher Chances of Immigration

- In USA, Canada, UK, Australia, Germany

Better Job and Business Opportunities

- Especially in USA and Germany

Global Internship Opportunities

- Popularly in Taiwan, Italy, Thailand, China, New Zealand

Higher Remuneration

- ROI in Less than a Year

Scope of Employment in Research Cell

- Especially in Australia, UK, and European Countries

Ranging from fun learning to amazing job opportunities, let's discuss a few of its benefits:

Global Popularity and Competition in STEM Education

STEM is highly popular in the international education system, especially after considering that 75% of the total jobs require STEM graduates. As such the influential universities offer a variety of courses in the fields of Science, Technology, Engineering, and Mathematics. However, on a global account, it has been observed that the majority of students do not stick to pursuing their interest in science as they grow. With changes in the global economic scenario and business environment, new courses such as STEM MBA in USA are gaining significant traction.

Many students switch their interest in pursuing Science and end up taking Commerce and Humanities after completing high school. Surprisingly, more than 50% of Science students change their domain after completing their twelfth grade. This reduces the competition of STEM applicants on the bachelor's level alone to a great extent.

In addition to this, there is a huge share of graduates who prefer not to complement their bachelor's degree with a master's qualification. This further reduces the competition, making it easier for master's applicants to add on to their academic profiles, thereby becoming more employable STEM professionals.

The depletion in students as they progress to higher education shows how minimal the competition gets moving forward. Most of the students who drop their plans of pursuing STEM are unaware of how much STEM Education has to offer.

Students who change their streams after the twelfth do not know how many stems Graduates have to relish. The ones who drop off after their bachelor's degree in STEM are the ones, who prefer finding jobs rather than continuing their education. Needless to say, STEM Graduates with a master's degree have an edge over STEM undergraduates in the global industry.

Eligibility for Pursuing STEM Courses Abroad

The basic eligibility to pursue STEM courses abroad requires one to crack the mandatory standardised tests such as the GRE/GMAT for master's courses and the SAT for bachelor's courses. Besides, you also need to clear the English language tests. Check out which exam to take: IELTS vs TOEFL vs PTE Academic.

Additionally, you should have a basic qualification in the exact field of Science, Technology, Engineering, and Mathematics that you aspire to pursue at a higher level. Your previous course curriculum should have exposed you to the domain you aspire to study.

If you have already finalised a course or a country and you seek guidance with the university shortlisting or the application process, you may connect with our education counsellors here. Our experts are trained to help Indian aspirants with the entire application process ranging from profile evaluation to visa filing. They would assist you in realising your dream of flying to your study destination with their effective free-of-cost counselling services.

STEM Careers

After pursuing STEM courses, there will be a plethora of opportunities which will open up for international students. There is a gaining popularity for STEM courses and the returns it brings. Check out the popular STEM career options for international students after graduation in the table below:

STEM Careers

Web Developer	Mechanical Engineer	Cybersecurity Analyst	Chemist
Database Administrator	Biologist	Civil Engineer	Architect
Computer Science	Mathematics	Data Science	Nuclear Engineer
Physicists	IT Manager	Actuary	Mathematician

These are some of the most popular STEM career options which can be taken up after graduation or postgraduation abroad.

CHAPTER-12

BEYOND CRISPR: EMERGING TECHNOLOGIES IN BIOTECHNOLOGY

The biotech industry is constantly evolving, and new technologies are emerging all the time that have the potential to revolutionize the way we diagnose, treat and prevent diseases. In this post, we will take a closer look at some of the emerging technologies that are shaping the future of biotech and their potential impact.

- CRISPR: CRISPR (Clustered Regularly Interspaced Short Palindromic Repeats) is a revolutionary gene-editing tool that allows scientists to make precise changes to DNA. This technology has the potential to treat and cure a wide range of genetic diseases, and is also being used to develop new treatments for cancer and other diseases.
- Artificial Intelligence (AI) and Machine Learning (ML): AI and ML are being used in the biotech industry to analyze large amounts of data, such as genetic sequences and medical records, to identify new patterns and insights that can help to improve the effectiveness of treatments. This technology can also be used to develop new drugs, identify new drug targets and predict the efficacy of treatments.
- Quantum Computing: Quantum computing is a new technology that has the potential to revolutionize the way we process and analyze large amounts of data. In the biotech industry, this technology could be used to design new drugs and analyze large amounts of genetic data to identify new treatments for diseases.
- Nanotechnology: Nanotechnology is the manipulation of materials on a molecular and atomic scale. In biotech, it can be used to create new medical devices, such as nanoparticles that can target specific cells in the body. This technology can also be used to develop new drugs and therapies that

can target specific cells and tissues, which could lead to more effective treatments with fewer side effects.

- Synthetic Biology: Synthetic biology is the design and construction of new biological parts, devices, and systems that do not exist in the natural world. This technology has the potential to revolutionize the way we produce drugs, vaccines, and other biological products.
- 3D printing: 3D printing technology can be used to produce complex biological structures, such as replacement organs and tissues. This technology could also be used to produce personalized medical devices and prosthetics.

All these technologies are at different stages of development, but they all have the potential to revolutionize the way we diagnose, treat and prevent diseases. They also create new opportunities for researchers, scientists and healthcare professionals, as well as new business opportunities.

Biotechnology comprises the study and manipulation of living organisms with the goal of creating new products and processes that can improve our health and/or standards of living. It's a field that can lead to revolutionary new developments in everything from healthcare to food production to waste management.

Major happenings in the biotech industry can have a significant impact on humanity, the Earth and the animal kingdom, so it's not surprising that headlines from the field are of immense interest to tech experts and the general public alike. Below, 18 members of Forbes Technology Council discuss some of the amazing new and emerging biotech tools and developments everyone should know about and why they're so exciting.

Brain Mapping

In 2021, Google and Harvard announced they had mapped the equivalent of a 1-millionth section of the human brain. The resulting map took up 1.4 petabytes of disk space, comparable in size to three decades' worth of satellite images of Earth. Once technology advances and allows us to map larger areas of the human brain, we will better understand and be able to help those with traumatic injuries or neurological disorders and diseases.

Autonomous Therapeutic Systems

Autonomous therapeutic systems are one of the most significant future medical technologies. These systems take over patient care from (human) providers by analyzing, determining and autonomously controlling conditions and treatments. Thus, we need to have a precise simulation of a patient's medical condition—the patient's bio digital twin. This should reduce human error and medical care costs.

AlphaFold

AlphaFold is a protein folding program developed by Google's DeepMind. It's a major breakthrough in the field of protein folding. By understanding how proteins fold, we can better understand how they work and how they can be manipulated to improve human health. Additionally, we can more quickly design better drugs that will bind to their targets with greater specificity and affinity.

Cellular Anti-Aging Research

One of the biggest areas of biotech research right now is anti-aging at the cellular level. Researchers believe that we can actually reverse aging, which could not only help people live longer, but could also provide cures or treatments for diseases that are incurable today. There are ethical issues to be worked out, but the potential to change life as we know it is exciting.

CRISPR-Based Gene Editing

One thrilling biotech breakthrough is CRISPR-based gene editing. This technique allows precise modifications to DNA, offering potential cures for genetic disorders. For the general public, the implications are vast, including everything from addressing inheritable diseases to revolutionizing agriculture. Imagine a future where we could eliminate conditions such as cystic fibrosis or grow drought-resistant crops.

Microbiome Manipulation

Manipulation of the microbiome is exciting everyone. The study of the numerous microbes that reside within our bodies indicates that their impact on our health extends beyond digestion. Leveraging this understanding has the potential to pave the way for personalized treatments targeting a range of conditions, from obesity to mental health, thereby opening novel pathways toward enhanced well-being.

Living Medicines

An exciting but less talked-about area in biotech is living medicines. Imagine taking a pill filled with good bacteria that's been programmed to fight your specific illness. These bacteria could sense what's wrong in your body and release medicine only when needed. This could be a game-changer for treating chronic conditions. But, as with any new tech, it also raises questions about long-term safety.

Lab-Grown Organs

One of the most exciting developments in biotech has been the creation of lab-grown organs. Using a patient's own cells, scientists can now cultivate organs in the lab that could potentially replace damaged ones, eliminating the need for organ donors and reducing the risk of organ rejection. This could revolutionize transplant medicine.

Epigenetics And Digital Therapeutics

Epigenetics and digital therapeutics represent groundbreaking frontiers in biotech that would be of mass interest. Epigenetics offers transformative potential for personalized, precision medicine, tailoring treatments to individual genetic profiles. Concurrently, digital therapeutics herald a new era of holistic wellness, integrating technology for enhanced healing outcomes.

Sophisticated Wearables

The convergence of artificial intelligence and biotech has given rise to highly sophisticated wearables that do more than just count steps or monitor your heart rate. These devices are becoming capable of diagnosing conditions ranging from sleep apnea to cardiac arrhythmias—all in real time. The beauty of this trend is how accessible it is; you don't need to be in a hospital to be closely monitored.

Bioluminescent Imaging

Bioluminescent imaging, explored since the '90s, is gaining renewed traction. It allows for real-time visualization of cellular processes in living organisms by making them glow! Imagine watching how a drug travels and affects different parts of a live organism. This could offer invaluable insights into drug effects and disease progression.

Brain-Computer Interfaces

While still a decade away, brain-computer interfaces have the potential to revolutionize the way we learn, work and communicate. The advances in non-intrusive BCIs by a company called Neurable are going to do for cognitive monitoring what the Apple Watch did for cardio health. Consumers will no longer need electrodes and swim caps to monitor their brain health—just Neurable-compatible headphones.

Lab-Grown Meats

Lab-grown or cultured meat is produced directly from animal cells without traditional animal farming. This can potentially address issues related to animal welfare, environmental sustainability and food security. It could significantly reduce the environmental impact of meat production while providing a more ethical and efficient way to meet the growing global demand for protein.

Organoid Intelligence

Organoid intelligence is a fascinating field, albeit far-reaching and difficult for a non-scientist to imagine as a reality. While AI aims to make computers more brain-like, OI research works to make a 3D brain cell culture more "computer-like" by giving it more inputs and outputs to "think." Envision a future of brain-machine interfaces and biological computing.

3D Bioprinting

3D bioprinting enables scientists to craft three-dimensional biological structures using living cells. This breakthrough could revolutionize organ transplants by offering patient-specific organs, drastically reducing waitlists. Additionally, it holds immense potential in reshaping personalized medicine and precise drug testing.

Plastic-Eating Bacteria

Scientists have recently engineered bacteria to "eat" plastic waste, addressing one of the planet's most pressing environmental concerns. These modified bacteria can degrade plastics much faster than natural processes, potentially revolutionizing waste management and significantly reducing pollution. This breakthrough shows how solutions inspired by nature can rise to meet global challenges.

Generative AI

For biotech companies, generative AI can improve the pace of innovation and improve efficiencies while providing better guardrails for privacy, security and compliance. Generative models trained on biotech content can be augmented with specific private project or company content, automated to carry out validation checks to ensure content compliance (such as not including any personally identifiable information), and verified against specified checklists.

Quantum Technology

Quantum may still be nascent, but it's already showing incredible potential to transform biotech in areas such as drug discovery. Hybrid approaches, such as quantum machine learning, are being explored now for their ability to analyze biological data, identify potential drug targets and rapidly predict drug interactions. This could unlock new drug treatments for complex diseases such as cancer.

RNA EDITING AND OTHER GENOME EDITING TOOLS

Genetic modification techniques cover a wide range of studies, including the generation of transgenic animals, functional analysis of genes, model development for diseases, or drug development. The delivery of certain proteins such as monoclonal antibodies, enzymes, and growth hormones has been suffering from several obstacles because of their large size. These difficulties encouraged scientists to explore alternative approaches, leading to the progress in gene editing. The distinguished efforts and enormous experimentation have now been able to introduce methodologies that can change the genetic constitution of the living cell.

The genome editing strategies have evolved during the last three decades, and nowadays, four types of "programmable" nucleases are available in this field: meganucleases, zinc finger nucleases, transcription activator-like effector nucleases, and the clustered regularly interspaced short palindromic repeats (CRISPR)/CRISPR associated protein 9 (Cas9) (CRISPR/Cas-9) system.

Each group has its own characteristics necessary for researchers to select the most suitable method for gene editing tool for a range of applications. Genome engineering/editing technology will revolutionize the creation of precisely manipulated genomes of cells or organisms in order to modify a specific characteristic. Of the potential applications are those in human health and agriculture. Introducing constructs into target cells or organisms is the key step in genome engineering.

RNA editing in Trypanosoma

The term "RNA editing" was first coined in 1986 to describe the addition of four nucleotide bases, not present in the genomic DNA, in the RNA sequence of the *Trypanosoma* mitochondrial cytochrome oxidase (cox) subunit II gene. These edits extensively modify the reading frame of the target mRNAs, and sometimes editing can result in two entirely different protein products.

Editing is important for maintaining mitochondrial electron transport and oxidative phosphorylation via the production of functional mRNAs. The molecular mechanism regulating RNA editing in Trypanosoma brucei is highly complex. A guide RNA (gRNA) is used to direct the insertion and deletion of uridines in the target mitochondrial mRNA. A family of RNA editing core complexes (RECCs) is involved in the cycles of U insertion and deletion.

Additionally, another dynamic multi-protein complex, designated as Mitochondrial RNA Binding complex (MRB1), was later considered as an essential component of RNA editing in Trypanosoma. The historical context, physiological role and molecular mechanism of RNA editing in Trypanosoma is well summarized in a recent review [5].

RNA editing in mammals

RNA editing in mammals was first reported in 1987, and one year after its discovery in Trypanosoma. The RNA editing was discovered only in vertebrates and it involves cytosine-to-uridine (C-to-U) editing. RNA editing targets the nuclear encoded mRNA for apolipoprotein B and it affects innate immunity as well as lipid metabolism and muscle development. RNA editing is predominantly catalyzed by AID/APOBEC proteins, encoded by a family of 10 genes (APOBEC1, 2, 3 A–D, 3 F–H, and 4) in the human genome. Various auxiliary factors required for the target mRNA recognition are also known.

Another important RNA editing category in mammalian cells is adenosine-to-inosine (A-to-I) RNA editing. Inosine is translated to guanosine; therefore, A-to-I editing can alter the amino acid sequences of proteins and can affect other transcriptional processes such as alternative splicing and miRNA binding. The prominent target mRNAs encode neurotransmitter receptors and ion channels in animals.

The A-to-I RNA editing is performed by adenosine deaminases acting on RNA (ADAR), a family of enzymes that recognize specific RNA duplex structure as the substrate.

Recent scientific advancements and applications of next-generation sequencing technology have enhanced the understanding of RNA editing and its effects. While insertions and deletions have not yet been reported, base conversion is a widespread phenomenon in mammals.

RNA editing in plants

Plant RNA editing was first reported in 1989 as a C-to-U conversion in mitochondrial transcripts. There are approximately 500 and 30 RNA editing events in the mitochondria and chloroplasts, respectively, recorded in a diverse list of plants such as Arabidopsis thaliana, *Oryza sativa, Brassica napus Beta vulgaris, Vitis vinifera,* and *Nicotiana tabacum*. Recent analysis showed that numerous nuclear transcripts are also differentially edited. Most C-to-U substitutions are found in the protein-coding regions safeguarding evolutionary codons.

Therefore, the RNA editing should be considered a corrective mechanism at the post-transcriptional level against C-to-T (or T-to-C) changes, which possibly support less favorable changes. RNA editing events have also been identified in introns and untranslated regions; these events affect mRNA stability or splicing as the secondary structure or regulatory element is modified. In the case of some lower plant families, in addition to the extensive C-to-U, "reverse" U-to-C RNA editing has also been identified in plastids and mitochondria.

Recent studies have demonstrated the molecular mechanism by which the target sequence recognition and the catalysis of C-to-U and U-to-C editing of organellar RNAs are conducted by pentatricopeptide repeat (PPR) and the auxiliary DYW domain, respectively; this is described in more detail later.

The nuclear transcriptome of *A. thaliana* revealed the presence of A-to-I RNA editing events, possibly caused by a plant homolog of ADAR, similar to the mammalian RNA editing. Additionally, recent transcriptomic studies have reported nine different RNA editing patterns in nuclear encoded RNAs of *Arabidopsis* and *Salvia;* these events are all intra-base substitutions (C-to-U, U-to-C, A-to-I, A-to-C, A-to-U, G-to-A, G-to-C, U-to-A, and U-to-G). The molecular mechanism of these peculiar RNA editing of nuclear encoded RNA is completely unknown.

PPR protein for recognition of RNA editing sites in plant organelles

Several lines of recent research have elucidated that a specific class of PPR proteins with a putative deaminase domain is responsible for C-to-U and U-to-C RNA editing in plant mitochondria and chloroplasts. In *Arabidopsis*, the *PPR* gene is nuclear encoded and comprises a large family of approximately 500 members. PPR proteins are characterized by the presence of signature tandem repeats of a 35-amino acid PPR motif, which is responsible for specific RNA binding properties.

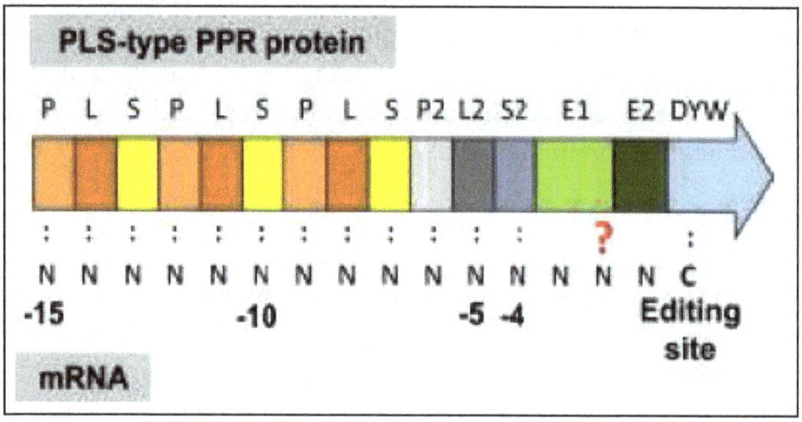

Fig. 1. Schematic illustration of the PLS-type PPR protein RNA binding activity for recognition of target sites in mRNA. The PLS-type PPR protein consists of PPR and PPR-like (P, L, S, P2, L2, S2) repeats and extended E (E1, E2) and DYW motifs. The last PPR motif (S2) recognizes the nucleotide at position -4 towards the editing site.

PPR proteins are classified into the P and PLS subfamilies. The P subfamily comprises 35 canonical amino acid PPR motifs. The PLS subfamily comprises the canonical and short (S; 31 amino acids) and long (L; 36 amino acids) PPR motifs, of which approximately 193 members have an extended E domain, while 87 others have both E and DYW domains in *Arabidopsis*.

Genetic investigations in model plants, including *Arabidopsis, Oryza sativa*, and *Physcomitrium patens* (moss), have revealed that the recognition of specific editing sites in plant organellar RNA editing is governed by the function of PPR proteins of the PLS subfamily [30]. The PPR array, typically 10–20 PPR motifs, recognizes the upstream nucleotide sequence via a 1 to 1 motif to nucleotide correspondence with the last PPR motif located four nucleotides before the editable C residue (Fig. 1). Sequence-specific RNA binding is determined by the RNA recognition code of the respective PPR motif.

The additional C-terminal DYW domain of the PPR family of proteins is known to have a conserved zinc-binding motif (HXE and CXXC), whose sequence is similar to that of cytidine deaminases, which are editing enzymes in mammals.

Therefore, the DYW domains are considered to have an active site for C-to-U editing. This hypothesis was verified by a recent report on the reconstitution of C-to-U RNA editing in *Escherichia coli* by the expression of a single DYW-containing PPR protein from *P. patens*, as well as using *in vitro* RNA editing experiments. Furthermore, structural analysis of the DYW domain has verified that it has a typical cytidine deaminase conformation, with a Zn ion-associated catalytic center.

The function of the remaining E domain, which connects the PPR array and the DYW domain, is still unknown. E1 and E2 domains can have PPR- or TPR-like structures, comprising of two alpha helixes. The E domains are expected to play an important role in RNA editing, as indicated by the fact that some E1 domains appear to interact with RNA in a sequence-specific manner. It has also been shown that the E1 domain is required for interaction with auxiliary factors of multiple organellar RNA editing factor (MORF) as discussed below.

In mammals, codon bias is an important factor for RNA editing. Other studies have reported a 3-fold higher frequency of RNA editing at the second nucleotide position of the codon than at the first and third nucleotide positions. This inconsistency may be due to the differences in the preferences of editing enzymes for the adjacent nucleotides; in the case of adenosine deaminase, editing is greatly affected by neighboring nucleotides.

These observations suggest that the E domain is involved in the appropriate binding of the PPR array and DYW domain for proper positioning of the catalytic reaction and/or sequence recognition in the vicinity of the RNA editing site (Fig. 1). The molecular mechanism of U-to-C editing, observed only in a limited plant species including hornwort, is largely unknown.

Sequence-based analysis of hornwort species has demonstrated that their PPR proteins have unique C-terminal DYW-like domains with characteristic signatures that are not found in flowering plants where only C-to-U RNA editing occurs. Proteins in this group have a distinct PG box sequence ("KPAxA") and are categorized into two subtypes, depending on which "DRH" or "GRP" amino acid motifs are substituted for the eponymous DYW tripeptide at the end of a "typical" DYW domain.

Recently, Ichinose et.al. successfully developed a U-to-C RNA editing factor based on DYW:KP proteins that is functional in bacteria and human cells.

However, the amino acid sequences of DYW motifs are very similar between C-to-U RNA editing and U-to-C RNA editing. In any case, the molecular mechanism of "reverse" U-to-C RNA editing should be studied further.

Applications of RNA editing

Advanced techniques of gene editing systems such as TALE and CRISPR- Cas9 allow DNA editing, which is the direct, irreversible conversion of one base pair to another at a target genomic sequence without inducing a double-stranded DNA break. These approaches have become popular for creating point mutations in numerous organisms and are expected to have applications in disease treatment.

However, incorrect and unexpected base editing in genomic DNA leads to the potential risk of cancers, fetal abnormalities, or other unexpected disorders. In contrast, RNA is generally expressed only in specific cell types or at certain times. Hence, fewer off-target editing side effects are expected when using an editing system targeting RNA than during genome sequence editing. Furthermore, incorrect RNA editing does not affect fetal development because the genome sequence is not affected, and mutated RNAs are quickly degraded if treatment is discontinued. Therefore, RNA editing is more flexible and safer than genome editing for clinical uses.

Artificial RNA editing systems

Researchers performed site-directed RNA editing using catalytically inactive Cas13 (dCas13) for directing the deamination of adenosine to inosine by ADAR2 in mammalian cell mRNA. This system is known as RNA editing for programmable A to I replacement (REPAIR). Artificial A-to-I RNA editing was performed using the ADAR family of enzymes tethered to gRNAs complementary to the target sequence. Later, the artificial cytidine deaminase system was expanded to induce C-to-U editing, derived from ADAR2DD enzymes fused with catalytically inactive dRanCas13b (dCas13) and used for programmable RNA editing for specific C-to-U exchange (RESCUE) in mammalian cells, exhibiting an editing efficiency of up to 76%. Additional systems were developed by Researchers who used a SNAP-tag for tethering, Montiel-Gonzalez et al. who used the Lambda-N protein and box B element RNA, and the Tsukahara group who used the MS2 system.

Recently, another study involving the MS2 system for C-to-U RNA editing similar to the A-to-I editase 48,49was published, which utilized members of the apolipoprotein B mRNA editing (APOBEC) family. The MS2-tagged system restored the original sequence of the G-to-A mutation via A-to-I editing using ADAR1 deaminase.

APOBEC 1 was bound to the MS2 coat protein, and the gRNA was bound to the MS2 stem-loop. Thus, the binding of the coat protein and stem-loop could be achieved, allowing the gRNA to guide the deaminase to the specific location/ site and perform editing at the targeted nucleotide sequence. This artificial RNA editing system successfully converted up to 21% C to U at the mRNA level, restoring the wild-type sequence.

Gene Stacking: A Step Forward from Gene Editing

Gene stacking emerged as a favorable insight for providing organisms with multiple characters. Previously, stacked products were developed by random gene insertions at multiple locations within organism's genome. Stacked products proved to be constructive but were producing off target effects. Pyramiding two or more genes at a precise location not only reduced the effects of off target gene activity but it provided a mechanism of removal of the whole gene cassette when the constructed stack is no longer working in future.

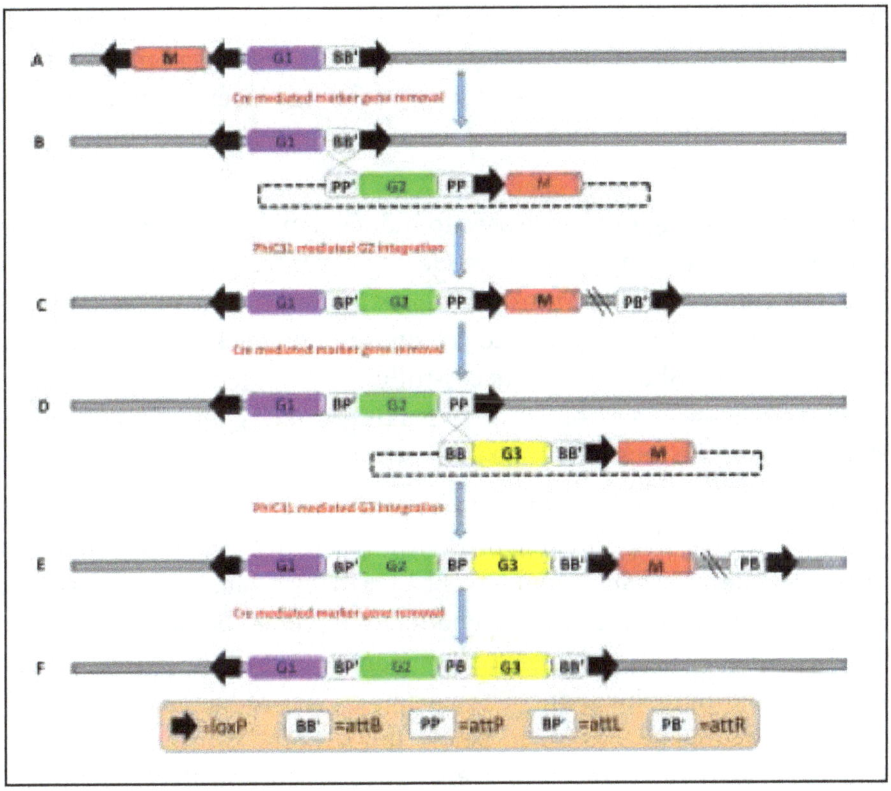

Fig. 2. Gene stsacking strategy.

Gene stacking strategy became more favorable by introducing molecular scissors with the addition of organism's health beneficial gene. The stacked products will be able to combat multiple diseases with additional benefit of promoting health of the organisms.

Recombinase-mediated gene stacking (RMCE) is a new transformation operation in which site-specific gene integration events can co-occur with randomly integrated transgenes (Fig. 2).

Site-specific recombination systems for eukaryotic genome engineering are equally beneficial for plants and animals, but the technology is used more in plants than animals due to consumer dislike of genetically modified crops. The objective was to produce marker-free transgenic plants and to clear a genetic burden that is useless to the plant. Site-specific gene engineering also reduced ambiguous position effects. Crops are improved through addition, deletion or inversion site-specifically, without disrupting other loci.

Site specific gene stacking means accumulating different trait genes in a single plant and at a single locus. Only a handful of commercial stacked products with pyramided traits were available in 2005. By 2010, many biotech companies had developed stacked trait products with pronounced insect and herbicide tolerance. A new method for combining transgenes into a single genome used specific independent loci for recombinase-mediated deletion, insertion and inversion of transgenes. Recombinase-mediated cassette exchange also helped increase gene stacking in crop plants.

Genome engineering uses nucleases to induce double-strand breaks at precise positions on a chromosome. A meganuclease was introduced at an endogenous specific locus contiguous to a transgenic locus of insect control region in cotton. Targeted insertion of a transgene into cotton is made practical by a DNA cleavage and repair pathway. About 2% of successfully transformed embryonic calli were recovered as precise insertion events, making multi gene stacking easier by demonstrating successful single gene integration.

A commercial stacked maize hybrid was confirmed by transcript analysis of the transgene together with proteomic profiling. The maize hybrid was more insect- and herbicide-tolerant than a single-event hybrid with same genetic background. The stacked hybrid was already genetically diverse before transformation. Protein enrichment analysis showed alteration in two major pathways: energy/carbohydrate and detoxification metabolism. A 34% reduction in transcription of the stacked transgene was found below single-event hybrid genotypes. These scientists concluded that expression of endogenous genes is affected by stacking of two or more genes in genetically modified maize. Significant alterations in protein expression change energy/carbohydrate metabolism in stacked trait products. The new world of "omics" could provide insight on security evaluation of stacked-trait products.

Most commercial operations use classical breeding to introgress transgenes from a laboratory cultivar into commercial varieties suitable for different growing regions. Prior to vegetative cloning, genetic crosses were performed for varietal selection even in vegetatively propagated crops. In a diploid species, if (1/4) (x) is the probability that "x" transgenic loci will assort into a genome and "y" is the number of non-transgenic loci that breeders need to accumulate in the same genome, the probability of a getting both into a breeding stack is (1/4) (x + y).

Site-specific integration of new transgenes at a preexisting locus is required to add new DNA sequences into transgenic varieties without generating a new segregating locus. Recombinase-mediated gene stacking and Bxb1 integrase from mycobacteriophage for site-specific integration between attP and attB recognition sites was used for gene stacking of tobacco. A frequency of about 10% site-specific integration events was observed, with one recognition site spared for the second round of integration. Gene silencing was observed in a third site-specific integrant. About 3% of the transformed clones showed accurate structure and reproducible expression of three consecutive gene-stacking events.

NANOTECHNOLOGY IN BIOMEDICINE

Nanotechnology is used in biomedicine for a variety of applications, including:

- Diagnostics: Nanotechnology devices can help detect cancer and infectious diseases early.
- Therapeutics: Nanotechnology can be used for drug discovery, drug delivery, and gene/protein delivery.
- Regenerative medicine: Nanotechnology can be used for sensitive medical procedures, such as gene therapy, dentistry, and oncology.
- Biomolecules detection: Nanotechnology can be used for tissue engineering and biomolecules detection.
- Imaging and diagnostics: Quantum dots (QDs) have unique photoelectric properties that make them useful as imaging and diagnostic agents.

The science of nanotechnology involves the development of new materials with dimensions on the order of tens to a few hundreds of nanometers.

Bigger than what we normally consider molecules and smaller than bulk materials, nanomaterials have emergent properties that are a direct result of their size. These new materials are proving to have a broad range of applications in medicine and basic biological research. Included in nanotechnology research are:

Nanostructured materials for biological sensing

Efforts towards characterizing the entire genomes and proteomes of humans and other organisms have unleashed a flood of new protein and nucleic acid targets on the biomedical research community, and have made the development of new ways of detecting and quantitating these targets rapidly and simply an urgent need.

Researchers are using nanostructured materials as substrates for the production of new biomedical sensors, which may prove useful as components of medical diagnostic systems.

Nanoporous membranes

Researchers at Rochester recently discovered a novel silicon membrane with well-defined, tunable nanometer-sized pores. Because this membrane is also exceptionally thin, it has the potential to dramatically improve the efficiency and speed of size-based biofiltration.

Researchers are examining the basic properties of this material, as well as ways in which it can be incorporated into biomedical devices.

Nanoparticle-based drug delivery

Delivery of therapeutic agents to target sites in the body is a continuing challenge. Polymer nanoparticles are proving to be one useful method for encapsulating therapeutics and delivering them in an effective manner.

Projects in this area involve the creation of new polymer nanomaterials, and examining their ability to deliver small-molecule or biopolymeric agents.

Imaging, transport, and toxicity properties of semiconductor nano crystals

Semiconductor nanocrystals, or quantum dots, are proving to be very useful as new reagents for biomedical imaging. Nanocrystals are also finding their way into numerous consumer products. However, little is yet known about the interactions of these materials with biological tissues, and their eventual fate in the body.

Projects center on the development and application of new nanocrystal-based reagents, as well as on evaluating their transport properties through tissues.

Nanobiomechanics

In recent years, through efforts in single molecule force spectroscopy, the role of mechanics at the molecular and cellular level have been shown to play a critical role in biology.

Researchers employ the atomic force microscope, optical tweezers, magnetic tweezers, bioforce probe and microcantilevers to study the mechanics which serve to influence life at the sub-cellular level.

Genomics and Nanotechnology

Goal no. 1 of the 2000 NNI forecasts an important role for nanotech in enabling inexpensive sequencing. The rapid advancement in next-generation genome sequencing (NGS) technologies has been a triumph of modern biotechnology, so that the $1000 genome is now a reality. In fact, the NGS tools have advanced to the point that the technology is effectively a given, and the intellectual effort within a genomics study now is centered on the scientific design and interpretation of population-based studies. However, NGS has major deficiencies.

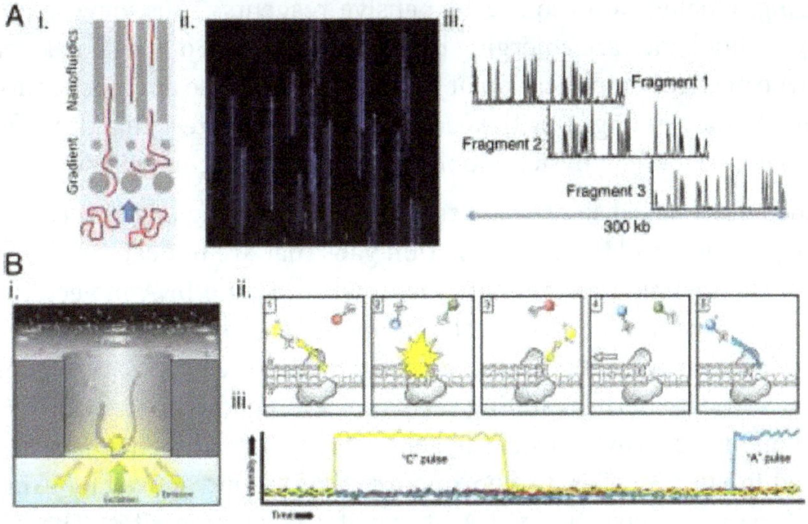

Fig. 3 Nanotechnologies for genome mapping and genome sequencing. (A) Nanochannelbased genome mapping. (i) The microchip is designed to untangle the DNA and guide long fragments into nanochannels with diameters <100 nm. (ii) The DNA fragments are fluorescently labeled at specific sites to provide a spatial map of those sites. (iii) Different fragments are lined up according to the spatially resolved fluorescent signatures to provide the genome map, shown here for a 300-kb segment. Adapted from ref. 9, with permission from Macmillan Publishers Ltd: Nature Biotechnology, copyright 2012. (B) The Pacific Biosciences nanopore-based genomesequencing platform, in which single-molecule, real-time sequencing data are obtained from a DNA polymerase that is isolated within a 100-nm-diameter pore in an aluminum film. (i) The pore serves as a zero mode waveguide for optical analysis of uninterrupted template-directed DNA synthesis using four distinguishable fluorescently labeled dNTPs. (ii) The temporal order of enzymatic incorporation of the dNTPs into a growing DNA strand is illustrated. At each step a fluorescent dNTP is incorporated, generating a fluorescent signal that is collected efficiently within the nanowaveguide. The fluorophore then is cleaved and diffuses out of the waveguide. (iii) An optical readout for a single channel is illustrated.

First, most NGS methods are limited to short reads, although alternative technologies are emerging that permit longer reads. Analyzing an NGS dataset is, by analogy, similar to reading a book in which short sentence fragments are placed in random order relative to how the text actually was written. To make it understandable, the book must be reassembled correctly from those short fragments. Second, NGS requires amplification.

This requirement, combined with short reads, means that the identification of many natural or disease-associated molecular lesions, such as repeat-rich regions, gene amplifications and deletions, or chromosomal translocations, can be challenging. Finally, NGS requires expensive reagents. Thus, new sequencing or mapping technologies are emerging based on long reads of single DNA molecules, without the need for amplification or even reagents. These approaches are enabled by nanotechnologies (Fig. 3) with specific design parameters that draw from many years of polymer-based physics, biophysics, and physical chemistry.

Nanochannel-based single-molecule genome mapping (Fig. 3*A*) is used to help assemble genomes and fill in information gaps that are not captured by NGS. The origins of the technique were methods in which single DNA molecules, 102–104 kb long, were stretched out on a microscope slide and cleaved using site-specific enzymes. In this way a 400-kb DNA molecule might be cut into 20–40 components. The physical length of the resulting segments, averaged over many DNA molecules similarly analyzed, provided a crude genome map. For nanochannels, the concept illustrated in Fig. 3*A* relies on fluorophores that are introduced by using nicking endonucleases that cut the dsDNA at specific restriction sites. These locations provide a map that includes deletions, amplifications, and translocations and provides guides for NGS genome assembly. Of course, many physical chemistry issues, such as the interplay of the persistence length of the DNA molecules, the ionic strength of the solvent, the conditions for uncoiling the DNA, and the dimensions of the nanochannel, have all provided a scientific foundation for this nanotechnology.

Long, accurate sequencing reads of unamplified DNA would negate the need for mapping. Nanopore-based sequencing is such a method, although it has been technically challenging to develop and only very recently has permitted some sequence determination of an actual genome. The basic idea dates back some 20 y and draws from fundamental work in surface science, molecular biology, nanofabrication, and electrochemistry.

In one manifestation, two aqueous electrolyte solutions are connected through a single protein nanopore, such as such as *Mycobacterium* smegmatis porin A (MspA). A bias is applied, and current is measured as DNA is drawn through a bilayer-entrained pore protein at a rate that is controlled through the use of molecular motors such as F 29 DNA polymerase (DNAP). A variant developed by Pacific Biosciences is shown in Fig. 3*B*. Four distinct fluorophores are used to read out the action of the polymerase, which is isolated within a nanofabricated waveguide.

The polymerase, which also is used for whole-genome amplification, permits long, relatively unbiased reads. Each polymerase permits a read rate of about four to five bases per minute, with thousands of single polymerases in nanowaveguides used in parallel. A third variant, which was released by Oxford Nanopore Technologies as the MinION nanopore sequencing product, is about the size of a

USB memory stick. Early literature reports on the MinION imply that it may not yet be ready for wholesale sequencing, but it does have some compelling applications.

Although genomics has led the -omics revolution, nanotechnologies (and microfluidics) are playing increasingly important roles in reducing -omics technologies to the level of a single cell, and I turn to this area next.

Single-Cell -Omics in Biology and Biomedicine

Single-cell biology holds the promise of unraveling the heterogeneity that often confounds the interpretation of biological or biomedical measurements. For example, single-cell analysis is increasingly applied toward understanding immune responses triggered during various cancer immunotherapies or for unraveling the functional behaviors that emerge from heterogeneous healthy or diseased tissues. Aside from flow- and mass-cytometry methods, virtually every single-cell -omics platform involves some level of microfluidics and/or nanotech. Specifically, a single cell has only a certain number of copies of any given analyte. By isolating a single cell within a nanoliter or subnanoliter volume, those copy numbers can correspond to detectable concentrations, and the concentrations of contaminants are minimized. Microchip methods also can uniquely permit analysis of very small tissue samples.

Single-cell methods have been reviewed recently, so only a general technology overview plus a brief discussion of what is learned uniquely through single-cell assays is given here. Single cells are, from a physico-chemical point of view, finite systems. That is, quantitative measurements of transcripts, proteins, metabolites, and so forth in different single cells will yield different copy numbers of those analytes. Thus, the most useful single-cell methods are designed to assay a panel of analytes from statistically significant numbers of single cells. Added value that is uniquely provided by such single-cell analyses includes

Histograms of the abundance distributions of a given analyte across a population of cells help define outliers, such as cells that overexpress a particular protein. This information can be particularly important for immunology applications. For certain analytes, the abundance distributions also may be interpreted as the fluctuations of the cell (similar to positional fluctuations in a Brownian particle), thus providing a bridge to statistical thermodynamics models.

Correlations (and anticorrelations) between the measured analytes are statistical relationships that provide important clues regarding the structure of gene or protein signaling networks and how that structure is altered by some perturbation. Correlations derived from single-cell assays are not the same as those derived from bulk assays. In a bulk assay, the average levels of two analytes may both be repressed by a drug and thus exhibit correlated behavior.

For single-cell analyses, statistical correlations between any two analytes are determined through an *x,y* scatter plot of the assayed values for each single cell. The average analyte abundance is not a factor in such plots (Fig. 4*A*).

Detailed lineage tracing is enabled, permitting diseased tissues or cells to be traced back to the originating healthy tissue or cell types or circulating tumor cells to be traced back to the originating lesion.

Some recent studies capture multiple levels of biological information (i.e., proteins and metabolites) from the same single cells (Fig. 4A). Such studies have many applications, ranging from associating a particular T-cell receptor gene with a T-cell antigen to associating the functional behaviors of cells [e.g., cellular motility and glycolytic rate, among others] with genetics, drug response, and so forth.

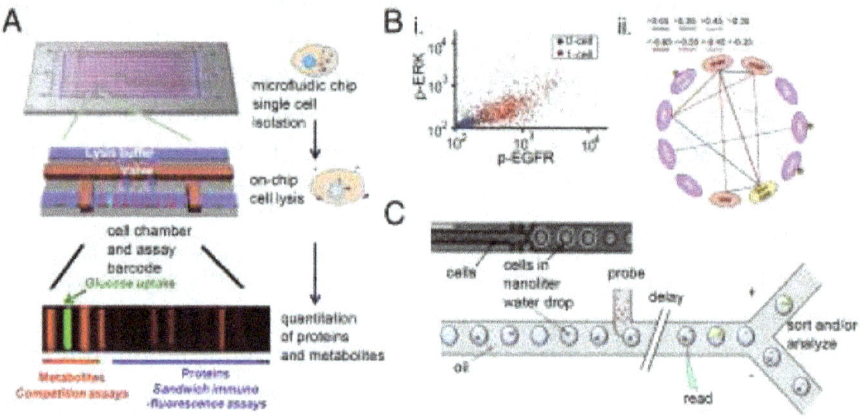

Fig. 4 Microfluidics and nanotech tools for single-cell analysis. (A) Illustration of a single-cell barcode chip (SCBC) in which individual cells are isolated within nanoliter or smaller volume microchambers within a microfluidics chip mounted on a microscope slide. The glass slide is patterned with a high-density barcode for protein and/or metabolite assays from isolated single cells. Cells are lysed using the valved microchamber structure shown in the middle drawing, and the contents are captured on specific locations within the barcode array. The fluorescence intensities of the developed barcode stripes are related to calibration curves to yield the level of the specific analytes. (B) Sample of single-cell data taken on an SCBC. (i) The scatter plot shows the correlated levels of two phosphoproteins as measured from single cells (red dots) or zero-cell chambers (blue dots). (ii) A protein–protein correlation matrix from a multiplex SCBC protein and metabolite assay.

Several platforms permit an -omics analysis of defined cell populations (i.e., zero, one, two, three, or more cells). Such platforms provide unique insights into cellular behaviors, such as how pairwise cell interactions influence bulk cellular architectures or quantitative analysis of immune cell–disease cell interactions.

A technologically distinct class of microchip designs permits individual cells to be isolated within droplets with a volume 0.05 pL to 1 nL, and those droplets are entrained within a fluidic flow that can be coupled with an optical analysis platform (Fig. 4*B*). The chemical composition of individual droplets can be manipulated to permit rapid and high-throughput screening assays.

Applications related to screening antimicrobial or antiviral agents or for various binding assays have been reported with very significant throughput advantages over traditional assays. For a technique known as Drop-seq, individual cells are similarly entrained within nanodrops on microfluidic chips. Each droplet also contains a microparticle that is encoded with a unique DNA barcode address. Cells are lysed within the droplet, the mRNAs are captured on the barcoded microbead, and then all microbeads are analyzed in parallel. The barcode address allows the transcriptome analysis to be reassociated with individual cells.

Inorganic Nanoparticles and Related Nanomaterials in Biomedicine

Metal nanoparticles (NPs) and semiconductor quantum dots (QDs) and related materials, such as single-wall carbon nanotubes (SWNTs), are increasingly used for preclinical in vitro and in vivo (imaging) diagnostics. Therapeutic applications have been slower to develop. The basic concept is that the size, shape, and composition of the inorganic core provide a useful physical property that enables a colorimetric, electrochemical, Raman, or other class of assays. The NP surface chemistry is tailored for the specific assay, including biomolecular recognition, solubility, and other characteristics that translate into a very large matrix of materials properties. For in vitro applications, there is substantial flexibility in exploiting this matrix. For in vivo applications, the task is much more challenging.

Each of the elements of this properties matrix [which includes the Stokes radius, the surface charge, the density of surface chemical groups, and the elastic modulus, among others] can strongly influence in vivo pharmacokinetics (PK) (i.e., circulation times, uptake within specific cells or within specific organs or disease sites, clearance route, and toxicology). In particular, understanding which basic properties of nano-materials can lead to nonspecific and potentially harmful biological interactions in vivo is a very active field of study.

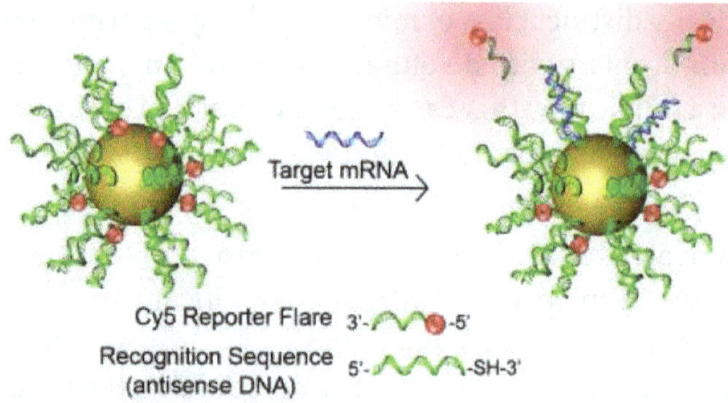

Fig. 5. The gold NP-based nanoflare construct is used for detecting specific mRNAs in live cells. The gold NPs are coated with a dense layer of DNA, which promotes cell penetration. The DNA shown comprises a fluorescent reporter (the Cy5 flare), which is nonfluorescent when bound to the Au NP. This nanoflare is hybridized with an antisense DNA. When the nanoflare encounters the target mRNA, the flare is released, thus activating fluorescence within the cell and permitting live-cell sorting based on the expression of a specific gene.

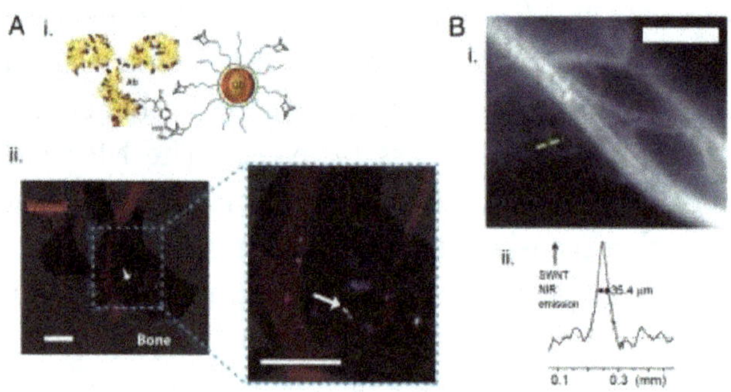

Fig. 6. emiconductor QD and SWNT in vivo imaging probes. (A) Semiconductor QD imaging of stem cells in bone marrow. (i) Polyimidazole incorporating norbornene provides a biocompatible surface functionalization for highly luminescent semiconductor core-shell QDs as well as a chemical handle for preparing QD-antibody conjugates, as shown (not to scale). Lyscine (amine-presenting) residues on the antibody are highlighted in red. The QD-antibody particles exhibit a moderate negative surface charge, which is generally favorable for improved in vivo circulation. (ii) Use of the QD-antibody particles as in vivo imaging probes for single-cell imaging in the bone marrow of a live murine model, viewed through a calvarial window. The arrow points to single Sca-1+c-Kit+ cell, which is a late-stage hematopoietic stem cell. The red and green cells represent Sca-1+ and c-Kit+ cells, respectively. Adapted from ref. 89. (B) Polymer-stabilized SWNTs used as in vivo near-infrared (NIR) fluorescent probes of vasculature. (i) Near-IR fluorescence of mouse vasculature. (ii) Fluorescence intensity taken along the dashed green line drawn on the image in i is plotted and reveals an imaging resolution of a few tens of micrometers.

In any case, this matrix of materials properties for optimization represents a sort of "opportunity hurricane" for the scientist or engineer. Efforts to elucidate how best to optimize NPs for specific tasks comprise much of the basic science of these NPs. Organizations such as the National Cancer Institute's Nanotechnology Characterization Laboratory have played important roles in guiding such efforts. Although approved clinical applications are beginning to appear, the bulk of the science is still maturing through mouse model studies. In following paragraphs, we provide a very brief overview of this underlying science and highlight illustrative examples in Figs. 5 and 6.

Noble-metal NPs have been used for several years for the point-of-care detection of blood-based biomarkers from droplets of blood; gold NPs provide the colorimetric agents for analyte detection. The basic exploited physical property is the surface plasmon resonance (SPR), which is in the visible or near-visible part of the spectrum for noble-metal NPs. The SPR is a collective resonance that carries a very high oscillator strength, with a peak wavelength, line shape, and intensity that strongly depend upon NP size, shape, and local dielectric environment. The basic optical properties of colloidal gold were described more than 100 y ago by Mie using Maxwell's equations.

Current models can capture the remarkable linear and nonlinear optical properties achievable through modern synthetic methods in which surface chemistry, local chemical environment, and NP size and shape are controlled. For in vitro biological applications, synthetic methods for surface-loading high-density coverages of DNA have enabled efficient cell delivery for various in-cell, live-cell assays (Fig. 5).

Other surface chemistries have enabled a myriad of electrochemical, chemical, and optical platforms for sensing and in vitro diagnostics. For in vivo applications, that synthetic control has been harnessed for minimizing toxicity [exhibited by some very small gold NPs] and for targeting delivery of gold nanorods to specific tissues. Gold nanorods have shown promise in animal models for photothermal therapy applications. In such a therapy, the particles are delivered to the disease site, and laser light is used to penetrate into the tissue, excite the NPs, and transfer thermal energy to—and thus kill—the diseased tissue. This concept has had only limited success in human trials and so may need more time to mature.

Remarkable advances in nanomaterials synthesis have occurred for a broad variety of NPs, including magnetic NPs. These advances have enabled applications of ferromagnetic and superparamagnetic NPs for both in vitro and in vivo (imaging) diagnostic assays. For in vitro assays, biofunctionalized magnetic NPs have a history of use for magnet-assisted cell sorting. A recent variant of that application is the use of antibody-coated magnetic NPs for detecting (using a custom-built microchip-NMR tool) and sorting microvesicles released by certain solid tumors. Microvesicles are 10- to 100-fold smaller than the tumor cells from which they are released but can carry many of the biomarkers that identify the originating cell as a diseased cell. For certain tumors, such as glioblastomas, microvesicles are released into the blood, although circulating tumor cells are not.

Fig. 6 illustrates two in vivo live-animal imaging applications of highly fluorescent semiconductor QDs (Fig. 6*A*) and near-IR fluorescing SWNTs (Fig. 6*B*). In both cases, the nanomaterial exhibits a compelling optical property that enables the imaging application and also is highly engineered, through surface chemistry control, for useful PK and pharmacodynamics (PD) properties. These types of probes, although not yet clinical tools, are providing powerful approaches for preclinical investigations.

Organic Nanoparticles for Drug Delivery

Goal no. 3 of the 2000 NNI, which describes the promise of NP drug-delivery systems, was largely inspired by the promising performance of liposomal delivery vehicles, at least four of which had received approval (one from outside the United States) before 2000. Liposomes are spherical vesicles with at least one lipid bilayer and can be used to carry hydrophobic drug molecules that associate within the membrane or hydrophilic molecules that are trapped in the core. The approval of the protein-drug conjugate Abraxane in 2005 gave further impetus to the field.

Nanotherapeutics such as liposomes are based on a concept in which many copies of a drug molecule are loaded into a delivery vehicle (the NP) that is optimized for improved PK and PD relative to the drug by itself. Nanotherapeutics have broad applications for many disease conditions; their use in oncology illustrates their general value. For almost all metastatic cancers, the frontline molecular treatments are small-molecule chemotherapies.

High systemic exposure of patients to those drugs frequently leads to dose-limiting toxicity. The nature of small-molecule drugs is that they exhibit short (<1 h) circulation times and indiscriminate tissue penetration, thus affecting both healthy and diseased tissue with a PK and PD profile that is defined largely by the nature of the drug rather than the needs of the patient. Early work on polymer–drug conjugates showed that large-molecular-weight particles, or macromolecules, could avoid such rapid clearance and, in fact, accumulate in tumor sites over a period of many hours.

This effect became known as the "enhanced permeability and retention" (EPR) effect. EPR is not a general characteristic of all nanotherapeutics; early liposomal formulations were found to clear rapidly from the blood and often to accumulate in the liver. The variations in PK and PD across different formulations suggested the compelling bioengineering challenge that drives the current science and translational progress in the field.

Organic nanotherapeutics under development today include liposomes with single or multiple bilayered membrane structures built from natural or synthetic lipids, dendrimer constructs, albumin-bound formulations, and polymeric nanoparticles engineered from biocompatible and biodegradable polymers. A common theme across these systems is that the NP delivery vehicle itself contains the engineering handles for controlling the drug PK and PD.

These handles include ligands for tumor targeting and mechanisms for triggered drug release based on pH or enzymatic signals as well as molecular or nanofabrication engineering to control surface charge (zeta potential), NP size, NP shape, and NP elastic modulus. Of course, as with inorganic NPs, the available design space is vast. A measure of progress in this field is the rapidly increasing number of NP formulations that have advanced into Phase I–III clinical trials.

In Figs. 5 and 6 we highlight two very different examples of organic NP preparations that reflect unique aspects of the state of the art and which are being translated through human trials. Fig. 5 illustrates the use of the particle replication in nonwetting templates (PRINT) technology, developed by DeSimone's group. This example highlights the role that the particle modulus can play in organ delivery. The highly versatile PRINT technology has been used to prepare NP vaccines that have been translated into an initial Phase I/IIa clinical trial for influenza.

Nanomechanical and Nanoelectronic Devices and Wellness Monitoring

Goals nos. 2 and 6 of the 2000 NNI, which emphasized remote sensor systems and in vivo devices for early disease detection and diagnosis, represented interrelated grand challenges with somewhat similar roadmaps to success. Recent influences on these goals are the proliferation of smart phones and other smart devices as well as the emerging emphasis on wellness. A relevant emerging wellness trend is that an individual is provided with regular measurements of his/her health status.

The individual can make lifestyle, exercise, and diet adjustments informed by those measurements so as to maximize health benefit. The technology challenge is to provide accurate, informative, and readily interpretable diagnostic measurements at low cost. The emerging science and technology of wearable health monitoring devices illustrates a dominant approach toward meeting this challenge. Nanotech does not play a role in the early-stage examples of such wearable devices, although it is clear that nanotech will play important roles in the near future. Current smart devices monitor a panel of basic body functions, such as heart rate, body temperature, and exercise but, with few exceptions, do not monitor specific biomarker analytes.

This current functionality is enabled by a menu of microelectromechanical machines (MEMs) such as microphones, accelerometers, and gyroscopes. The uptake of MEMs in consumer products was young in 2000, but of the three waves of proliferation initially envisioned for MEMs—automotive, consumer electronics, and home integration—the first two are now considered complete. A fourth wave, relevant to this discussion, is likely personal integration.

Recent advances in integrating electronics onto low-elastic modulus polymer substrates and other moldable materials that are conformal, biocompatible, and, in some instances, even self-powered are opening up a large menu of wearable devices based on mechanical, electronic, or electrochemical signal transduction. A large fraction of these devices contain some nanotechnology, with a scientific basis that can be traced back to early demonstrations of nanotube or nanowire sensors and the integration of nanowires and nanotubes onto plastic substrates.

The net result is that biomarker-rich liquids, such as sweat or saliva, can be used for continuous multiplex monitoring of health status-relevant analytes. An example of a nanotech-enabled tooth "decal" sensor. This device straddles the line between an implant and a wearable device. Although the presented data are from a benchtop experiment, they clearly illustrate the potential for real-time monitoring of both bodily functions (breathing) and analytes (bacteria in saliva).

INTEGRATION OF AL AND BIOTECHNOLOGY

Artificial intelligence in biotechnology can speed up drug discovery, deliver analytics, accurately diagnose medical conditions, edit gene structures, develop personalized medicine, and do much more to help mankind. Here, we'll discuss the role and importance of AI in biotech industries.

The role of AI in biotechnology is gaining momentum in recent times. The biotech industry is redefining its processes using AL and ML technology to get better, faster, and more accurate results. Be it the pharma industry, healthcare, animal husbandry, or agriculture, AI and biotech are setting the stage for greater advancements and innovations.

AI in biotechnology

AI in agricultural biotechnology

Biotechnology firms are now leveraging AI/ML solutions to develop autonomous robots that handle important agricultural tasks such as harvesting crops at a much faster pace than humans. Computer Vision and DL algorithms are leveraged to process and analyze the data captured by drones. This helps in monitoring crop and soil health. ML algorithms help in tracking and predicting various environmental changes including weather changes that impact the crop yield. Digital transformation is also having a strong impact on the field of smart agriculture. Numerous isolated, often non-interoperable solutions exist in digital ecosystems in agriculture. This is where an "Agricultural Data Space", such as that used in the Fraunhofer lighthouse project "Cognitive Agriculture" (COGNAC), can offer great added value. One application example is the evaluation of ecological and economic sustainability via the nutrient cycle. In agriculture, a balanced and appropriate nutrient cycle is at the core of efficient, productive and sustainable production of crop and livestock products. In addition to documentation, e.g., for regulatory monitoring, the focus is increasingly on optimizing the nutrient cycle.

Typical examples include dairy farming with arable and grassland management. Various suppliers of sensor systems record soil, plant and weather data. On dairy farms, conclusions about nutrient inputs and outputs can be made by collecting data on feeding along with the milk yield. However, to optimize nutrient cycling, the relevant data must be complete and of sufficient quality.

This is often the key problem. Important components here are interoperability, uniform ontologies, and the cognitive processing of the data. Missing data must, for example, be interpolated or modelled accordingly. By representing the nutrient cycle in the form of a digital twin, the farmer can obtain information about current nutrient balance and thus identify possible problem areas.

AI in agriculture can provide a solution for food security by adapting agricultural management to a changing climate. This includes identification of resistant crops that are more resilient to environmental changes and extremes such as drought periods. This would allow maintaining crop yields under abiotic stresses, which can severely effect crop productivity. Extreme temperatures can lower wheat yields by 6% per C.

As Rubisco activity and hence the photosynthetic process is sensitive to high temperature and stops at temperatures above 35 °C. Stress physiology of crops effected by water and nutrient limitation can be directly addressed by advances in remote sensing beyond solutions that are already used in agriculture for biomass estimation, crop type classification, and mapping of soil characteristics. The use of AI together with low-cost multi-channel sensors as well as remote sensing to collect big data requires infrastructure with data security.

These technologies can also be applied to identify new crop phenotypes that are more efficient in resource use and resistant to highly variable climate conditions. Phenotyping has become a key discipline of plant sciences in the last decade. Sensor developments and "big data" technologies have driven phenotyping as a main field of application in breeding to meet better the challenges of global change with increasing pressure from abiotic and biotic stresses.

As change in climate goes along with increased pressure of pests and disease, it is crucial that plant breeding programs also cover natural defence mechanisms of crops including wild relatives which often show a more diverse rhizomicrobiome and root exudates. Wild crop relatives should be considered in breeding programs, as they have a proven pool of genetic diversity to enhance adaptive capacity of agricultural systems to limiting resource availability and disease pressures. While climate change and severe events such as droughts and heavy precipitation causes plant stress with serious consequences for food production (yield potential and stability) and in particular also decreases plant fitness, identification of efficient natural defence mechanisms are thus highly needed.

However, the stress from low doses of herbicide for example, however can also have a positive effect on plant fitness through hormesis. New imaging technologies combined with AI are at the nexus to process big data and support data interpretation for climate change effects. Besides image-based phenotyping in plant breeding, imaging technologies have also been proposed to identify the effect of biostimulants which are of increasing interest to enhance plant fitness.

Combining data from image-based phenotyping with information on stress responses on a molecular level, such as genomic variation, gene and protein expression, and metabolite biosynthesis from stressors and their doses to develop hormesis management protocols could be highly rewarding to align different types of information sources. The potential of big data leverage has increased the significance of AI in agricultural disciplines. Data availability increases exponentially with technologies that are already widely applied in agriculture, from proximal to remote sensing, such as image-based phenotyping platforms in greenhouses, unmanned aircraft systems (UAS) at the field scale and satellite based remote sensing up to the landscape and global scale. To make use of the information, advances in computer vision algorithms are critical, while offering unprecedented opportunities to advance our system understanding.

For example, the data produced from UAS with appropriate sensors are of high accuracy and predictability for management decision, and generally more objective than traditional visual scoring methods or point sensors that show limited samples sizes or do not allow detecting in-field variations. The data generation is cost-efficient and hence can deliver not only fine spatial but also high temporal resolution on, e.g., plant growth.

This high-throughput technology and its benefits have been combined with other technologies such as plant phenotyping in the field, yield and biomass estimation, crop disease monitoring, water stress of crops, as well as weed mapping. Crop characteristics such as vegetation indices, canopy height and cover, or selection of genomes are also possible at high temporal resolution.

The huge benefit of UAS based phenotyping is its combination with AI, as the huge datasets created increase the performance and robustness of models via better training conditions. For the application of the genomics tools available to improve crops, precise phenotypic measurements at the field level are required.

The application of molecular biology tools, in particular the genetic manipulation of DNA to improve animal or plant traits, is termed molecular breeding. The tools include molecular marker assisted or genomic selection, and gene manipulation or genetic engineering. Plant tissue culture technique is a useful technique for plant propagation at a commercial scale. In recent years it has been applied for

- Rapid production of plants regardless of season,
- Production of (heat, drought, salinity) tolerant varieties,
- Disease resistant plants
- For conservation of endangered species, and
- Genetic transformation among others.

These applications aim to improve crop performance in agriculture and consist of a wide array of techniques, such as, tissues, organs and in-vitro regeneration of plants, aseptic growth of cells, molecular genetics, genome analyses, transfer of genes, and recombinant technologies. Among the benefits of these technologies used in agriculture are, genetic preservation of desirable genetics, uniform plant growth, genetic enhancement for increased plant efficiency, and year-round production regardless of season.

The combined use of tissue culture with AI and other optimization algorithms has been demonstrated as a field of technology for optimization of production efficiency. Plant tissue culture, which is based on "totipotenti", the ability of a stem cell to differentiate into all kind of cell types, provides the basis for "micropropagation". In this process plants are grown in vessels containing culture media from different explants. This in vitro culture contains nutrients and growth regulators. The smaller growth of plants in vitro compared to in vivo, leads to the term "micro-propagation".

In vitro culture is among the most significant technologies for propagation and breeding of various crop species, enabling different methods such as shoot multiplication, or production of plants from cells and tissues via the development of somatic embryos or adventitious shoots. The particular nutrient demands of different plant cells and tissues differ with plant species and hence the improvement of culture media is a time-consuming process that needs an enormous number of media compositions. In this case AI models are highly beneficial in overcoming the problem of complex interactions with multiple factors of in vitro culture which cannot be solved with an unrealistic quantity of treatments and traditional statistics.

These AI models can simulate and forecast development and growth of plant tissue in vitro under various conditions to optimize the media, with realistic number of treatments. The potential of AI models to cope with different in vitro systems and results of the cultures have turned them into a common method applied among plant tissue scientists.

Pest and disease control, via intelligent and precise application of plant protection agents, is expected to enhance sustainability in conventional agricultural systems. Imaging of plant disorders in real time, and the use of UAS and AI, allow a targeted and automated spraying of pesticides and fertilizers in high precision, and reduced risks of contamination for crops, animals, humans and other environmental resources (i.e., water bodies). Beside identification of disorders of plants/crops, targeting their maturity and timing of harvest is also a field of computer vision and AI in precision agriculture. Already this technology is at a stage, where computer vision models can outperform human observation accuracy .

After harvest, AI assisted computer vision with imaging algorithms can be used for sorting and grading of agricultural products, by identifying disease and defects, or products that do not meet the demands regarding size, shape and colour. In addition to external values, as well as the internal ones, the quality of the products can be assessed with AI machine learning and image data. By automating this process, AI can overcome labour shortages that are already being faced in many industrialized countries.

Coming back to the ground, AI can also be of excellent use in soil health monitoring. On the one hand, computer vision is able to characterize efficiently soil organic matter and texture at large spatial scales, providing a source of information to improve existing soil maps. The identification of soil health parameters is usually a time and cost demanding effort.

Therefore, identifying key soil variables underlying soil health functions for larger scale monitoring is of high importance to define effective monitoring schemes. Laboratory data as well as low cost handheld devices can be used to train algorithms to identify soil health or potentials for improvement.

A digital hub could focus on developing tools and algorithms to better measure and process data characterizing agricultural ecosystem complexity based on the interaction of microorganisms, plants and animals and feedback to humans and their health. The focus is resilient biodiversity in production systems.

In particular the role of fungi and bacteria in soils frames the design and the bandwidth of driving processes of the system. High resolution data sets with biostatistical-mathematical modelling are the tools to identify key species for dynamic buffering or stabilising ecosystems.

Soil health maps for different land use areas are a key challenge in advancing towards sustainable management, as they require precise spatial data acquisition and thus improved usage of remote sensing information beyond the current state of the art. Beyond novel remote and proximal sensors with higher spatial resolution and extended spectral bands, better learning algorithms to infer on belowground processes from data fusion are highly needed. Multimodal data processing to merge soil biodiversity data from the increasing databases with other types of environmental information (e.g., soil maps, meteorological data) can advance our understanding and monitoring capabilities of soil health and underlying natural and human drivers. Implementing these data in the evaluation of soil health and ecosystem service measures could provide a novel approach for environmental risk assessment in terms of biodiversity and losses in ecosystem functions in agricultural systems.

Digitalization in agriculture can improve data collection and recording on the *status quo* of soil health and future application of regenerative agricultural practices. Soil attributes and functions, such as soil organic matter, pore volume, aggregate stability, water holding capacity, activity of microorganisms, and nutrient (mainly nitrogen) availability, have positive impacts on the environment and crop yield. These soil functions are strongly affected by soil management practices (e.g., tillage, mechanical weed control) and crop rotations (i.e. legumes, grass-legume, field grass). Positive effects of crops on soils by providing nutrients and as a source of organic matter are often complex and not considered due to a lack of immediate responses. Nevertheless, sustainably managed soils can play an important role in climate change mitigation by storing stable soil organic carbon and reducing greenhouse gas emissions to the atmosphere. This has caused a rethinking of agricultural practices, as soil fertility and soil health is the most important resource. Soil health monitoring assessment, using soil microbiome-based diagnostics and AI models proved most accurate when trained with the highest taxonomic resolution. This is becoming increasingly important to observe impacts of management and to ensure improvements in crop productivity and sustainable agricultural systems. However, there is still a lack for a universal parameters or simple measures that enable high throughput analyses for soil quality or health.

Soil health or quality requires optimal physicochemical characteristics, which are less sensitive to soil degradation; and soil biological characteristics which respond quickly, but often are also sensitive to seasonal changes.

Soil microbiotas produce enzymes and actively contribute to the formation of biophysical structures that are essential for soil functions, and thereby drive "quality", "health" and "fertility" of soils. This integral information from soil microbial data can be a strategy identifying an integrated measure for soil health, via machine learning algorithms, that is currently lacking.

In particular, it is crucial to also capture sites specific factors that often limit current model predictions, and to predict soil health for a wide scale. For a robust and universally applicable soil health indices a collaborative effort among environmental, biological and computer science disciplines is needed for sustainable ecosystems and agricultural management practices.

Nowadays awareness of the factors that influence human health is increasing and the concept termed "one health" considers the environment of a human being and the functioning of the surrounding ecosystems as a pre-requisite for healthy communities. These ecosystem functions include the provision of fresh water, clean air, food security and medicine. Extinction of species above and belowground is a fundamental problem, as our soils are a potential source for new antibiotics in human medical applications.

Hence, restoring and sustaining healthy soils and their biodiversity may have numerous benefits inducing preserving future opportunities for human health. AI can assist in identifying key drivers for ecosystem functions and how they can be regulated via measures to be implemented in land use and in particular agricultural systems. The loss of key taxa is one concern, but AI is also able to identify the complex interplay of the food-web belowground, as environmental disturbances, not only cause a reduction in the abundance of some organisms this has also consequences for others, as there are multiple interactions within food webs.

A reservoir of genetic resources for crops, livestock and soil biota is only provided in biodiverse ecosystems that are fundamental for a nutritious variety, which are an essential determent for health, i.e., via micronutrient availability. There is a large use of traditional medicine by 60% of the world's population, and this originates from plant for medical use from wild populations and cultivation. Manifold communities depend on natural products collected from ecosystems not only for medical treatment, but also for cultural activities.

AI in forest biotechnology

Wood is an increasingly important resource for humanity and natural forests are of enormous ecological value. However, these slow-growing forests are unable to meet current demand, resulting in loss and degradation of forest resources. This is where forest biotechnology, especially genetic engineering, can help. This is very important because plantation forests, for example, are urgently needed to sustainably meet global demand for wood. There are many potential applications for AI, including:

- *Predictive modelling*: AI can be used to analyze data from satellite imagery, drone imagery, and other sources to predict the growth and yield of different species of trees in different locations. This can help to optimize the planting and management of forests for maximum productivity.
- *Disease and pest management*: AI can be used to analyze data on the presence and spread of diseases and pests in forests, as well as to predict their likely impact on the health and productivity of trees. This can help to identify areas that are at risk and to implement preventative measures to protect forests.
- *Environmental monitoring*: AI can be used to analyze data from sensors and other sources to monitor the health of forests and identify potential environmental impacts, e.g., wildfire. This can help to identify areas that are at risk and to implement measures to protect forests, also.
- *Resource management*: AI can be used to optimize the use of resources, such as water and nutrients, in forests to maximize productivity and minimize waste.
- *Inventory management*: AI can be used to optimize the management of forests for different purposes, such as timber production, conservation, and recreation. This can involve the use of AI to analyze data on the location, age, and species of trees, as well as the availability of resources and the demand for different products and services.

AI in medical biotechnology

The European In Vitro Diagnostics Regulation (IVDR) explicitly includes software and thus AI algorithms in its requirements. This poses significant challenges for in vitro diagnostics (IVD) companies that use AI for data analysis and decision support. However, if the ethical and legal issues are taken into account and addressed well, we see enormous potential for AI to revolutionize medical biotechnology by enabling the faster, more accurate and more cost-effective identification and development beyond new drugs. Some specific ways in which AI can be used in medical biotechnology include:

- *Drug target identification*: AI can be used to analyze data from various sources, such as genomic data and protein-protein interaction data, to identify potential therapeutic targets for the treatment of diseases. This can involve the use of machine learning algorithms to identify patterns and correlations that may not be apparent to humans.

- *Drug screening*: AI can be used to analyze data on the activity of potential drugs against different targets to identify those that are most likely to be effective. This can involve the use of ML algorithms to predict the likelihood of a particular drug being effective based on its characteristics and the characteristics of the target.
- *Image screening*: AI can be used to analyze medical images, such as CT scans and MRI images, to identify abnormalities and diagnose diseases. This can involve the use of DL algorithms to automatically segment and classify structures in medical images.
- *Predictive modelling*: AI can be used to analyze data from various sources, such as electronic health records and wearable devices, to make predictions about an individual's health. This can include the use of machine learning algorithms to predict the likelihood of an individual developing a particular disease or the likelihood of a particular treatment being effective.

AI in Animal biotechnology

Livestock production systems have intensified in recent years in terms of productivity per animal or unit of land or labour. This intensification of various livestock farming practices has also led to social concerns in terms of consumer acceptance, concerning food and nutrition security, food safety, sustainability, animal welfare, animal health and human health.

Livestock is globally the largest user of land resources, as almost 80% of the total agricultural land is dedicated to the production of feed and pasture, despite the large proportion of land, as well as the energy (and water), used to produce animal protein, by converting crops which are dedicated to feed livestock by 50–60%, for feed production on declining arable land per person. Along the whole value chain, livestock related direct and indirect emissions are 16.5% of the total, or 8.1 GtCO2 equivalents according to the FAO.

In addition, climate change, with more severe extreme weather events such as droughts and heavy precipitations, may result in adverse effects on animal health and welfare, as well as increased GHG emissions, reduced quality and quantity of feed and food, and hence human health. To tackle these challenges, agricultural systems must produce more with the use of fewer resources and focus on the reduction of food waste with more closed production chains. Emerging technologies such as "Precision Livestock Farming" can be beneficial for animal health and well-being together with information on environmental and economic aspects in agricultural production. Sensor-based monitoring in animal husbandry can provide optimal quantitative input data for life cycle analyses. A combination with communication and information technologies build a fundament for agriculture 4.0.

Life cycle analyses (LCA) of agricultural production systems are of crucial importance considering the environment, natural resources and human health which are classified as areas of protection in life cycle impact assessment (LCIA). Product chain related life cycle analyses also include the evaluation of precision life stock farming technologies. These technologies provide an opportunity for the animal biotechnology sector, by improved performance from replacement of manual labour with intelligent labour techniquesenhancing production sustainability and animal welfare by reducing costs and environmental impact.

Animal farming involves tracing of animals that allows tracking the whole production process to be made available even to consumers and stakeholders via data collection and evaluation. Complete data tracking to be evaluated via LCA includes sustainable crop production for feed, tracing real-time positioning and health monitoring of animals, as well as the transportation and food processing and storage, resulting in a full tracking from farm to fork, for consumer health and safety, and could improve their awareness of behaviours. In particular, the data on environmental and health conditions will ensure animal welfare and well-being.

Tracing (animal) products, creates big data which could facilitate more closed cycles, by reducing inputs and saving resources and costs, as well as reductions in GHG emissions (including manure management) that feed back onto the global climate, considering in particular that carbon taxes would increase beef prices by more than 100%without considering actual effects of the energy crisis that accelerate costs for fossil fuel and fertilizer as well.

AI in bioinformatics

While ML is already well established in medical research integrating multi omic approaches for system biology, there are still challenges in environmental sciences. For example the use of soil metaproteomics and the link to other omic data, or even the lack of this information, are consuming computational power and time, as the size of general databases is increasing dramatically.

Here, DL algorithms may be a solution to save resources, as ML is particularly helpful for the prediction of large datasets, and human-in-the-loop can increase the explanatory power by excluding hits that are not plausible for the studied ecosystem. The combination of omics data with bioinformatics and ML will enable moving from data that are of explanatory nature to application in fields such as medicine, but also to agriculture and forestry.

One example is breeding for improved crops via soil rhizomicrobiome selection where a combined approach with AI in bioinformatics can enhance detection of genotypes with improved biotic and/or abiotic stress resistance (e.g., pest/herbivore resistance, water and nutrient use efficiency) via association with a specific rhizosphere community to promote plant growth/health and thereby reduce the input of agrochemicals. This underlines that AI provides multiple combined applications for breeding programs and are a potential biotechnology to adjust to the current production needs.

Another example where crop-soil microbial interactions are targeted by using large sequencing data is the effort of targeted design of microbial products, such as biostimulants, biofertilizers and biopesticides, to tackle functions such as improved nutrient uptake or plant immune system. Also for revealing untargeted effects of biopesticides on changing the soil microbial community composition and their functions (such as N-fixation) sequencing data are a potential source for understanding the relevant changes in microbial structure to underlying observed functional effects, under a given combination of climatic, soil type and crop species.

Finally, in global change research, large datasets are crucial on global (soil) biodiversity, and the drivers for biodiversity losses and ecosystem functioning are critical to sustain stable ecosystem health. The implementation of computomics, with modern high-throughput omic measurement platforms, is fundamental to unravel environmental system understanding and discover keystone taxa that are crucial in sustaining ecosystem functions that are vital for human life and wellbeing

Diagnosing and Treating Mental Illness

The growing cases of mental illness have been a cause of concern in the last few years. Mental illness affects around 13% of the world's population. The Covid-19 pandemic has led to an alarming increase in depression and anxiety. People with depressive symptoms went from 193 million to 246 million (28% increase), while anxiety disorders showed a 25% increase (298 million to 374 million).

Artificial intelligence can help doctors and scientists detect early signs of mental illness and use preventive medication or treatment to help patients feel better. Treating mental illness is hard because the symptoms exhibited by patients are varied and not universally applicable. However, advanced AI programs can observe the brain waves of patients to recommend the best antidepressant therapies for each patient.

AI algorithms are being used to study and identify the behavioral loops in addicts to determine the chances of relapse so that physicians can take the necessary action. Furthermore, AI can help create personalized medicine that will be more effective in treating the patient. However, there is a lot more research to be done in this field to help people deal with mental illness and lead better lives.

Creating 3D Protein Structures

Proteins are one of the four most important macromolecules essential to building life. The scientific community has been using X-ray crystallography and nuclear magnetic resonance to identify protein structures. They have added 187,000 identified structures to the database. However, the process is slow and laborious. Many more proteins are yet to be identified.

A UK-based team of AI researchers created AlphaFold, a machine-learning platform that uses the existing database to predict protein structures and build 3D models. The tool takes only a small fraction of the time takes originally to complete the job, that too with greater accuracy. This can help biotech companies identify more proteins quickly and use the vital ones in developing effective drugs. Diseases like cystic fibrosis and muscular dystrophy can be treated with the help of this technology.

Gene Coding Identification

Machine learning in Biotech can be researchers build powerful models to study and understand human genomics. The next-gen technology used by biotech companies helps sequence a gene in less time and creates an alternative method to homology-based sequence analysis.

CRISPR (clustered regularly interspaced short palindromic repeats) can electively modify and edit the DNA or genes of living organisms. This allows scientists to identify and destroy the DNS from bacteriophages that cause infections and diseases. The technology can also help create personalized medicine based on the genome of the patient. This can be a game-changer in treating hereditary diseases and identifying the probability of a patient inheriting a family disease or illness.

Lab Assistants

AI in biotech is also being used to simplify tedious tasks. Artificial intelligence programs are acting as lab assistants and managing backend work as well as complex tasks like data analytics. H2O.ai is a fast, accurate, and open-source big data analytics platform that allows people to use its statistical models to identify patterns in datasets. Even though the company caters to many industries, many other AI platforms cater exclusively to the biotech, health, and pharma industries.

Another role of AI as a lab assistant is the use of robotic devices in research centers, healthcare units, and hospitals. Though this area is still being explored by the biotech industry, we can soon expect robotic assistants to have a prominent role.

Speed Up Innovation and Analytics

Artificial intelligence is now a part of research and development in the biotech industry. AI tools can process large datasets in a quick time and deliver faster analytics. The initial stages, where scientists have to repeatedly conduct lab experiments and record their findings, can be hastened using AI. AI tools can make the recordings and come up with possible variations to the experiments and the subsequent results.

The process of discovering and analyzing extensive information doesn't have to take years. Artificial intelligence and machine learning can accurately analyze the data and deliver insights in days or hours. From identifying patterns to detecting errors in datasets, all repetitive tasks can be automated with AI technology. Scientists and researchers can directly use the insights in their experiments.

New Vaccines and Drugs

With new viruses and their mutant versions affecting people around the world, it has become important to speed up the process of developing vaccines. The biotech industry had to race against time to create vaccines for the Covid-19 pandemic. The entire process can be hastened by relying on advanced AI and ML algorithms to study molecular data and identify the right composition as an antidote to the virus.

Earlier, scientists had to manually perform most processes, and it took around five to ten years to create a vaccine or new drug for a virus. However, by using AI in biopharma research, companies can develop new vaccines in two or three years.

Advancement in Industrial Biotech

AI is used in the field of biopolymer replacements, molecular designing, and robotics. For example, AI tools can create 3D images of molecules and edit the structures to create new chemical compositions. This makes it easier for laboratories to develop new and better chemicals for industrial use. AI, ML, and IoT are already a part of industries and help with day-to-day operations.

Many companies that deal with textiles, fuels, car components, chemicals, and biotech manufacturing gain an advantage by streamlining their processes using artificial intelligence. Automated robots can replace humans in hazardous zones to minimize the risk of accidents.

Agricultural Biotech

AI and biotech have a great role in the agricultural industry. Genetically modified and hybrid plants are not new. Researchers say that genetically modified plants provide better and greater yields while consuming fewer resources. However, successfully creating such a seed or plan takes years of work. Scientists work with many procedures like tissue culture, molecular breeding, and micropropagation to get the results.

Biotech companies in the agriculture sector use AI and ML tools to speed up genetic engineering. Computer vision and deep learning technologies are used to monitor soil and crop health in farmlands. The tools can also be used to track environmental changes and predict rainfall, storms, and weather conditions to help farmers decide the best crop to plant for the season. Furthermore, robots can help farmers with hard jobs like harvesting crops through automated machines.

Animal Biotech

Artificial intelligence in the animal biotech industry can help companies use molecular biology to modify the genes and traits of animals and create mixed or cross-breed versions for agricultural and pharmacological applications. AI is also useful in selective breeding, where animals of specific qualities are bred to give birth to offspring with similar attributes.

Machine learning algorithms can be used to process large datasets with genomic data about different animal species and ways to choose animals for selective breeding. This helps the animal industry breed healthier and stronger offspring that are less prone to diseases or illness.

Personalized Medicines

Biopharma companies take personalization one step ahead by developing medicines based on a patient's condition or symptoms. While the actual process is a lot more complex and involves the use of several AI tools, it is quite possible to find medication for lesser-known or hereditary diseases that affect only a small portion of the community.

Here, Artificial Intelligence and Machine Learning are not only used in discovering new drugs but play a crucial role in understanding how diseases develop in humans, the various stages of disease development, the corresponding symptoms, and ways to effectively treat the conditions based on the parameters of each case.

CHAPTER-13

RESOURCES FOR FURTHER EXPLORATION

CRISPR-based approaches have quickly become a favored method to perturb genes to uncover their functions. Here, we review the key considerations in the design of genome editing experiments, and survey the tools and resources currently available to assist users of this technology.

RISPR-Cas is an adaptive immune system existing in most bacteria and archaea, preventing them from being infected by phages, viruses and other foreign genetic elements. It is composed of CRISPR repeat-spacer arrays, which can be further transcribed into CRISPR RNA (crRNA) and *trans*-activating CRISPR RNA (tracrRNA), and a set of CRISPR-associated (cas) genes which encode Cas proteins with endonuclease activity. When the prokaryotes are invaded by foreign genetic elements, the foreign DNA can be cut into short fragments by Cas proteins, then the DNA fragments will be integrated into the CRISPR array as new spacers. Once the same invader invades again, crRNA will quickly recognize and pair with the foreign DNA, which guides Cas protein to cleave target sequences of foreign DNA, thereby protecting the host.

INNOVATIONS OF CRISPR-CAS SYSTEMS

CRISPR-Cas systems have become the most favorite genome editing tool in the molecular biology laboratory since they were confirmed to have genome editing capabilities in 2012. They have made numerous achievements in the field of correcting pathogenic mutations, searching for essential genes for cancer immunotherapy, and solving key problems in organ xenotransplantation. Unfortunately, there are still some limitations which need to solve in CRISPR-Cas systems, such as potential off-target effects, limited genome-targeting scope restricted by PAM sequences, and low efficiency and specificity. Therefore, many research teams have been trying to improve this tool.

Dead-Cas9 system

By introducing two point mutations, H840A and D10A, into HNH and RuvC nuclease domain, researchers have obtained a nuclease dead Cas9 (dCas9). The dCas9 lacks DNA cleavage activity, but DNA binding activity is not affected. Then, by fusing transcriptional activators or repressors to dCas9, the CRISPR-dCas9 system can be used to activate (CRISPRa) or inhibit (CRISPRi) transcription of target genes. Additionally, dCas9 can be fused to various effector domains, which enables sequence-specific recruitment of fluorescent proteins for genome imaging and epigenetic modifiers for epigenetic modification. Furthermore, this system is easy to operate and allows simultaneous manipulation of multiple genes within a cell.

Base editing system

In order to improve the efficiency of site-directed mutagenesis, base editing systems containing dCas9 coupled with cytosine deaminase (cytidine base editor, CBE) or adenosine deaminase (adenine base editor, ABE) have been developed. It can introduce C·G to T·A or A·T to G·C point mutations into the editing window of the sgRNA target sites without double-stranded DNA cleavage. Since base editing systems avoid the generation of random insertions or deletions to a great extent, the results of gene mutation are more predictive. However, owing to the restriction of base editing window, base editing systems are not suitable for any target sequence in the genome. Accordingly, C-rich sequences, for example, would produce a lot of off-target mutations.

Therefore, researchers have always been trying to develop and optimize novel base editing systems to overcome this drawback. At present, base editing systems have been widely used in various cell lines, human embryos, bacteria, plants and animals for efficient site-directed mutagenesis, which may have broad application prospects in basic research, biotechnology and gene therapy. In theory, 3956 gene variants existing in Clin var database could be repaired by base substitution of C-T or G-A.

Cas9 variant system

An NGG PAM at the 32 end of the target DNA site is essential for the recognization and cleavage of the target gene by Cas9 protein. Besides classical NGG PAM sites, other PAM sites such as NGA and NAG also exist, but their efficiency of genome editing is not high. However, such PAM sites only exist in about one-sixteenth of the human genome, thereby largely restricting the targetable genomic loci. For this purpose, several Cas9 variants have been developed to expand PAM compatibility.

In 2018, researchers developed xCas9 by phage-assisted continuous evolution (PACE), which can recognize multiple PAMs (NG, GAA, GAT, etc.). In the latter half of the same year, researchers developed SpCas9-NG, which can recognize relaxed

NG PAMs. In 2020, researchers developed three new SpCas9 variants recognizing non-G PAMs, such as NRRH, NRCH and NRTH PAMs. Later in the same year, Walton *et al.* developed a SpCas9 variant named SpG, which is capable of targeting an expanded set of NGN PAMs. Subsequently, they optimized the SpG system and developed a near-PAMless variant named SpRY, which is capable of editing nearly all PAMs (NRN and NYN PAMs).

By using these Cas9 variants, researchers have repaired some previously inaccessible disease-relevant genetic variants. However, there are still some drawbacks in these variants, such as low efficiency and cleavage activity. Therefore, they should be further improved by molecular engineering in order to expand the applications of SpCas9 in disease-relevant genome editing.

RNA editing system

In addition to editing DNA, CRISPR-Cas systems can also edit RNA. Class 2 Type VI CRISPR-Cas13 systems contain a single RNA-guided Cas13 protein with ribonuclease activity, which can bind to target single-stranded RNA (ssRNA) and specifically cleave the target. To date, four Cas13 proteins have been identified: Cas13a (also known as C2c2), Cas13b, Cas13c and Cas13d. They have successfully been applied in RNA knockdown, transcript labeling, splicing regulation and virus detection. Later, researchers developed two RNA base edting systems (REPAIR system, enables A-to-I (G) replacement; RESCUE system, enables C-to-U replacement) by fusing catalytically inactivated Cas13 (dCas13) with the adenine/cytidine deaminase domain of ADAR2 (adenosine deaminase acting on RNA type 2).

Compared with DNA editing, RNA editing has the advantages of high efficiency and high specificity. Furthermore, it can make temporary, reversible genetic edits to the genome, avoiding the potential risks and ethical issues caused by permanent genome editing. At present, RNA editing has been widely used for pre-clinical studies of various diseases, which opens a new era for RNA level research, diagnosis and treatment.

Prime editing system

Recently, researchers developed a novel genome editing technology, named prime editing, which can mediate targeted insertions, deletions and all 12 types of base substitutions without double-strand breaks or donor DNA templates.

This system contains a catalytically impaired Cas9 fused to a reverse transcriptase and a prime editing guide RNA (pegRNA) with functions of specifying the target site and encoding the desired edit. After Cas9 cleaves the target site, the reverse transcriptase uses pegRNA as a template for reverse transcription, and then, new genetic information can be written into the target site. Prime editing can effectively improve the efficiency and accuracy of genome editing, and significantly expand the scope of genome editing in biological and therapeutic research. In theory, it is possible to correct up to 89% known disease-causing gene mutations. Nevertheless, as a novel genome editing technique, more research is still needed to further understand and improve prime editing system.

APPLICATIONS OF CRISPR-CAS SYSTEMS IN HUMAN DISEASE RESEARCH

Applications of CRISPR-Cas systems in establishing animal and cell models of human diseases

So far, as a rapid and efficient genome editing tool, CRISPR-Cas systems have been extensively used in a variety of species, including bacteria, yeast, tobacco, Arabidopsis, sorghum, rice, Caenorhabditis elegans, Drosophila, zebrafish, Xenopus laevis, mouse, rat, rabbit, dog, sheep, pig and monkey, as well as various human cell lines, such as tumor cells, adult cells and stem cells. In medical field, the most important application of CRISPR-Cas systems is to establish genetically modified animal and cell models of many human diseases, including gene knockout models, exogenous gene knock-in models, and site directed mutagenesis models.

Establishing animal models of human diseases

Animal models are crucial tools for understanding gene function, exploring pathogenesis of human diseases and developing new drugs. However, traditional methods for generating animal models are complex, costly and time-consuming, which severely limit the application of animal models in basic medical research and preclinical studies. Since the discovery of CRISPR-Cas systems, a series of genetically modified animal models have successfully been generated in a highly efficient manner.

Among numerous model animals, mice are widely used for scientific studies and recognized as the most important model animals in human disease research. So far, researchers have successfully generated many genetically modified mouse models, such as cancer, cardiovascular disease, cardiomyopathy, Huntington's disease, albino, deafness, hemophilia B, obesity, urea cycle disorder and muscular dystrophy.

Nevertheless, owing to the great species differences between humans and rodents, they can't provide effective assessment and long-term follow-up for research and treatment of human diseases. Therefore, the application of larger model animals, such as rabbits, pigs and non-human primates, is becoming more and more widespread. With the development of CRISPR-Cas systems, generating larger animal models for human diseases has become a reality, which greatly enriches the disease model resource bank.

Applications of CRISPR-Cas systems in disease diagnosis

With the development of CRISPR-Cas systems and the discovery of novel Cas enzymes (Cas12, Cas13, etc.), CRISPR-based molecular diagnostic technology is rapidly developing and has been selected as one of the world's top ten science and technology advancements in 2018.

Unlike Cas9, Cas13 enzymes possess a 'collateral cleavage' activity, which can induce cleavage of nearby non-target RNAs after cleavage of target sequence. Based on the 'collateral cleavage' activity of Cas13, researchers developed a Cas13a-based *in vitro* nucleic acid detection platform, named SHERLOCK (Specific High Sensitivity Enzymatic Reporter UnLOCKing).

It is composed of Cas13a, sgRNA targeting specific RNA sequences and fluorescent RNA reporters. After Cas13a protein recognizes and cleaves the target RNA, it will cut the report RNA and release the detectable fluorescence signal, so as to achieve the purpose of diagnosis. Researchers have used this method to detect viruses, distinguish pathogenic bacteria, genotype human DNA and identify tumor DNA mutations. Later, researchers improved SHERLOCK system and renamed it as SHERLOCKv2, which can detect four virus at the same time.

In addition to Cas13, Cas12 enzymes are also found to possess collateral cleavage activity. Researchers developed a nucleic acid detection system based on Cas12a (also known as Cpf1), named DETECTR (DNA endonuclease-targeted CRISPR *trans* reporter). DETECTR has been used to detect cervical cancer associated HPV subtypes (HPV16 and HPV18) in either virus-infected human cell lines or clinical patient samples. Furthermore, researchers are trying to use the newly discovered Cas14 and CasX proteins in molecular diagnosis, which may further enrich the relevant techniques of CRISPR-based molecular diagnosis.

CRISPR-based molecular diagnostic technology has incomparable advantages over traditional molecular diagnostic methods, such as high sensitivity and single-base specificity, which is suitable for early screening of cancer, detection of cancer susceptibility genes and pathogenic genes. Meanwhile, CRISPR diagnostics is inexpensive, simple, fast, without special instrument, and is suitable for field quick detection and detection in less-developed areas. At present, many companies are trying to develop CRISPR diagnostic kits for family use, to detect HIV, rabies, Toxoplasma gondi, etc.

Applications of CRISPR-Cas systems in genome-scale screening

CRISPR-Cas9 system enables genome-wide high-throughput screening, making it a powerful tool for functional genomic screening. The high efficiency of genome editing with CRISPR-Cas9 system makes it possible to edit multiple targets in parallel, thus a mixed cell population with gene mutation can be produced, and the relationship between genotypes and phenotypes could be confirmed by these mutant cells. CRISPR-Cas9 library screening can be divided into two categories: positive selection and negative selection. It has been utilized to identify genes associated with cancer cell survival, drug resistance and virus infection in various models. Compared with RNAi-based screening, high-throughput CRISPR-Cas9 library screening has the advantages of higher transfection efficiency, minimal off-target effects and higher data reproducibility. At present, scientists have constructed human and mouse genome-wide sgRNA libraries, and they have been increasingly improved according to different requirements. In the future, CRISPR-Cas9-based high-throughput screening technology will definitely get unprecedented development and application.

Applications of CRISPR-Cas systems in gene therapy

Gene therapy refers to the introduction of foreign genes into target cells to treat specific diseases caused by mutated or defective genes. Target cells of gene therapy are mainly divided into two categories: somatic cells and germ line cells. However, since germ line gene therapy is complicated in technique as well as involves ethical and security issues, today gene therapy is limited to somatic cell gene therapy. Traditional gene therapy is usually carried out by homologous recombination or lentiviral delivery.

Nevertheless, the efficiency of homologous recombination is low, and lentiviral vectors are randomly inserted into the recipient genome, which may bring potential security risks to clinical applications. Currently, with the rapid development of CRISPR-Cas systems, they have been widely applied in gene therapy for treating various of human diseases, monogenic diseases, infectious diseases, cancer, etc. Furthermore, some CRISPR-mediated genome-editing therapies have already reached the stage of clinical testing.

Monogenic diseases

Monogenic diseases refer to the genetic diseases caused by mutations of a single allele or a pair of alleles on a pair of homologous chromosomes. There are more than 6600 known monogenic diseases around the world, â-thalassaemia, sickle cell disease (SCD), hemophilia B (HB), retinitis pigmentosa (RP), leber congenital amaurosis type 10 (LCA10), duchenne muscular dystrophy (DMD), hutchinson-gilford progeria syndrome (HGPS), hereditary tyrosinemia (HT), cystic fibrosis (CF), etc. Most of the monogenic diseases are rare diseases lacking of effective treatment, which will greatly affect the life quality of patients. Nowadays, many animal models of monogenic diseases have been treated with CRISPR-mediated gene therapy.

The incorporation of CRISPR/*Cas9* into clinical settings has promising applications for the future treatment of genetic disorders, human immunodeficiency virus (HIV) infection, and various malignant diseases. Although considerable research is still needed to optimize the targeted delivery of Cas9 molecules into cells and prevent unwanted off-target effects, CRISPR/*Cas9* technology has the potential to revolutionize the treatment of many debilitating health conditions.

The need for targeted drugs

Considerable scientific advancements over the past several decades have contributed to the development of novel treatment options that target specific molecular perturbations in different diseases.

Within oncology, the combination of targeted treatments with traditional therapies like chemotherapy, radiation, and/or immunotherapy has significantly increased survival rates while also reducing the risk of undesirable side effects.

Although many targeted anticancer therapies (TATs) are associated with less severe toxicity as compared to standard chemotherapy treatments, they are still associated with their unique adverse effects, some of which include rash, cardiac dysfunction, hypertension, and bleeding.

This necessitates the involvement of toxicity specialists early on in the drug development process to allow drug developers to gain a thorough understanding of a compound's toxicity and determine how to circumvent it in a timely manner.

In addition to the limitations associated with current targeted therapies, the identification and subsequent development of new targeted agents is also a tedious and expensive process. In fact, current estimates indicate that the time to develop a new anticancer drug is between 12-15 years, with costs reaching up to $2 billion USD.

CRISPR/Cas9 in drug discovery

Within the pharmaceutical industry, drug screening has historically been used to identify chemically synthesized small molecule drugs capable of inducing desired effects in a cell-based disease model. Drug screening often involves large-scale and high-throughput assays that allow researchers to test and compare the in vitro efficacy of numerous molecules simultaneously.

CRISPR technology can be incorporated into drug screening assays for both full and specified genome applications. Often, CRISPR is used to differentially express specific genetic sequences that are relevant to the disease being studied. This creates an *in vitro* model in which researchers can then evaluate the efficacy and toxicity of various compounds before initiating further *in vivo* studies.

In addition to this type of hypothesis-driven screen, CRISPR can also be applied to genome-wide screens in which all known genes are targeted in an effort to identify potentially unknown genes involved in the disease phenotype of interest. A restricted CRISPR screen can also be used for drug discovery purposes, in which a specific subset of genes that are relevant to the disease phenotype are explored.

PRECISION MEDICINE AND PERSONALIZED THERAPIES

Precision medicine incorporates individual differences in patients' genes, as well as their environments and lifestyles, in order to identify specific treatment options that are most likely to be effective in treating their disease.

Within oncology, for example, cancer patients each have their own unique genetic and epigenetic patterns. Even cancer patients with the same type of cancer will exhibit different transcriptomes, proteomes, and metabolomes, thus causing patients to often respond to the same therapeutic regimen differently.

CRISPR technologies have already been incorporated into the personalized screening of cancers in an effort to identify the presence of certain genes that may contribute to tumor growth, drug resistance, epithelial-to-mesenchymal transition, cancer origins, metabolic adaptations, and metastasis. CRISPR screens have also been performed on patient tissues to identify individualized drug targets and model how they might respond to different therapeutic regimens.

A vast amount of research has also been dedicated to the development of CRISPR therapeutics; however, the efficacy and safety of these treatments are primarily dependent upon their mode of delivery. In various preclinical cancer models, CIRSPR-Cas components have been successfully delivered in the form of plasmids and nanoparticles, as well as both non-viral and viral vectors.

DECIPHERING CELL-SPECIFIC GENE EXPRESSION USING CRISPRI IN IPSC-DERIVED NEURONS

Deploying CRISPR technology in iPSCs has been notoriously challenging, as Martin Kampmann, of the Kampmann lab at the University of California San Francisco, says: "CRISPR introduces DNA breaks, which can be toxic for iPSCs, since these cells have a highly active DNA damage response." To overcome the issue of toxicity, as a postdoc in the lab of Professor Jonathan Weissman, Kampmann co-invented a tool known as CRISPR interference (CRISPRi), where the DNA cutting ability of CRISPR/Cas9 is disabled. The "dead" Cas9 (or, dCas9) is still recruited to DNA as directed by a single guide RNA. It can therefore operate as a recruitment platform to target protein domains of interest to specific places in the genome.

CRISPRi permits gene repression at the transcription level, as opposed to RNAi which controls genes at the mRNA level. This allows researchers to repress certain genes within stem cells and decipher their function. Kampmann explains: "For CRISPRi, we target a transcriptional repressor domain (the KRAB domain) to the transcription start site of genes to repress their expression. This knockdown approach is highly effective and lacks the notorious off-target effects of RNAi-based gene knockdown."

In a study published just last month, Kampmann's laboratory adopted a CRISPRi-based platform to conduct genetic screens in human neurons derived from iPSCs: "CRISPR-based genetic screens can reveal mechanisms by which these mutations cause cellular defects, and uncover cellular pathways we can target to correct those defects. Such pathways are potential therapeutic targets."

We expressed the CRISPRi machinery (dCas9-KRAB) from a safe-harbor locus in the genome, where it is not silenced during neuronal differentiation. We also developed a CRISPRi construct with degrons, stability of which is controlled by small molecules. This way, we can induce CRISPRi knockdown of genes of interest at different times during neuronal differentiation.

Previous CRISPR-based screens have focused on cancer cell lines or stem cells rather than healthy human cells, thus limiting potential insights into the cell-type-specific roles of human genes. The researchers opted to screen in iPSC-derived neurons as genomic screens have revealed mechanisms of selective vulnerability in neurodegenerative diseases, and convergent mechanisms in neuropsychiatric disorders.

The large-scale CRISPRi screen uncovered genes that were essential for both neurons and iPSCs yet caused different transcriptomic phenotypes when knocked down. "For me, one of the most exciting findings was the broadly expressed genes that we think of as housekeeping genes had different functions in iPSCs versus neurons. This may explain why mutations in housekeeping genes can affect different cell types and tissues in the body in very different ways," says Kampmann. For example, knockdown of the E1 ubiquitin activating enzyme, UBA1, resulted in neuron-specific induction of a large number of genes, including endoplasmic reticulum chaperone *HSPA5* and *HSP90B1*.

These results suggest that comprised UBA1 triggers a proteotoxic stress response in neurons but not iPSCs - aligning with the suggested role of UBA1 in several neuro-degenerative diseases. The authors note: "Parallel genetic screens across the full gamut of human cell types may systematically uncover context-specific roles of human genes, leading to a deeper mechanistic understanding of how they control human biology and disease."

DEVELOPING AND TESTING CELL-BASED THERAPIES FOR HUMAN DISEASE USING CRISPR

Several laboratories across the globe are in an apparent race to develop the first clinically relevant, efficacious and safe cell-based therapy utilizing CRISPR gene-editing technology.

Whilst a plethora of literature demonstrates the efficacy of CRISPR in editing the genome of mammalian cells *in vitro*, for *in vivo* application, particularly in humans, rigorous long-term testing of safety outcomes is required. This month, researchers from the laboratory of Hongkui Deng, a Professor at Peking University in Beijing, published a paper in *The New England Journal of Medicine.* The paper outlined their world-first study in which they transplanted allogenic CRISPR-edited stem cells into a human patient with HIV.

The rationale for the study stems back to the "Berlin patient", referring to Timothy Ray Brown who is one of very few individuals in the world that has been cured of HIV. Brown received a bone marrow transplant from an individual that carries a mutant form of the *CCR5* gene. Under normal conditions, the *CCR5* gene encodes a receptor on the surface of white blood cells. This receptor effectively provides a passageway for the HIV to enter cells. In individuals with two copies of the *CCR5* mutation, the receptor is distorted and restricts strains of HIV from entering cells.

Researchers used CRISPR to genetically edit donor hematopoietic stem and progenitor cells (HSPCs) to carry either a *CCR5* insertion or deletion. They were able to achieve this with an efficiency of 17.8%, indicated by genetic sequencing. The CRISPR-edited HSPCs were then transplanted into an HIV patient who also had leukaemia and required a bone-marrow transplant, with the goal being to eradicate HIV.

The study was designed to assess the safety and feasibility of the transplantation of CRISPR–Cas9–modified HSPCs into HIV-1–positive patients with hematologic cancer. The success of genome editing in human hematopoietic stem and progenitor cells was evaluated in three aspects including editing persistence, specificity and efficiency in long-term engrafting HSPCs. Long-term monitoring of the HIV patient found that, 19 months after transplantation, the CRISPR-edited stem cells were alive - however, they only comprised five to eight percent of total stem cells. Thus, the patient is still infected with HIV.

Despite the seemingly low efficiency in long-term survival, the researchers were encouraged by the results from the safety assessment aspect of the study: "Previously reported hematopoietic stem and progenitor cells-based gene therapies were less effective because of random integration of exogenous DNA into the genome, which sometimes induced acute immune responses or neoplasia," Deng says. "The apparent absence of clinical adverse events from gene editing and off-target effects in this study provides preliminary support for the safety of this gene-editing approach."

To further clarify the anti-HIV effect of CCR5-ablated HSPCs, it will be essential to increase the gene-editing efficiency of our CRISPR–Cas9 system and improve the transplantation protocol.

The marrying of CRISPR gene-editing and stem cell research isn't *just* bolstering therapeutic developments in HIV. An ongoing clinical trial is evaluating the safety and efficacy of autologous CRISPR-Cas9 modified CD34+ HSPCs for the treatment of transfusion-dependent â-thalassemia, a genetic blood disorder that causes hemoglobin deficiency.

The therapeutic approach – known as CTX001 – involves extracting a patient's HSPCs and using CRISPR-Cas9 to modify the cells at the erythroid lineage-specific enhancer of the *BCL11A* gene. The genetically modified cells are then infused back into the patient's body, where they produce large numbers of red blood cells that contain fetal hemoglobin. Currently no results are available, but reports confirm that participants have been recruited on to the trial.

GLOSSARY

Adenine: a purine nucleobase that binds with thyamine in DNA and with uracil in RNA.

Adult stem cell: An undifferentiated cell found in a differentiated tissue in an adult organism that can renew itself and can differentiate to yield specialized cell types of the tissue in which it is found.

Allele: A variant form of a gene at a particular locus on a chromosome. Different alleles produce variations in inherited characteristics.

Aneuploidy: The presence of an abnormal number of chromosomes in a cell.

Assisted reproductive technology (ART): A fertility treatment or procedure that involves laboratory handling of gametes (eggs and sperm) or embryos. Examples of ART include in vitro fertilization (IVF) and intracytoplasmic sperm injection (ICSI).

Autologous transplant: Transplanted tissue derived from the intended recipient of the transplant. Such a transplant helps to avoid complications of immune rejection.

Blastocyst: A preimplantation embryo in placental mammals (about 5 days after fertilization in humans) of 50-150 cells. The blastocyst consists of a sphere made up of an outer layer of cells (the trophectoderm), a fluid-filled cavity (the blastocoel or blastocyst cavity), and a cluster of cells in the interior (the inner cell mass). Cells from the inner cell mass, if grown in culture, can give rise to embryonic stem cell lines.

Cas9 (CRISPR Associated Protein 9): A specialized enzyme known as a nuclease that has the ability to cut DNA sequences. Cas9 makes up part of the "toolkit" for the CRISPR/Cas9 method of genome editing.

Chimera: An organism composed of cells derived from at least two genetically different individuals.

Choriocarcinoma: A type of tumor that originates from the trophoblast, the precursor of the placenta, and invades the uterine wall.

Chromatin: The complex of DNA and proteins that forms chromosomes. Some of the proteins are structural, helping to organize and protect the DNA, while others are regulatory, acting to control whether genes are active or not, and to promote DNA replication or repair.

Chromosome: A thread-like structure that contains a single length of DNA, usually carrying many hundreds of genes. This is packaged with proteins to form chromatin. The DNA within the complete cellular set of chromosomes (23 pairs in humans) comprises two copies of the genome, one from each parent. The chromosomes usually reside in the nucleus of a cell, except during cell division when the nuclear membrane breaks down and the chromosomes become condensed and can be visualized as discrete entities.

Cleavage: The process of cell division in the very early embryo before it becomes a blastocyst. Also used to describe breaking or cutting DNA.

Clinical application: The use of a biomedical reagent, procedure, or device to treat a clinical condition.

Clinical trial: A supervised and monitored experimental test in patients of a newly developed clinical application to ensure minimization of risk and optimization of efficacy. Clinical trials are required before a treatment is approved for general use.

CRISPR (Clustered Regularly-Interspaced Short Palindromic Repeats): A naturally occurring mechanism found in bacteria that involves the retention of fragments of foreign DNA, providing the bacteria with some immunity to viruses. The system is sometimes referred to as CRISPR/Cas9 to denote the entire gene-editing platform in which RNA homologous with the targeted gene is combined with Cas9.

CRISPRa: CRISPR activation, using a guide RNA and nuclease-deficient or nuclease-dead Cas9 (dCas9) linked to one or more activation domains to increase transcription of a target gene.

CRISPRr/CRISPRi: CRISPR repression, or CRISPR interference, using a dCas9 or dCas9-repressor with a guide RNA to decrease transcription of a target gene.

Cultured cell: A cell maintained in a tissue culture allowing expansion of its numbers.

dCas9 (Nuclease-deficient Cas9 or nuclease-dead Cas9): This can still bind DNA, together with a guide RNA, but not cut it. It is often linked to a transcription factor, chromatin-modifying enzyme, or fluorescent protein to mediate alterations to gene expression or to mark specific sites.

Deontology ethics: A normative theory regarding which choices are morally required, forbidden, or permitted.

Deoxyribonucleic acid (DNA): A two-stranded molecule, arranged as a double helix, that contains the genetic instructions used in the development, functioning, and reproduction of all known living organisms.

Differentiation: The process whereby an unspecialized early embryonic cell acquires the features of a specialized cell, such as a heart, liver, or muscle cell (IOM, 2005, p. 116).

Diploid: Cells that contain a full set of DNA: half from each parent. In humans, diploid cells contain 46 chromosomes (in 23 pairs).

Divergence (evolutionary): During evolution, variations occur in the sequences of genes; if these variations confer some advantage natural selection increases their prevalence. Different selective pressures select for different variations so that the prevalence of different gene variants diverges in different populations.

Dominant: A pattern of inheritance of a gene or trait in which a single copy of a particular allele (gene variant) confers a function independent of the nature of the second copy of the gene in a diploid cell of an organism.

Double-strand break (DSB): A break in the DNA double helix in which both strands are cut, as distinct from a single-strand break or "nick."

Ectoderm: The outermost of the three primitive germ layers of the embryo; it gives rise to skin, nerves, and brain.

Ectopic: Found in an unusual location, such as an ectopic pregnancy outside the uterus.

Embryo: An animal in the early stages of growth and differentiation that are characterized by cleavage (cell division of the fertilized egg), differentiation of fundamental cell types and tissues, and the formation of primitive organs and organ systems; the developing human individual from the time of implantation to the end of the eighth week after conception, after which stage it becomes known as a fetus.

Embryonic germ (EG) cell: A pluripotent stem cell that migrates during early development to the future gonads to form the progenitors of egg or sperm cells. The properties of EG cells are similar to those of embryonic stem cells, but may differ in the DNA methylation of some imprinted regions.

Embryonic stem (ES) cell: A primitive (undifferentiated) cell from the embryo that has the potential to become a wide variety of specialized cell types (i.e., is pluripotent). It is derived from the inner cell mass of the blastocyst. An embryonic stem cell is not an embryo; by itself, it cannot produce the necessary cell types, such as trophectoderm cells, so as to give rise to a complete organism. Embryonic stem cells can be maintained as pluripotent cells in culture and induced to differentiate into many different cell types.

Endoderm: Innermost of the three primitive germ layers of the embryo; it later gives rise to the lungs, liver, and digestive organs.

Endogenous: Originating from within a cell or an organism.

Endometrium: The inner epithelial lining of the uterus into which embryos implant.

Endonuclease: An enzyme that breaks down a nucleotide chain into two or more shorter chains by cleaving at internal phosphodiester bonds.

Enhancement: Improving a condition or trait beyond a typical or normal level.

Enucleated cell: A cell whose nucleus has been removed.

Enucleation: A process whereby the nuclear material of a cell is removed, leaving only the cytoplasm. When applied to an egg, can involve the removal of the maternal chromosomes, when they are not surrounded by a nuclear membrane.

Enzyme: A protein that acts as a biological catalyst, speeding up chemical reactions.

Epiblast: A specific layer of cells in an early vertebrate embryo that gives rise to the entire embryo other than yolk sac and placenta. Epiblast cells are pluripotent and can give rise to embryonic stem cells.

Epigenetic effects: Changes in gene expression that occur without changing the DNA sequence of a gene; for example, in the epigenetic effect called genomic imprinting, chemical molecules called methyl groups attach to DNA and alter the gene's expression.

Epigenome: A set of chemical modifications to the DNA of the genome and to proteins that bind to DNA in the chromosomes to affect whether and how genes are expressed.

Ex vivo: Latin: "out of the living"; outside an organism.

Exogenous: Introduced or originating from outside a cell or an organism.

Fertilization: The process whereby male and female gametes (sperm and egg) unite.

FokI: The nuclease from which the cleavage domain has been abstracted and joined to zinc finger (ZF) or transcription activator-like effector (TALE) DNA-binding domains. The FokI cleavage domain cuts only one strand of the DNA (a nick), so a pair of ZFNs or TALENs is required to create double-strand breaks. The FokI cleavage domain has also been linked to nuclease-deficient Cas9 (dCas9), and this fusion must also dimerize to cut DNA.

Gain of function: A type of mutation that results in an altered gene product that possesses a new molecular function or a new pattern of gene expression.

Gamete: A reproductive cell (egg or sperm). Gametes are haploid (having only half the number of chromosomes found in somatic cells: 23 in humans), so that when two gametes unite at fertilization, the resulting one-cell embryo (zygote) has the full number of chromosomes (46 in humans).

Gastrulation: The procedure by which an animal embryo at an early stage of development produces the three primary germ layers: ectoderm, mesoderm, and endoderm.

Gene: A functional unit of heredity that is a segment of DNA in a specific site on a chromosome. A gene typically directs the formation of a protein or RNA molecule.

Gene drive: A system of biased inheritance in which the ability of a particular genetic sequence to pass from a parent to its offspring through sexual reproduction is enhanced. Gene drive technology actively copies a sequence on one chromosome to its partner chromosome, so that the organism carries two copies of the intentionally modified gene. This process ensures that all of an organism's offspring and subsequent generations will inherit the edited genome and related trait(s). Thus, the result of a gene drive is the preferential increase of a specific genotype from one generation to the next, and potentially throughout a population.

Gene editing: A technique that allows researchers to alter the DNA of cells or organisms to insert, delete, or modify a gene or gene sequences to silence, enhance, or otherwise change the gene's characteristics.

Gene expression: The process by which RNA and proteins are made from the instructions encoded in genes. Gene expression is controlled by proteins and RNA molecules that bind to the genome or to the RNA copy and regulate their levels of production and those of their products. Alterations in gene expression change the functions of cells, tissues, organs, or whole organisms and sometimes result in observable characteristics associated with a particular gene.

Gene targeting: A procedure used to produce an alteration in a specific gene.

Gene therapy: Introduction of exogenous genes into cells with the goal of ameliorating a disease condition.

Gene transfer: Any process often used to describe the transfer of genes into cells: as used in gene therapy.

Genetic element: A segment of the DNA in a genome that has some particular property conferred by its sequence, such as a gene encoding a protein or RNA: more often used to refer to sequences that are not such genes but may control gene expression or genome organization.

Genome: The complete set of DNA that makes up an organism. In humans, the genome is organized into 23 pairs of homologous chromosomes.

Genome editing: The process by which the genome sequence is changed through the intervention of a DNA break or other DNA modification.

Genotype: Genetic constitution of an individual.

Germ cell (or germline cell): A cell at any point in the lineage of cells that will give rise to sperm or eggs. The germline is this lineage of cells. Eggs and sperm fuse during sexual reproduction to create an embryo. In so doing, the germline continues into the next generation.

Germ layer: In early development, the embryo differentiates into three distinct germ layers (ectoderm, endoderm, and mesoderm), each of which gives rise to different parts of the developing organism.

Gestation: The period of development of an organism from fertilization of the egg until birth.

Governance: The process of exercising oversight through traditions (standards of practice) or regulations by which individuals and communities are held accountable. Governance often involves such policy tools as professional standards of practice and codes of conduct; formal guidelines, agreements, and treaties; and legislation or other governmental regulation.

Guide molecule: A protein or short section of RNA used to guide the genome-editing machinery to the desired location in the DNA sequence.

Guide RNA (gRNA): Short segments of RNA used to direct the DNA-cutting enzyme to the target location in the genome. gRNA segments contain the region of homology to the target sequence (usually 20 bases), and a sequence that interacts with the nuclease (e.g., Cas9). gRNAs used in genome editing are synthetic and do not occur in nature.

Haploid: Refers to a cell (usually a gamete or its immediate precursor) having only one set of chromosomes (23 in humans). In contrast, body cells (somatic cells) are diploid, having two sets of chromosomes (46 in humans).

Heritable genetic change: Modifications to genes that could be passed down through generations.

Heterozygous: Having two different variants (alleles) of a specific gene on the two homologous chromosomes of a cell or an organism.

Homologous recombination: Recombining of two like DNA molecules, including a process by which gene targeting produces an alteration in a specific gene.

Homology-directed repair (HDR): A natural repair process used to repair broken DNA, which relies on a DNA "template" with homology to the broken stretch of DNA. This usually occurs during or after DNA synthesis, which provides this template. In genome editing via HDR, the DNA template is synthesised or made by recombinant DNA techniques, and usually contains regions of exact homology to the target locus at each end, with the desired alteration contained within the middle.

Homozygous: Having the same variant (allele) of a specific gene on both homologous chromosomes of a cell or an organism.

Human Fertilisation and Embryology Authority (HFEA): The United Kingdom's independent regulator overseeing the use of germ cells and embryos in fertility treatment and research. It also stands for the Human Fertilisation and Embryology Act, the law under which the Authority operates and which it upholds.

Implantation: The process by which an embryo becomes attached to the inside of the uterus (7-14 days in humans).

In utero: Latin: "in the uterus."

In vitro: Latin: "in glass"; in a laboratory dish or test tube; in an artificial environment.

In vitro fertilization (IVF): An assisted reproduction technique in which fertilization is accomplished outside the body.

In vivo: Latin: "in the living"; in a natural environment, usually in the body of the subject. This term is often also used to refer to events in "living" cells in culture.

Indel: An insertion or deletion of DNA sequence. Small indels (e.g., one to four base pairs) are often associated with nonhomologous end joining. These often result in the disruption of a gene by shifting the open reading frame and/or creating premature stop codons.

Induced pluripotent stem (iPS) cell: A cell induced by the introduction or activation of genes conferring pluripotency and stem celllike properties. Thus, cells already committed to a particular fate (e.g., skin) can be induced to become pluripotent. This is useful in regenerative medicine because the iPS cells can be introduced back into the donor of the original cells with much less risk of transplant rejection.

Insertional mutagenesis: The alteration of the sequence of a gene by the insertion of exogenous sequence such as by integration of viral sequences.

Institutional review board (IRB): An administrative body in an institution (such as a hospital or a university) established to protect the rights and welfare of human research subjects recruited to participate in research activities conducted under the auspices of that institution. The IRB has the authority to approve, require modifications in, or disapprove research activities in its jurisdiction, as specified by both federal regulations and local institutional policy (NRC, 2002, p. 266).

Lentivirus: A subclass of retroviruses, viruses whose genome are made of RNA but during viral replication becomes copied into a DNA form that can integrate into the DNA genome of a cell. Often used as carriers of genes (vectors) to introduce genes into cells.

Ligase: An enzyme that catalyzes joining of two pieces of DNA.

Loss of function: A type of mutation in which the altered gene product lacks the molecular function of the wild-type gene.

Meganuclease: A special type of enzyme that binds to and cuts DNA at specific DNA sequences of a length that occurs at few sites in the genome. These are natural enzymes (and their synthetic derivatives) that catalyze DNA rearrangement events via DNA cleavage.. They can be used in genome editing for both nonhomologous end joining and homology directed repair–mediated alterations. It was the study of these that first revealed the basic mechanisms of DNA cleavage and the DNA repair processes on which genome editing depends.

Mesoderm: The middle layer of the embryo, which consists of a group of cells derived from the inner cell mass of the blastocyst; it is formed at gastrulation and is the precursor to blood, bone, muscle, and connective tissue.

Mitochondrial transfer (or mitochondrial replacement): Novel procedures designed to prevent the maternal transmission of mitochondrial DNA (mtDNA) diseases.

Mitochondrion (plural, Mitochondria): A cellular structure in the cytoplasm that provides energy to the cell. Each cell contains many mitochondria. In humans, a single mitochondrion contains 37 genes on a circular mitochondrial DNA, compared with about 35,000 genes contained in the nuclear DNA.

Mosaicism: Variation among cells, such that the cells are not all the same: for example, in an embryo when not all the cells are edited.

Multipotent stem cells: Stem cells from the embryo, fetus, or adult, whose progeny are of multiple differentiated cell types and usually, but not necessarily, all of a particular tissue, organ, or physiological system.

Murine: Derived from mice.

Mutation: A change in a DNA sequence. Mutations can occur spontaneously during cell division or can be triggered by environmental stresses, such as sunlight, radiation, and chemicals.

Nickase: A nuclease that cuts only one strand of the DNA double helix.

Nonhomologous end joining (NHEJ): A natural repair process used to join the two ends of a broken DNA strand back together. This is prone to errors where short indels (usually of two to four base pairs of DNA) are introduced.

Normative theory: A theory of how people should make decisions, as opposed to how they actually do or will make decisions.

Nuclease: An enzyme that can cut through DNA or RNA strands.

Off-target effect: A direct or indirect, unintended, short- or long-term consequence of an intervention on an organism other than the intended effect on that organism.

Off-target event (or off-target cleavage): when a genome-editing nuclease cuts DNA at a location other than the one for which it was targeted. This can occur because the off-target sequence is similar to but not identical with the intended target sequence.

Oocyte: Developing egg; usually a large and immobile cell.

Phenotype: Observable properties of an organism that are influenced by both its genotype and its environment.

Plasmid: A self-replicating circular DNA molecule. A plasmid can be engineered to carry and express genes of interest in target cells.

Pluripotent stem cell (PSC): A stem cell that includes in its progeny all cell types that can be found in a postimplantation embryo, fetus, or developed organism.

Population: All of the individuals of a given species within a defined ecological area.

Preclinical research: Research conducted to investigate potential clinical applications but not involving humans. For example, research on molecules, cells, tissues, or animals.

Precursor cell or Progenitor cell: In fetal or adult tissues, it is a partially committed but not fully differentiated cell that divides and gives rise to differentiated cells.

Preimplantation genetic diagnosis (PGD): Before an in vitro–fertilized embryo is implanted in a woman's uterus, it can be screened for specific genetic mutations that are known to cause particular genetic diseases or for chromosomal abnormalities. One or more cells are removed from the preimplantation embryo for testing and the surviving embryo that is implanted is one that is not carrying the genetic abnormality.

Prenatal diagnosis: Detection of abnormalities and disease conditions while a fetus is developing in the uterus. Many techniques for prenatal diagnosis, such as chorionic villus sampling and amniocentesis, require sampling placental tissue or fetal cells found in the amniotic fluid or fetomaternal circulation. Others, such as ultrasonography, can be performed without cell or tissue samples.

Primitive streak: An elongated band of cells that forms along the axis of an embryo early in gastrulation by the movement of lateral cells toward the axis and that develops a groove along its midline through which cells move to the interior of the embryo to form the mesoderm.

Pronucleus: The haploid nucleus of an oocyte or sperm, either prior to fertilization or immediately after fertilization, before the sperm and egg nuclei have fused into a single diploid nucleus.

Protein: A large complex molecule made up of one or more chains of amino acids. Proteins perform a wide variety of activities in the cell.

Recessive: A recessive allele of a gene is one whose effects are masked by the second allele present in a diploid cell or organism, which is referred to as dominant.

Recombinant DNA: A recombinant DNA molecule is made up of DNA sequences that have been artificially modified or joined together so that the new genetic sequence differs from naturally occurring genetic material.

Recombinant DNA Advisory Committee (RAC): Oversees and reviews proposals for research funded by the National Institutes of Health (NIH) or similar projects conducted at institutions funded by NIH that involve recombinant or synthetic DNA, such as gene therapy.

Recombination: The process, natural or engineered, in which two pieces of DNA undergo breakage and reunion to generate a new combination of DNA segments.

Regenerative medicine: Medical treatments that seek to replace defective, damaged, or missing tissue by engineered cells, tissues, or implants, often involving stem cells.

Restriction enzyme: An enzyme from bacteria that is used to cut DNA at defined sequences, used in DNA analysis and in joining DNA fragments through the cut ends.

Retrovirus: A virus whose genome is made of RNA but during viral replication becomes copied into a DNA form that can integrate into the DNA genome of a cell. Often used as carriers of genes (vectors) to introduce genes into cells. A subset of retroviruses is called lentiviruses.

Risk: The probability of an effect on a specific endpoint or a set of endpoints due to a specific set of a stressor or stressors. An effect can be beneficial or harmful.

Risk assessment: The process by which all available evidence on the probability of effects is collected, evaluated, and interpreted to estimate the probability of the sum total of effects.

RNA (ribonucleic acid): A chemical that is similar in structure to DNA. One of its main functions is to translate the genetic code of DNA into structural proteins.

RNP (ribonuclear protein complex): Many types exist within cells, and this is a general term encompassing all of these, but in the context of genome editing it is often used to refer to a guide RNA molecule combined with a DNA-cutting enzyme such as Cas9.

Selective advantage: Some variants of genes provide a trait that confers a survival or a reproductive advantage that can be selected by natural selection and therefore increases in prevalence in a population.

Single guide RNA (sgRNA): A short piece of RNA that binds to a nuclease such as Cas9 and also to a specific DNA sequence to guide the nuclease to a specific location in the genome. This term is synonymous with guide RNA (vide infra) in most usages.

Somatic cell: Any cell of a plant or animal other than a reproductive cell or reproductive cell precursor. Latin: soma = body.

Somatic cell nuclear transfer (SCNT): The transfer of a cell nucleus from a somatic cell into an egg (oocyte) whose nucleus has been removed.

Spermatogonial stem cells: The self-replicating precursors of sperm cells.

Stem cell: A nonspecialized cell that has the capacity to divide indefinitely in culture and to differentiate into more mature cells with specialized functions.

Stem cell therapy: The use of stem cells in regenerative medicine to replace defective, damaged, or missing tissue.

Syncytiotrophoblast cell: A cell derived from trophectodermal cells from the early mammalian embryo that fuse (into multinucleate syncitia) and contribute to the structure and function of the placentae.

Synthetic biology: The development of living cells from separate genetic components, using engineering principles to build desired functions into living organisms.

Synthetic DNA: DNA molecules that are chemically or by other means synthesized or amplified; they may be chemically or otherwise modified but can base pair, or be recombined with, naturally occurring DNA molecules.

T cells: Types of white blood cells that are of crucial importance in the immune system. They cooperate with other immune cells in killing infected or cancerous cells but can also participate in inflammation or in autoimmunity when they become activated against an organism's own cells or tissues.

Target sequence: Specific sequence of DNA bases within the genome that is the target of genome-editing tools. For CRISPR/Cas9 methods this will be a 20 nucleotide sequence that the gRNAs are designed to recognize (i.e., they will contain a complementary sequence of the same length).

Therapy (or therapeutic intervention): The treatment or prevention of disease or disability.

Tissue culture: The growth of cells or tissue segments in vitro in an artificial medium for experimental research.

Totipotent cell: A stem cell that has unlimited developmental capability. The totipotent cells of the very early embryo (an embryo prior to the blastocyst stage) have the capacity to differentiate into extraembryonic tissues, membranes, the embryo, and all postembryonic tissues and organs.

Transcription: Making an RNA copy from a gene or other DNA sequence. Transcription is the first step in gene expression.

Transcription Activator-Like Effector Nuclease (TALEN): A class of engineered restriction enzymes generated by the fusion of a transcription activator-like effector DNA-binding domain (that binds to a specific DNA sequence) to a DNA-cleavage domain (nuclease) to be used as a genome-editing tool. TALENs followed zinc finger nucleases and preceded CRISPR/Cas9 as genome-editing tools.

Transcription factor: A protein that binds to control regions (enhancers and promoters) of genes to activate or repress their transcription (or expression).

Transfection: A method by which experimental DNA may be introduced into a cell.

Transgene: A gene or genetic material that has been introduced into a cell or organism. Transgenes can be integrated at random, or targeted to a specific site by homologous recombination or by genome editing using methods of homology-directed repair.

Transgenic organism: An organism into which one or more genes from another species (transgenes) have been transferred or otherwise artificially introduced.

Transhumanism: A class of philosophies of life that seek the continuation and acceleration of the evolution of intelligent life beyond its currently human form and human limitations by means of science and technology, guided by life-promoting principles and values.

Translation: The process of forming a protein molecule from information contained in a messenger RNA: a step in gene expression following transcription (copying of RNA from DNA).

Trophectoderm: The outer layer of the developing blastocyst that will ultimately form the embryonic side of the placenta.

Undifferentiated: Not having developed into a specialized cell or tissue type.

Unipotent stem cell: A stem cell that both divides and gives rise to a single mature cell type, such as a spermatogenic stem cell, which only gives rise to sperm. Alternatively called a progenitor.

Utilitarianism: The morally right action as the action that produces the most "good."

Variant: Genes have many variants in a population that can differ somewhat in function, some being advantageous and some being deleterious or nonfunctional.

Vector: A vehicle that transfers a gene into a new site (analogous to insect vectors that transfer a virus or parasite into a new animal host). Vectors used in molecular cell biology and genetic engineering include plasmids and modified viruses engineered to carry and express genes of interest in target cells. The most clinically relevant viral vectors for gene transfer include retroviral, lentiviral, adenoviral, and adeno-associated viral vectors.

Virtue ethics: A focus on moral character as opposed to duties (deontology) or consequences (consequentialism).

Wild type (noun); Wild-type (adjective): The "normal" type of an organism or a gene.

X-inactivation: The process in which one X chromosome of the two present in a female mammalian cell is inactivated so that only the genes of one X chromosome are expressed.

Zinc finger: A small protein structure based on naturally occurring transcription factors that bind to defined DNA sequences to control the activity of nearby genes. Zinc fingers can be custom engineered to target a specific section of the DNA sequence for use in genome engineering.

Zinc finger nuclease (ZFN): A class of engineered enzymes generated by the fusion of zinc finger DNA-binding domains to a DNA-cleavage enzyme (usually FokI) that can be used as a genome-editing tool. One of the first and a reliable method of genome editing.

Zygote: The one-cell embryo formed by the union of sperm and egg at fertilization.

BIBLIOGRAPHY

Abdallah NA, Prakash CS, McHughen AG. Genome editing for crop improvement: challenges and opportunities. GM Crops & Food. 2015; 6(4): 183-205.

Adetunji CO, Kremer RJ, Makanjuola R, Sarin NB. Application of molecular biotechnology to manage biotic stress affecting crop enhancement and sustainable agriculture. Advances in Agronomy. 2021; 168: 39-81.

Ahmad MI, Farooq S, Alhamoud Y, Li C, Zhang H. Soy Leghemoglobin: A review of its structure, production, safety aspects, and food applications. Trends in Food Science & Technology. 2023; 104199.

Ali Q, Yu C, Hussain A, Ali M, Ahmar S, Sohail MA, Zhou L. Genome engineering technology for durable disease resistance: Recent progress and future outlooks for sustainable agriculture. Frontiers in Plant Science. 2022; 13: 860281.

Anzalone, Andrew V et al. "Search-and-replace genome editing without double-strand breaks or donor DNA." Nature vol. 576, 7785 (2019): 149-157. doi: 10.1038/s41586-019-1711-4

Araki, M., & Ishii, T. (2014). International regulatory landscape and integration of corrective genome editing into in vitro fertilization. Reprod Biol Endocrinol, 12, 108. doi:10.1186/1477-7827-12-108

Bailey-Serres J, Parker JE, Ainsworth EA, Oldroyd GE, Schroeder JI. Genetic strategies for improving crop yields. Nature. 2019; 575(7781): 109-118.

Baranski R, Klimek-Chodacka M, Lukasiewicz A. Approved genetically modified (GM) horticultural plants: a 25-year perspective. Folia Horticulturae. 2019; 31(1): 3-49.

Barrett CB. Overcoming global food security challenges through science and solidarity. American Journal of Agricultural Economics. 2021; 103(2): 422-447.

Baweja P, Kumar S, Kumar G. Fertilizers and pesticides: Their impact on soil health and environment. Soil health. 2020; 265-285.

Borrás L, Caballero-Rothar NN, Saenz E, Segui M, Gerde JA. Challenges and opportunities of hard endosperm food grade maize sourced from South America to Europe. European Journal of Agronomy. 2022; 140: 126596.

Borrelli VM, Brambilla V, Rogowsky P, Marocco A, Lanubile A. The enhancement of plant disease resistance using CRISPR/Cas9 technology. Frontiers in plant science. 2018; 9: 1245.

Boutin J, Cappellen D, Rosier J, Amintas S, Dabernat S, Bedel A, Moreau-Gaudry F. ON-target adverse events of CRISPR-Cas9 nuclease: more chaotic than expected. The CRISPR Journal. 2022; 5(1): 19-30.

Bucci G, Bentivoglio D, Finco A. Precision agriculture as a driver for sustainable farming systems: state of art in literature and research. Calitatea. 2018; 19(S1): 114-121.

Callaway, E. (2016). UK scientists gain licence to edit genes in human embryos. Nature News, 530(7588), 18. doi:doi:10.1038/nature.2016.19270

Caradus JR. Processes for regulating genetically modified and gene edited plants. GM Crops & Food. 2023; 1-41.

Ceasar SA, Rajan V, Prykhozhij SV, Berman JN, Ignacimuthu S. Insert, remove or replace: A highly advanced genome editing system using CRISPR/Cas9. Biochimica et Biophysica Acta (BBA)-Molecular Cell Research. 2016; 1863(9): 2333-2344.

Chen, Janice S et al. "CRISPR-Cas12a target binding unleashes indiscriminate single-stranded DNase activity." Science (New York, N.Y.) vol. 360, 6387 (2018): 436-439. doi: 10.1126/science.aar6245

Chen, Kok-Siong et al. "Bifunctional cancer cell-based vaccine concomitantly drives direct tumor killing and antitumor immunity." Science translational medicine vol. 15, 677 (2023): eabo4778. doi: 10.1126/scitranslmed.abo4778

Cook N, Greenaway A, Kirk N. Annotated bibliography: Potential ethical issues and unintended effects of CRISPR gene editing and gene drives for invasive species control. Lincoln, New Zealand: Landcare Research; 2017.

Cook, Peter J, and Andrea Ventura. "Cancer diagnosis and immunotherapy in the age of CRISPR." Genes, chromosomes & cancer vol. 58, 4 (2019): 233-243. doi: 10.1002/gcc.22702

Courtier Orgogozo V, Morizot B, Boëte C. Agricultural pest control with CRISPR based gene drive: time for public debate: should we use gene drive for pest control?. EMBO reports. 2017; 18(6): 878-880.

Cyranoski, D., & Reardon, S. (2017). Chinese scientists genetically modify human embryos. Nature News. doi:doi:10.1038/nature.2015.17378

Delhove J, Osenk I, Prichard I, Donnelley M. Public acceptability of gene therapy and gene editing for human use: a systematic review. Human gene therapy. 2020; 31(1-2): 20-46.

Duncan E, Glaros A, Ross DZ, Nost E. New but for whom? Discourses of innovation in precision agriculture. Agriculture and Human Values. 2021; 38: 1181-1199.

EFSA Panel on Genetically Modified Organisms (GMO), Naegeli H, Bresson JL, Dalmay T, Dewhurst IC, Epstein MM, Firbank LG. Adequacy and sufficiency evaluation of existing EFSA guidelines for the molecular characterisation, environmental risk assessment and post market environmental monitoring of genetically modified insects containing engineered gene drives. EFSA Journal. 2020; 18(11): e06297.

Filippova J, Matveeva A, Zhuravlev E, Stepanov G. Guide RNA modification as a way to improve CRISPR/Cas9-based genome-editing systems. Biochimie. 2019; 167: 49-60.

Gascuel Q, Diretto G, Monforte AJ, Fortes AM, Granell A. Use of natural diversity and biotechnology to increase the quality and nutritional content of tomato and grape. Frontiers in plant science. 2017; 8: 652.

Glick BR, Patten CL. Molecular biotechnology: principles and applications of recombinant DNA. John Wiley & Sons; 2022.

Gootenberg, Jonathan S et al. "Multiplexed and portable nucleic acid detection platform with Cas13, Cas12a, and Csm6." Science (New York, N.Y.) vol. 360, 6387 (2018): 439-444. doi: 10.1126/science.aaq0179

Gupta S, Kumar A, Patel R, Kumar V. Genetically modified crop regulations: scope and opportunity using the CRISPR-Cas9 genome editing approach. Molecular Biology Reports. 2021; 48(5): 4851-4863.

Hampton, T. (2016). Ethical and Societal Questions Loom Large as Gene Editing Moves Closer to the Clinic. JAMA, 315(6), 546-548. doi:10.1001/jama.2015.19150

Heckl, Dirk et al. "Generation of mouse models of myeloid malignancy with combinatorial genetic lesions using CRISPR-Cas9 genome editing." Nature biotechnology vol. 32, 9 (2014): 941-6. doi: 10.1038/nbt.2951

Hinge VR, Chavhan RL, Kale SP, Suprasanna P, Kadam US. Engineering resistance against viruses in field crops using CRISPR-Cas9. Current Genomics. 2021; 22(3): 214.

Hirel B, Tétu T, Lea PJ, Dubois F. Improving nitrogen use efficiency in crops for sustainable agriculture. Sustainability. 2011; 3(9): 1452-1485.

Huang Y, Brown M. Advancing to the next generation of precision agriculture. Agriculture & food systems to. 2019; 2050: 285â.

Ishii, T. (2017). Germ line genome editing in clinics: the approaches, objectives and global society. Brief Funct Genomics, 16(1), 46-56. doi:10.1093/bfgp/elv053

Ismail AM, Singh US, Singh S, Dar MH, Mackill DJ. The contribution of submergence-tolerant (Sub1) rice varieties to food security in flood-prone rainfed lowland areas in Asia. Field Crops Research. 2013; 152: 83-93.

Israel, Mason A et al. Probing sporadic and familial Alzheimer's disease using induced pluripotent stem cells. Nature vol. 482, 7384 216-20. 25 Jan. (2012).

Javaid M, Haleem A, Khan IH, Suman R. Understanding the potential applications of Artificial Intelligence in Agriculture Sector. Advanced Agrochem. 2023; 2(1): 15-30.

Jiang, Ling et al. "Ultrasensitive CRISPR/Cas13a-Mediated Photoelectrochemical Biosensors for Specific and Direct Assay of miRNA-21." Analytical chemistry, 10.1021/acs.analchem.2c03945. 5Jan. 2023, doi: 10.1021/acs.analchem.2c03945

Jørgensen MH. Agricultural field production in an 'Industry 4.0'concept; 2018.

Kaul T, Sony SK, Raman NM, Eswaran M, Verma R, Prakash AT, Kaul R. How crisp is CRISPR? CRISPR-Cas-mediated crop improvement with special focus on nutritional traits. Advancement in crop improvement techniques. 2020: 159-197.

Kellner, Max J et al. "SHERLOCK: nucleic acid detection with CRISPR nucleases." Nature protocols vol. 14, 10 (2019): 2986-3012. doi: 10.1038/s41596-019-0210-2

KhokharVoytas A, Shahbaz M, Maqsood MF, Zulfiqar U, Naz N, Iqbal UZ, AlShaqhaa MA. Genetic modification strategies for enhancing plant resilience to abiotic stresses in the context of climate change. Functional & Integrative Genomics. 2023; 23(3): 283.

Kumlehn J, Pietralla J, Hensel G, Pacher M, Puchta H. The CRISPR/Cas revolution continues: from efficient gene editing for crop breeding to plant synthetic biology. Journal of Integrative Plant Biology. 2018; 60(12): 1127-1153.

Kuriakose SV, Pushker R, Hyde EM. Data-driven decisions for accelerated plant breeding. Accelerated Plant Breeding, Cereal Crops. 2020; 1: 89-119.

Kusch S, Panstruga R. mlo-based resistance: an apparently universal "weapon" to defeat powdery mildew disease. Molecular Plant-Microbe Interactions. 2017; 30(3): 179-189.

Kuzma J, Grieger K, Cimadori I, Cummings CL, Loschin N, Wei W. Parameters, practices, and preferences for regulatory review of emerging biotechnology products in food and agriculture. Frontiers in Bioengineering and Biotechnology. 2023; 11.

Landrum AR, Hallman WK, Jamieson KH. Examining the impact of expert voices: communicating the scientific consensus on genetically-modified organisms. Environmental Communication. 2019; 13(1): 51-70.

Lanphier, E., Urnov, F., Haecker, S. E., Werner, M., & Smolenski, J. (2015). Don't edit the human germ line. Nature News, 519(7544), 410. doi:10.1038/519410a

Lenaerts B, Collard BC, Demont M. Improving global food security through accelerated plant breeding. Plant Science. 2019; 287: 110207.

Li S, Lin D, Zhang Y, Deng M, Chen Y, Lv B, Gao C. Genome-edited powdery mildew resistance in wheat without growth penalties. Nature. 2022; 602(7897): 455-460.

Li Z, Taylor J, Frewer L, Zhao C, Yang G, Liu Z., Sun Z. A comparative review on the state and advancement of Site-Specific Crop Management in the UK and China. Frontiers of Agricultural Science and Engineering; 2019.

Liu, Wenyi et al. "Applications and challenges of CRISPR-Cas gene-editing to disease treatment in clinics." Precision clinical medicine vol. 4, 3 179-191. 10 Jul. 2021, doi: 10.1093/pcmedi/pbab014

MacDonald KI. Landscapes of Diversity: Development and Vulnerability to Food Insecurity in Subsistence Agroecosystems of Northern Pakistan; 2010.

Mall T, Gupta M, Dhadialla TS, Rodrigo S. Overview of biotechnology-derived herbicide tolerance and insect resistance traits in plant agriculture. Transgenic Plants: methods and protocols. 2019; 313-342.

Massel K, Lam Y, Wong AC, Hickey LT, Borrell AK, Godwin ID. Hotter, drier, CRISPR: the latest edit on climate change. Theoretical and Applied Genetics. 2021; 134: 1691-1709.

McCarthy JJ, McLeod HL, Ginsburg GS. Genomic medicine: a decade of successes, challenges, and opportunities. Science translational medicine. 2013; 5(189): 189sr4-189sr4.

Meena BR, Jatav HS, Dudwal BL, Kumawat P, Meena SS, Singh VK, Sathyanarayana E. Fertilizer Recommendations by Using Different Geospatial Technologies in Precision Farming or Nanotechnology. Ecosyst. Serv. 2022; 14: 241-257.

Monast JJ. Editing nature: Reconceptualizing biotechnology governance. BCL Rev. 2018; 59: 2377.

Munteanu ITT. Evolution of crispr system and their applicability to genome editing in bacteria. Journal of Experimental and Molecular Biology. 2022; 23(1): 46-53.

National Academies of Sciences, E., Medicine, . (2017). Human Genome Editing: Science, Ethics, and Governance. Washington, DC: The National Academies Press.

Nordberg A, Minssen T, Holm S, Horst M, Mortensen K, Møller BL. Cutting edges and weaving threads in the gene editing (ß) evolution: reconciling scientific progress with legal, ethical, and social concerns. Journal of Law and the Biosciences. 2018; 5(1): 35-83.

Novak, Lance C et al. "PICKLES v3: the updated database of pooled in vitro CRISPR knockout library essentiality screens." Nucleic acids research vol. 51, D1 (2023): D1117-D1121. doi: 10.1093/nar/gkac982

Pardee, Keith et al. "Rapid, Low-Cost Detection of Zika Virus Using Programmable Biomolecular Components." Cell vol. 165, 5 (2016): 1255-1266. doi: 10.1016/j.cell.2016.04.059

Park, A. (2016). UK Approves First Studies Using New Gene Editing Technique. Time Health.

Pei, Wen-Di et al. "Epigenome editing by CRISPR/Cas9 in clinical settings: possibilities and challenges." Briefings in functional genomics vol. 19, 3 (2020): 215-228. doi: 10.1093/bfgp/elz035

Pickar-Oliver A, Gersbach CA. The next generation of CRISPR–Cas technologies and applications. Nature reviews Molecular cell biology. 2019; 20(8): 490-507.

Rai KK. Integrating speed breeding with artificial intelligence for developing climate-smart crops. Molecular Biology Reports. 2022; 49(12): 11385-11402.

Rejeb A, Keogh JG, Zailani S, Treiblmaier H, Rejeb K. Blockchain technology in the food industry: A review of potentials, challenges and future research directions. Logistics. 2020; 4(4): 27.

Rode NO, Estoup A, Bourguet D, Courtier-Orgogozo V, Débarre F. Population management using gene drive: molecular design, models of spread dynamics and assessment of ecological risks. Conservation Genetics. 2019; 20: 671-690.

Savulescu, J., Pugh, J., Douglas, T., & Gyngell, C. (2015). The moral imperative to continue gene editing research on human embryos. Protein Cell, 6(7), 476-479.

Schnurr MA. Africa's gene revolution: Genetically modified crops and the future of African agriculture. McGill-Queen's Press-MQUP; 2019.

Scholthof KBG. Plant Sciences: A Brief History. A Companion to American Agricultural History. 2022; 175-187.

Shah F, Wu W. Soil and crop management strategies to ensure higher crop productivity within sustainable environments. Sustainability. 2019; 11(5): 1485.

Shahriar SA, Islam MN, Chun CNW, Rahim MA, Paul NC, Uddain J, Siddiquee S. Control of plant viral diseases by CRISPR/Cas9: resistance mechanisms, strategies and challenges in food crops. Plants. 2021; 10(7): 1264.

Shi J, Gao H, Wang H, Lafitte HR, Archibald RL, Yang M, Habben JE. ARGOS 8 variants generated by CRISPR Cas9 improve maize grain yield under field drought stress conditions. Plant biotechnology journal. 2017; 15(2): 207-216.

Singh AK. Precision agriculture in india–opportunities and challenges. Indian Journal of Fertilisers. 2022; 18(4): 308-331.

Steinbach A. Technology, Patents, and Plants: Are the Next Generation of GMOs Patentable?. Patents, and Plants: Are the Next Generation of GMOs Patentable; 2018.

Suprasanna P, Mirajkar SJ, Patade VY, Jain SM. Induced mutagenesis for improving plant abiotic stress tolerance. In Mutagenesis: exploring genetic diversity of crops (p. e30765). Wageningen Academic Publishers; 2014.

Syed MA, Ahmed MME, Debsharma SK, Jahan N, Afrin W, Biswas A, Akter MB. Breeding and Omics Approaches to Understand Abiotic Stress Response in Rice. In Omics Approach to Manage Abiotic Stress in Cereals (pp. 341-404). Singapore: Springer Nature Singapore; 2022.

Taylor J. Towards Incorporating Within-Field Variation into Spatial Agronomic Decision Processes (Doctoral dissertation, Université de Montpellier); 2019.

The Hinxton Group. (2015). Statement on Genome Editing Technologies and Human Germline Genetic Modification.

Tudi M, Daniel Ruan H, Wang L, Lyu J, Sadler R, Connell D, Phung DT. Agriculture development, pesticide application and its impact on the environment. International journal of environmental research and public health. 2021; 18(3): 1112.

Weersink A, Fraser E, Pannell D, Duncan E, Rotz S. Opportunities and challenges for big data in agricultural and environmental analysis. Annual Review of Resource Economics. 2018; 10: 19-37.

Xue, Hui-Ying et al. "CRISPR-Cas9 for medical genetic screens: applications and future perspectives." Journal of medical genetics vol. 53, 2 (2016): 91-7.

Yuan GN, Marquez GPB, Deng H, Iu A, Fabella M, Salonga RB, Cartagena JA. A review on urban agriculture: technology, socio-economy, and policy. Heliyon; 2022.

Zhang D, Zhang Z, Unver T, Zhang B. CRISPR/Cas: A powerful tool for gene function study and crop improvement. Journal of Advanced Research. 2021; 29: 207-221.

Zhang H, Li Y, Zhu JK. Developing naturally stress-resistant crops for a sustainable agriculture. Nature plants. 2018; 4(12): 989-996.

Zhao H, Wolt JD Risk associated with off-target plant genome editing and methods for its limitation. Emerging Topics in Life Sciences. 2017; 1(2): 231-240.

INDEX

P

Q

R

S

www.ingramcontent.com/pod-product-compliance
Lightning Source LLC
LaVergne TN
LVHW062215200926
848484LV00006B/1401